The IMA Volumes
in Mathematics
and Its Applications

Volume 8

Series Editors
Geroge R. Sell   Hans Weinberger

Institute for Mathematics and Its Applications
IMA

The **Institute for Mathematics and Its Applications** was established by a grant from the National Science Foundation to the University of Minnesota in 1982. The IMA seeks to encourage the development and study of fresh mathematical concepts and questions of concern to the other sciences by bringing together mathematicians and scientists from diverse fields in an atmosphere that will stimulate discussion and collaboration.

The IMA Volumes are intended to involve the broader scientific community in this process.

Hans Weinberger, Director
George R. Sell, Associate Director

IMA Programs

1982–1983 Statistical and Continuum Approaches to Phase Transition

1983–1984 Mathematical Models for the Economics of Decentralized Resource Allocation

1984–1985 Continuum Physics and Partial Differential Equations

1985–1986 Stochastic Differential Equations and Their Applications

1986–1987 Scientific Computation

1987–1988 Applied Combinatorics

1988–1989 Nonlinear Waves

Springer Lecture Notes from the IMA

The Mathematics and Physics of Disordered Media
　　Editors: Barry Hughes and Barry Ninham
　　(Lecture Notes in Mathematics, Volume 1035, 1983)

Orienting Polymers
　　Editor: J. L. Ericksen
　　(Lecture Notes in Mathematics, Volume 1063, 1984)

New Perspectives in Thermodynamics
　　Editor: James Serrin
　　(Springer-Verlag, 1986)

Models of Economic Dynamics
　　Editor: Hugo Sonnenschein
　　(Lecture Notes in Economics, Volume 264, 1986)

Harry Kesten
Editor

Percolation Theory and Ergodic Theory of Infinite Particle Systems

With 47 Illustrations

Springer Science+Business Media, LLC

Harry Kesten
Institute for Mathematics and
    Its Applications
Minneapolis, Minnesota 55455, USA

AMS Classification: 60K 35 82 A43

Library of Congress Cataloging in Publication Data
Percolation theory and ergodic theory of infinite
    particle systems.
    (The IMA volumes in mathematics and its
applications ; v. 8)
    Bibliography: p.
    1. Percolation (Statistical physics) 2. Ergodic
theory. I. Kesten, Harry, 1931–   .   II. Series
QC174.85.P45P48   1987   530.1′595   87-9512

9 8 7 6 5 4 3 2 1

ISBN 978-1-4613-8736-7   ISBN 978-1-4613-8734-3 (eBook)
DOI 10.1007/978-1-4613-8734-3

The IMA Volumes in Mathematics and Its Applications

Current Volumes:

Forthcoming Volumes:

CONTENTS

FOREWORD

This IMA Volume in Mathematics and its Applications

PERCOLATION THEORY AND ERGODIC THEORY OF INFINITE PARTICLE SYSTEMS

represents the proceedings of a workshop which was an integral part of the
1984-85 IMA program on STOCHASTIC DIFFERENTIAL EQUATIONS AND THEIR APPLICATIONS
We are grateful to the Scientific Committee:

>   Daniel Stroock (Chairman)
>   Wendell Fleming
>   Theodore Harris
>   Pierre-Louis Lions
>   Steven Orey
>   George Papanicolaou

for planning and implementing an exciting and stimulating year-long program.

We especially thank the Workshop Organizing Committee, Harry Kesten
(Chairman), Richard Holley, and Thomas Liggett for organizing a workshop which
brought together scientists and mathematicians in a variety of areas for a
fruitful exchange of ideas.

>   George R. Sell
>   Hans Weinberger

PREFACE

Percolation theory and interacting particle systems both have seen an explosive growth in the last decade. These subfields of probability theory are closely related to statistical mechanics and many of the publications on these subjects (especially on the former) appear in physics journals, with a great variability in the level of rigour. There is a certain similarity and overlap between the methods used in these two areas and, not surprisingly, they tend to attract the same probabilists. It seemed a good idea to organize a workshop on "Percolation Theory and Ergodic Theory of Infinite Particle Systems" in the framework of the special probability year at the Institute for Mathematics and its Applications in 1985-86. Such a workshop, dealing largely with rigorous results, was indeed held in February 1986. These proceedings contain contributions of almost all the speakers at this meeting. As usual some of the papers are not very close to the talk given by the same author. We hope that nevertheless these proceedings give the reader a good impression of the state of the art in 1986. It is an indication of the success of the gathering that several joint papers here report on work which was finished (or even begun) during the workshop.

I am grateful to S. Orey and D. Stroock who coordinated the probability year at the IMA for including percolation theory and infinite particle systems in the program. The organizing committee of the workshop consisted of T.E. Harris, R. Holley, T. Liggett and myself. I am much obliged to my fellow committee members for their help and support. I also wish to thank the directors of the IMA, H. Weinberger and G. Sell, for their hospitality and help in making the workshop a success. Last but not least I thank the staff of the IMA for taking care of practical matters in such a way that things ran smoothly, and Kaye Smith and Patricia V. Brick for their excellent typing and preparation of these proceedings.

Harry Kesten

Ithaca, N.Y., Oct 1986

RAPID CONVERGENCE TO EQUILIBRIUM OF STOCHASTIC ISING MODELS IN THE DOBRUSHIN SHLOSMAN REGIME

M. Aizenman[1]
Department of Mathematics
Rutgers University
New Brunswick, New Jersey 08903

R. Holley[2]
Department of Mathematics
University of Colorado
Boulder, Colorado 80309

Abstract

We show that, under the conditions of the Dobrushin Shlosman theorem for uniqueness of the Gibbs state, the reversible stochastic Ising model converges to equilibrium exponentially fast on the L^2 space of that Gibbs state. For stochastic Ising models with attractive interactions and under conditions which are somewhat stronger than Dobrushin's, we prove that the semi-group of the stochastic Ising model converges to equilibrium exponentially fast in the uniform norm. We also give a new, much shorter, proof of a theorem which says that if the semi-group of an attractive spin flip system converges to equilibrium faster than $1/t^d$ where d is the dimension of the underlying lattice, then the convergence must be exponentially fast.

0. Introduction

This paper resulted from our attempt to understand the relationship between the Dobrushin-Shlosman criteria (Theorem 1 below; see [1]) for rapid decay of correlations in the equilibrium Gibbs states, and the dynamical property of rapid (i.e. exponentially fast) convergence to equilibrium in stochastic Ising models.

For a description of the stochastic Ising models on Z^d associated with a potential $\{J_R : R \subset Z^d\}$ see chapter 4 in [5]. We use the same notation that is used there. We will always assume (even when

---
[1]Research supported in part by NSF Grant PHY-8301493A02
[2]Research supported in part by NSF Grant MCS-8310542.

we do not mention it) that the potential $\{J_R : R \subset Z^d\}$ is translation invariant (i.e. $J_{R+k} = J_R$ for all $R \subset Z^d$ and all $k \in Z^d$) and has finite range L (i.e. $|J_R| = 0$ if the diameter of R is greater than L).

If Λ is a subset of Z^d we let $E(\Lambda) = \{-1,1\}^\Lambda$ be the space of spin configurations on Λ, and $C(\Lambda) = \{$real valued functions on $E(\Lambda)\}$. We denote $E(Z^d)$ by E, and often think of $C(\Lambda)$ as a subset of $C(E)$ by means of the obvious projections. If Λ is a finite subset of Z^d, $\eta \in E$ and $\sigma \in E(\Lambda)$ we let

$$g_\Lambda(\sigma|\eta) = \exp(-\sum_{R \cap \Lambda \neq \varnothing} J_R \prod_{j \in R} \sigma\eta(j))/Z(\Lambda,\eta)$$

be the conditional Gibbs state on $E(\Lambda)$ with boundary conditions η. Here $\sigma\eta$ is the configuration which is equal to σ on Λ and equal to η on Λ^c, and $Z(\Lambda,\eta)$ is the normalizing constant needed to make $g_\Lambda(\cdot|\eta)$ have total mass 1.

Let $\Delta_k f(\eta) = f(\eta^k) - f(\eta)$, where $f \in C(E)$, and η^k is the configuration which is equal to η at all sites $j \in Z^d$ except for j=k, where it is equal to $-\eta(k)$. Finally let $\|f\| = \sup_\eta |f(\eta)|$.

The recent result of Dobrushin and Shlosman [1] may be conveniently reformulated as follows (this formulation was given by E. H. Lieb and M. A.).

Theorem 1: (Dobrushin-Shlosman) If there is some finite cube, $\Lambda_0 \subset Z^d$ for which there are constants $\alpha_{j,k}$, $j \in \Lambda_0$, $k \in \Lambda_0^c$ such that

$$\frac{1}{|\Lambda_0|} \sum_{j \in \Lambda_0} \sum_{k \in \Lambda_0^c} \alpha_{j,k} = \gamma < 1$$

and such that for all $f \in C(\Lambda_0)$ and all $\eta,\eta' \in E$

$$\left| \int f(\sigma) g_{\Lambda_0}(d\sigma|\eta) - \int f(\sigma) g_{\Lambda_0}(d\sigma|\eta') \right| \leq \sum_{j \in \Lambda_0} \sum_{k \in \Lambda_0^c} \|\Delta_j f\| \alpha_{j,k} |\eta(k) - \eta'(k)|/2,$$

then there is an $\epsilon > 0$ and a constant $C < \infty$ such that for all finite cubes $\Omega \subset Z^d$ and all $f \in C(\Omega)$

$$\sup_{\eta,\eta'} \left| \int_{E(\Omega)} f(\sigma) g_\Omega(d\sigma|\eta) - \int_{E(\Omega)} f(\sigma) g_\Omega(d\sigma|\eta') \right| \leq C|\partial\Omega| \sum_{k \in \Omega} \|\Delta_k f\| e^{-\epsilon \text{dist}(k,\Omega^c)}.$$

Note that the conclusion of Theorem 1 implies that there is a unique Gibbs state, g, for the given potential.

Our first theorem is the following (here and in what follows $\{T_t : t \geq 0\}$ denotes the semi-group of the stochastic Ising model).

Theorem 2: Under the hypotheses of Theorem 1, there is an $\epsilon > 0$ such that for all $f \in L^2(g)$,

$$(0.1) \qquad \|T_t f - \int f dg\|_{L^2(g)} \leq \|f - \int f dg\|_{L^2(g)} \, e^{-\epsilon t}.$$

What we would really like is a conclusion similar to that of Theorem 2 with the L^2 norm on the left side of (0.1) replaced by the uniform norm and $\|f - \int f dg\|_{L^2(g)}$ replaced on the right side by some other constant depending on f. We do not have such a result, but for attractive stochastic Ising models (see chapter 3 in [5]) we can come closer, as the following theorem shows.

Theorem 3: If the stochastic Ising model has attractive flip rates and if there is a finite cube Λ_0 and constants $\alpha_{j,k}$ $j\in\Lambda_0$, $k\in\Lambda_0^c$ with

$$\frac{1}{|\Lambda_0|} \sum_{j\in\Lambda_0} \sum_{k\in\Lambda_0^c} \alpha_{j,k} = \gamma < 1$$

and such that for all $k \in \Lambda_0^c$ all $f \in C(\Lambda_0)$ and all large enough cubes Ω

$$(0.2) \qquad \|\Delta_k \int f(\sigma) g_{\Lambda_0 \cap \Omega}(d\sigma|\eta)\| \leq \sum_{j\in\Lambda_0} \|\Delta_j f\|\alpha_{j,k},$$

then there is a constant $B < \infty$ and a $\lambda > 0$ such that

$$(0.3) \qquad \sup_{\eta,\eta'} |T_t f(\eta) - T_t f(\eta')| \leq B \sum_{k\in Z^d} \|\Delta_k f\| e^{-\lambda t}$$

Remark: It could still be the case that for certain nontrivial models the assumptions of Theorem 3 are satisfied whenever the Dobrushin-Shlosman criterion is met. However, the "Čzech models" discussed in ref. [6] (which are not attractive) seem to show that the assumptions in Theorem 3 are in general stronger that those of Theorem 1.

Once one sees how to prove Theorem 2, that proof together with the techniques in [2] and the results in [3] can be combined to yield a proof of Theorem 3. The proof of Theorem 2 is in section 1. Our efforts to understand the relationship between Theorem 1 and Theorem 3 led us to a greatly simplified proof of Theorem 4 below, which is the main result of [3].

Before stating that theorem we need some notation. For each $k \in Z^d$ let $\phi_k(\eta) = \frac{1}{2}\eta(k)$. Let $\overline{+1}$ be the configuration of spins which is identically one and $\overline{-1}$ be the configuration which is identically minus one.

Theorem 4: For finite range, translation invariant, attractive spin systems on Z^d, if

$$T_t \phi_0(\mp I) - T_t \phi_0(\overline{-I}) = o(t^{-d})$$

then there are constants $B < \infty$ and a $\lambda > 0$ such that (0.3) holds.

Section 2 contains the proof of Theorem 4 and related results. In section 3 we briefly sketch the proof of Theorem 3 based on Theorem 4 and the proof of Theorem 2. The ideas in that section are not new (see [3]).

1. Proof of Theorem 2

We define a semi-norm $|||f|||$ on $D \equiv \{ f \in C(E) : \sum_{k \in Z^d} ||\Delta_k f|| < \infty \}$ given by $|||f||| = \sum_{k \in Z^d} ||\Delta_k f||$ and note that if $f \in C(E)$ then $\sup_{\eta,\eta'} |f(\eta) - f(\eta')| \leq |||f|||$.

The infinitesimal generator of the semi-group $\{T_t : t \geq 0\}$ when restricted to D is given by

$$(1.1) \qquad Lf(\eta) = \sum_{k \in Z^d} c_k(\eta) \Delta_k f(\eta),$$

where the c_k's are positive functions called the flip rates which satisfy the detailed balance equations (see (4.2.2) in [5]).

Let $\overline{L}$ be the closure of L on $L^2(g)$. Then we must show that there is an $\epsilon > 0$ such that for all $f \in L^2(g)$,

$$(1.2) \qquad \int f \overline{L} f dg \leq -\epsilon ||f - \int f dg||_{L^2(g)}$$

We prove (1.2) by first considering an auxiliary process which we now define. Let Λ_0 be as in the hypotheses of Theorem 1. Note that by translation invariance we may as well assume that Λ_0 is centered at the origin. Denote $\Lambda_0 + k$ by Λ_k. Define an operator A on D by

$$A f(\eta) = \frac{1}{|\Lambda_0|} \sum_{k \in Z^d} \int_{E(\Lambda_k)} (f(\sigma\eta) - f(\eta)) g_{\Lambda_k}(d\sigma|\eta).$$

A generates a positive contraction semi-group, $S_t = e^{tA}$, on $C(E)$ (see [5]). Moreover A is essentially self-adjoint on $L^2(g)$ (see [2]). Let $R_\lambda = \int_0^\infty e^{-\lambda t} S_t dt$ be the resolvent of A on $C(E)$; let $\overline{S_t}$ be the extension of S_t from $C(E)$ to $L^2(g)$, and let $\overline{A}$ be the generator of $\{\overline{S_t} : t \geq 0\}$. Since D is dense in $C(E)$ and in $L^2(g)$, if we can show that

(1.3)
$$|||S_t f||| \leq e^{-(1-\gamma)t} |||f||| \quad \text{for all } f \in \mathbf{D},$$

then $\overline{A}$ has a gap of length at least $1 - \gamma$ between 0 and the rest of its spectrum (see Lemma (1.13) in [2]). The gap in the spectrum of $\overline{L}$ is at least as big as a positive multiple of the gap in the spectrum of $\overline{A}$ (see Lemma (1.15) in [2]). Thus the entire proof comes down to proving (1.3). Since $S_t = \lim_{n \to \infty} (\frac{n}{t} R_{\frac{n}{t}})^n$, (1.3) will follow from

(1.4)
$$(\lambda + 1 - \gamma)|||R_\lambda f||| \leq |||f||| \quad \text{for all sufficiently large } \lambda.$$

The fact that $|||R_\lambda f|||$ is finite if $f \in \mathbf{D}$ and λ is sufficiently large is a consequence of Theorem (1.3.9) in [5]. From now on we assume that λ is sufficiently large that we know a-priori that $|||R_\lambda f||| < \infty$ for all $f \in \mathbf{D}$. Then

(1.5)
$$\lambda R_\lambda f - A R_\lambda f = f,$$
and therefore,

(1.6)
$$\lambda \Delta_k R_\lambda f = \Delta_k f + \Delta_k A R_\lambda f.$$

We need to analyze $\Delta_k A R_\lambda f$.

To simplify the notation write $R_\lambda f = \phi$. Then

(1.7)
$$\Delta_k A \phi = \Delta_k \frac{1}{|\Lambda_0|} \sum_j \int_{E(\Lambda_j)} (\phi(\sigma\eta) - \phi(\eta)) g_{\Lambda_j}(d\sigma|\eta).$$

Now let $\overline{\eta}$ maximize $\Delta_k \phi(\eta)$. Then since $\Delta_k \phi(\eta) = -\Delta_k \phi(\eta^k)$, we have $\Delta_k \phi(\overline{\eta}) = ||\Delta_k \phi||$.

Note that if $k \in \Lambda_j$ (or equivalently $j \in \Lambda_k$) then $\Delta_k \int_{E(\Lambda_j)} \phi(\sigma\eta) g_{\Lambda_j}(d\sigma|\eta) = 0$, and hence

(1.8)
$$\Delta_k \int_{E(\Lambda_j)} (\phi(\sigma\overline{\eta}) - \phi(\overline{\eta})) g_{\Lambda_j}(d\sigma|\overline{\eta}) = -\Delta_k \phi(\overline{\eta}) = -||\Delta_k \phi||.$$

If $k \notin \Lambda_j$ then,

(1.9)
$$\Delta_k \int_{E(\Lambda_j)} (\phi(\sigma\overline{\eta}) - \phi(\overline{\eta})) g_{\Lambda_j}(d\sigma|\overline{\eta}) = \int_{E(\Lambda_j)} (\Delta_k \phi(\sigma\overline{\eta}) - \Delta_k \phi(\overline{\eta})) g_{\Lambda_j}(d\sigma|\overline{\eta}^k)$$
$$+ \int_{E(\Lambda_j)} \phi(\sigma\overline{\eta}) g_{\Lambda_j}(d\sigma|\overline{\eta}^k) - \int_{E(\Lambda_j)} \phi(\sigma\overline{\eta}) g_{\Lambda_j}(d\sigma|\overline{\eta}).$$
$$\leq \sum_{l \in \Lambda_j} ||\Delta_l \phi|| \alpha_{l-j, k-j}.$$

The last inequality in (1.9) follows from the fact that $\Delta_k \phi(\sigma\overline{\eta}) \leq \Delta_k \phi(\overline{\eta})$ for all σ and the hypotheses of Theorem 2.

Thus from (1.6), (1.7), (1.8), and (1.9) we have

$$(1.10) \qquad \lambda \|\Delta_k \phi\| = \lambda \Delta_k \phi(\overline{\eta}) \leq \|\Delta_k f\| - \|\Delta_k \phi\| + \frac{1}{|\Lambda_0|} \sum_{j \notin \Lambda_k} \sum_{l \in \Lambda_j} \|\Delta_l \phi\| \alpha_{l-j, k-j}.$$

Summing both sides of (1.10) over $k \in \mathbb{Z}^d$ we get

$$(\lambda + 1) \|\|R_\lambda f\|\| \leq \|\|f\|\| + \gamma \|\|R_\lambda f\|\|,$$

which is (1.4).

The reader should notice the similarity between the proof of (1.3) and the proof of Theorem (1.3.9) in [5] or the proof of Theorem (1.4) in [4].

2. Possible rates of convergence

Throughout this section we assume that we are dealing with a spin flip process which is attractive (see page 134 ff. in [5]). In this section the process need not be a stochastic Ising model (i.e. need not be time reversible). We want to show that convergence to equilibrium cannot be faster than t^{-d} without being exponentially fast (see Theorem (2.12) below).

One way to describe attractive spin flip systems is the following. The state space, E, can be made into a lattice by the ordering $\eta \leq \eta'$ if and only if $\eta(k) \leq \eta'(k)$ for all $k \in \mathbb{Z}^d$. Let $\mathbf{M}$ be the set of real valued increasing functions on this lattice. To say that the spin slip process is attractive is the same as saying that its semi-group $\{T_t : t \geq 0\}$ leaves $\mathbf{M}$ invariant.

(2.1) **Lemma:** If μ_1 and μ_2 are probability measures on E and if $\int f d\mu_1 \leq \int f d\mu_2$ for all $f \in \mathbf{M}$, (we write this as $\mu_1 <_s \mu_2$) then for all $f \in \mathbf{D}$

$$(2.2) \qquad | \int f d\mu_1 - \int f d\mu_2 | \leq \sum_{k \in \mathbb{Z}^d} \|\Delta_k f\| (\int \phi_k d\mu_2 - \int \phi_k d\mu_1)$$

Proof. $\mu_1 <_s \mu_2$ implies that there is a measure m on $E \times E$ such that $m(A \times E) = \mu_1(A)$ and $m(E \times A) = \mu_2(A)$ for all measurable $A \subset E$, and $m(\{(\eta, \sigma) : \eta(k) \leq \sigma(k) \text{ for all } k \in \mathbb{Z}^d\}) = 1$ (see Theorem (2.2.4) in [5]). Therefore

$$\left|\int f d\mu_1 - \int f d\mu_2\right| = \left|\int (f(\eta) - f(\sigma))m(d\eta,d\sigma)\right| \le \int \sum_k \|\Delta_k f\| \tfrac{1}{2}|\eta(k) - \sigma(k)|m(d\eta,d\sigma)$$

$$= \sum_k \|\Delta_k f\| \int |\phi_k(\eta) - \phi_k(\sigma)|m(d\eta,d\sigma) = \sum_k \|\Delta_k f\| \left(\int \phi_k d\mu_2 - \int \phi_k d\mu_1\right).$$

The last equality follows since $\phi_k(\eta) \le \phi_k(\sigma)$ a.s. m.

Now let $\{T_t : T \ge 0\}$ be the semi-group of an attractive spin flip system and define
$\xi^t_{j,k} = \|\Delta_k T_t \phi_j\|$, $j,k \in Z^d$, $t \ge 0$.

(2.3) Lemma: For all $f \in D$, $\|\Delta_k T_t f\| \le \sum_j \|\Delta_j f\|\xi^t_{j,k}$.

Proof. For $\eta \in E$ let $\mu_{t,\eta}$ be the measure such that $T_t f(\eta) = \int f d\mu_{t,\eta}$ for all $f \in C(E)$. Since either $\eta^k \le \eta$ or $\eta \le \eta^k$, we have either $\mu_{t,\eta} <_s \mu_{t,\eta^k}$ or $\mu_{t,\eta^k} <_s \mu_{t,\eta}$. (The process is attractive). Also

$$\Delta_k T_t f(\eta) = \int f d\mu_{t,\eta^k} - \int f d\mu_{t,\eta}.$$

Thus the lemma follows from Lemma (2.1).

(2.4) Lemma: For all $j,k \in Z^d$ and all $s,t \ge 0$, $\xi^{s+t}_{j,k} \le \sum_l \xi^s_{j,l}\, \xi^t_{l,k}$.

Proof. Apply Lemma (2.3) to the function $f(\eta) = T_s \phi_j(\eta)$.

For $\tau \ge 0$ set $\delta_\tau = \sum_k \xi^\tau_{0,k}$. By Theorem (1.3.9) in [5], $\delta_\tau < \infty$ for each τ.

(2.5) Lemma: For all $n \in Z^+$, all $\tau > 0$, and all $f \in D$,

$$\sup_{\eta,\sigma}|T_{n\tau} f(\eta) - T_{n\tau} f(\sigma)| \le \delta^n_\tau \, \||f\||.$$

Proof. $|T_t f(\eta) - T_t f(\sigma)| \le \sum_k \|\Delta_k T_t f\| \le \sum_k \sum_l \|\Delta_l f\|\xi^t_{l,k} = \||f\|| \delta_t$. If $t = n\tau$ then by Lemma (2.4) $\delta_t \le \delta^n_\tau$.

Now define

(2.6) $$\lambda_0 = -\inf_{\tau > 0}\frac{1}{\tau}\log(\delta_\tau) = -\lim_{\tau \to \infty}\frac{1}{\tau}\log(\delta_\tau).$$

The limit exists and equals the infimum in (2.6) since by Lemma (2.4) $\log(\delta_\tau)$ is subadditive.

(2.7) Lemma: For any $\lambda < \lambda_0$ there is a constant B_λ such that for all $f \in D$

(2.8)
$$\sup_{\eta,\sigma} |T_t f(\eta) - T_t f(\sigma)| \le B_\lambda |||f||| e^{-\lambda t}.$$

Proof. Since for any $s > 0$ and all $f \in D$, $\inf_\eta T_t f(\eta) \le \inf_\eta T_{t+s} f(\eta) \le \sup_\eta T_{t+s} f(\eta) \le \sup_\eta T_t f(\eta)$, this follows immediately from the definition of λ_0 and Lemma (2.5).

From (2.6) and Lemma (2.7) we see that in order to get rapid convergence of $T_t f(\eta)$ to a limiting value it suffices to prove that

(2.9)
$$\delta_\tau < 1 \quad \text{for some } \tau < \infty. \quad \text{(i.e. that } \lambda_0 > 0\text{)}$$

The next lemma is an easy corollary of Theorem (1.3.9) in [5]. The proof of a closely related result may be found in [4].

(2.10) Lemma: Set $\chi = \inf_\eta (c_0(\eta) + c_0(\eta^0))$ and $M = \sum_{k \ne 0} ||\Delta_k c_0||$, and denote the range of the interaction by L. Let γ solve $\gamma \log(\gamma) = 2e^{-1}$ and set $L_t = \gamma e M L t$. Then

(2.11)
$$\sum_{|k| > L_t} \xi_{0,k}^t \le e^{-(\chi+M)t}.$$

(2.12) Theorem: If for some $t < \infty$,

(2.13)
$$\sum_{|k| \le L_t} \xi_{0,k}^t < 1 - e^{-(\chi+M)t}$$

then there is a $\lambda > 0$ and a constant $B < \infty$ such that for all $f \in D$

(2.14)
$$\sup_{\eta,\sigma} |T_t f(\eta) - T_t f(\sigma)| \le B |||f||| e^{-\lambda t}.$$

In particular if $T_t \phi_0(\overline{+1}) - T_t \phi_0(\overline{-1}) = o(t^{-d})$ then (2.14) holds.

Proof. (2.11) and (2.13) imply that $\delta_t < 1$ and hence (2.14) follows. The final statement follows from the observation that

$$\Delta_k T_t \phi_0(\eta) = T_t \phi_0(\eta^k) - T_t \phi_0(\eta) \le T_t \phi_0(\overline{+1}) - T_t \phi_0(\overline{-1}),$$

and that $|\{k : |k| \le L_t\}| = O(t^d)$.

(2.15) Remark. The proof of the last statement of Theorem (2.12) shows that (2.14) implies (2.13).

Thus (2.13) is necessary and sufficient for rapid convergence of $T_t f$ to a limiting value for all $f \in D$. From this it is easily seen that within any set of attractive spin flip processes having a common finite bound on the range of their interactions, the set for which (2.14) holds is open in the topology of point-wise convergence of their flip rates.

3. Sketch of the proof of Theorem 3

The ideas behind the proof of Theorem 3 are not new, so we just outline the proof. The details can be found in [2], and [3].

By Theorem (2.12) it suffices to show that for some $A < \infty$, and some $\epsilon > 0$,

$$(3.1) \qquad T_t \phi_0(\mp I) - T_t \phi_0(\overline{-I}) \leq A e^{-\epsilon t^{1/d}}.$$

(In fact, a bound of the form $o(1/t^d)$ would do.)

For each cube Ω centered at the origin and each $\omega \in E$ let $T_t^{\Omega,\omega}$ be the semi-group generated by $L^{\Omega,\omega} = \sum_{k \in \Omega} c_k^{\Omega,\omega}(\cdot)\Delta_k$, where $c_k^{\Omega,\omega}(\eta) = c_k(\eta\omega)$ and $\eta\omega$ is the configuration which agrees with η inside Ω and agrees with ω outside Ω. Then

$$(3.2) \qquad \begin{aligned} T_t \phi_0(\mp I) - T_t \phi_0(\overline{-I}) &\leq T_t^{\Omega,+I}\phi_0(\mp I) - T_t^{\Omega,-I}\phi_0(\overline{-I}) \\ &= (T_t^{\Omega,+I}\phi_0(\mp I) - \int \phi_0(\sigma)g_\Omega(d\sigma|\mp I)) + (\int \phi_0(\sigma)g_\Omega(d\sigma|\mp I) - \int \phi_0(\sigma)g_\Omega(d\sigma|\overline{-I})) \\ &\quad + (\int \phi_0(\sigma)g_\Omega(d\sigma \overline{-I}) - T_t^{\Omega,-I}\phi_0(\overline{-I})). \end{aligned}$$

By Theorem 1

$$(3.3) \qquad \int \phi_0(\sigma)g_\Omega(d\sigma|\mp I) - \int \phi_0(\sigma)g_\Omega(d\sigma|\overline{-I}) \leq C|\partial\Omega|e^{-\mu \operatorname{dist}(0,\Omega^c)} \quad \text{for some } C < \infty \text{ and some } \mu > 0.$$

To bound the first and third terms we first note that there is an $\epsilon > 0$ such that for all Ω

$$(3.4) \qquad \|T_t^{\Omega,\pm I}\phi_0(\pm I) - \int \phi_0(\sigma)g_\Omega(d\sigma|\pm I)\|_{L^2_{g_\Omega}(\cdot|\pm I))} \leq \|\phi_0 - \int \phi_0(\sigma)g_\Omega(d\sigma|\pm I)\|_{L^2_{g_\Omega}(\cdot|\pm I))}e^{-\epsilon t}.$$

The proof of (3.4) is exactly the same as the proof of Theorem 2 except that we now use the auxiliary process with generator given as follows. If $j \in \Omega$ let $\Lambda_j = (\Lambda_0 + j) \cap \Omega$. Then for $\omega \in E$ define the generator of the auxiliary process by the formula

$$A^{\Omega,\omega}f(\eta) = \sum_{j \in \Omega} \int_{E(\Lambda_j)} (f(\sigma\eta) - f(\eta))g_{\Lambda_j}(d\sigma|\eta\omega).$$

The reason that we need the hypotheses in Theorem 3 instead of those in Theorem 2 is to handle the boundary terms that arise in the truncation caused by intersecting $\Lambda_0 + j$ with Ω.

Once (3.4) is proved we proceed as follows. Write $\overline{\phi_0} = \phi_0 - \int \phi_0(\sigma) g_\Omega(d\sigma|\omega)$. Then letting $I_\eta(\sigma)$ be one if $\eta = \sigma$ and zero otherwise, we have

$$(3.5) \qquad |T_t^{\Omega,\omega}\phi_0(\eta) - \int \phi_0(\sigma)g_\Omega(d\sigma|\omega)| = |T_t^{\Omega,\omega}\overline{\phi_0}(\eta)|$$

$$= |\int T_t^{\Omega,\omega}\overline{\phi_0}(\sigma)I_\eta(\sigma)g_\Omega(d\sigma|\omega)|/g_\Omega(\eta|\omega) \leq \|T_t^{\Omega,\omega}\overline{\phi_0}\|_{L^2(g_\Omega(\cdot|\omega))}\|I_\eta(\cdot)\|_{L^2(g_\Omega(\cdot|\omega))}/g_\Omega(\eta|\omega)$$

$$\leq \|\overline{\phi_0}\|_{L^2(g_\Omega(\cdot|\omega))}e^{-\epsilon t}/(g_\Omega(\eta|\omega))^{1/2} \leq e^{-\epsilon t}e^{\beta|\Omega|} \qquad \text{for some } \beta < \infty.$$

Setting ω, and η equal to $\overline{+I}$ and then equal to $\overline{-I}$ in (3.5) and combining the resulting inequalities with (3.3) we have, for all cubes, Ω, centered at the origin, and all $t \geq 0$

$$(3.6) \qquad T_t\phi_0(\overline{+I}) - T_t\phi_0(\overline{-I}) \leq C|\partial\Omega|e^{-\mu \text{dist}(0,\Omega^c)} + 2e^{-\epsilon t}e^{\beta|\Omega|}.$$

Letting $|\Omega| \sim \epsilon t/2\beta$ as $t \to \infty$ we get

$$T_t\phi_0(\overline{+I}) - T_t\phi_0(\overline{-I}) = O(e^{-\mu_0 t^{1/d}}) \qquad \text{for some } \mu_0 > 0.$$

This is just (3.1), which is what we wanted to show.

Acknowledgements

M.A. wishes to thank R. L. Dobrushin and E. H. Lieb for most pleasant and stimulating discussions of the "Dobrushin uniqueness theory". We also thank H. Weinberger for hospitality at the Institute for Mathematics and its Applications which enabled our collaboration.

References

[1] Dobrushin, R. L., and S. B. Shlosman, Constructive criterion for the uniqueness of Gibbs Field, Statistical Physics and Dynamical Systems, pp. 347-370, edited by J. Fritz, A. Jaffe, and D. Szasz, Birkhauser 1985 (Boston-Basel-Stuttgart)

[2] Holley R., Rapid convergence to equilibrium in one-dimensional stochastic Ising models, Ann. Prob. 13 (1985), 72-89.

[3] Holley R., Possible rates of convergence in finite range, attractive spin systems, Contemporary Math. 41 (1985), 215-234.

[4]   Holley R., and Stroock D. W., Applications of the Stochastic Ising model to the Gibbs states, Comm. Math. Phys. 48 (1976), 246-265.

[5]   Liggett, T.M., Interacting Particle Systems, Grundlehren der mathematischen Wissenschaften, 276 (1985) Berlin-Heidelberg-New York: Springer.

[6]   Shlosman S. B., Meditation on Čzech models: uniqueness, half-space non-uniqueness and analyticity properties, to appear.

UNIQUENESS OF THE INFINITE CLUSTER AND RELATED RESULTS IN PERCOLATION

M. Aizenman[1]

Departments of Mathematics and Physics
Rutgers University
New Brunswick, NJ 08903

H. Kesten[2]

Department of Mathematics
Cornell University
Ithaca, NY 14853
and
Institute for Mathematics and its Applications
University of Minnesota
Minneapolis, MN 55455

C.M. Newman[3]

Department of Mathematics
University of Arizona
Tucson, AZ 85721

Abstract

We present the following results for independent percolation:

a) continuous differentiability (in the natural parameters) of the free
    energy function (mean number of clusters per site),

b) uniqueness of the infinite cluster,

c) continuity of the connectivity functions.

As a corollary of a) and previous results of van den Berg and Keane, there follows

d) continuity of the percolation density, except possibly at the critical
    point.

---

[1] Research supported in part by NSF Grant PHY-8301493A0T2.
[2] Research supported in part by the NSF through a grant to Cornell University.
[3] Research supported in part by NSF Grant DMS-8514834.

These results are valid for both site and bond percolation on d-dimensional lattices with arbitrary d and for translation-invariant long range bond models (satisfying a natural irreducibility condition).

Main Results

In this paper we present some general results dealing with several related issues in percolation theory: a) differentiability of the free energy, b) uniqueness of the infinite cluster, c) continuity of the connectivity functions and d) continuity of the percolation density. Although our main results for bond percolation are precisely given in Theorems 1-3, the proofs are either only sketched here or are left out entirely. All the details may be found in [AKN]. Also in [AKN] are some results related to a) which are not presented here; these give sufficient conditions for twice differentiability of the free energy and some related inequalities involving the critical exponents commonly denoted α, γ and δ.

Although, as explained below, our results are valid for a very large class of percolation models, we begin by presenting them for the simplest case, nearest neighbor bond percolation on Z^d. This is the one parameter family of random graphs obtained by starting with the infinite graph with vertex set Z^d and an edge between every nearest neighbor pair, and then independently deleting edges with probability $1-p$.

Let N denote the number of (distinct) connected components (or clusters) of this random graph which are infinite. By ergodicity, N takes on some value in $\{0,1,2,\ldots,\infty\}$ with probability one. The critical parameter p_c is defined as $\sup \{p \epsilon[0,1]: \text{Prob}_p \ (N=0)=1\}$. Although $p_c = 1$ for d=1, $p_c \ \epsilon(0,1)$ for $d \geqslant 2$ [BH,H]. Our main result concerns the impossibility of any value other than 0 or 1 for N; as a corollary we obtain results about $\tau(x,y) := \text{Prob} \ (x$ and y belong to the same cluster) or more generally $\tau(x_1, x_2,\ldots,x_n) := \text{Prob} \ (x_1 ,\ldots, x_n$ all belong to the same cluster), and about $P_\infty: = \text{Prob} \ (0$ belongs to an infinite cluster).

Theorem 1. For standard nearest neighbor bond percolation on Z^d:

(i)   for any p, either there are no infinite clusters (with prob. 1) or there is exactly one infinite cluster (with prob. 1),

(ii) for any $x_1,...,x_n$ in Z^d, $\tau(x_1,..., x_n)$ is a continuous function of p in [0,1],

(iii) $P_\infty(p)$ is continuous everywhere on [0,1], except possibly at p_c.

Remark. Uniqueness of the infinite cluster was already known for d = 2 and $p > p_c$ [Ha;F] and for d > 2 and $p > p_c^\infty$, where p_c^∞ is a certain limit of "slab" critical values conjectured, but not yet proven, to coincide with p_c [ACCFR; K2, p.382]. See [AKR] for dependent percolation when d = 2. It was also known that rather generally, N = 0 or 1 or ∞ [NS1]. For d=2, where $p_c=\frac{1}{2}$ [K1], P_∞ has been proven continuous on [0,1] [R; K2, pp. 122,123].

Part iii) of Theorem 1 is an immediate corollary of part i) and the result of [BK] that uniqueness strictly above p_c implies continuity of P_∞. Part i) is an almost immediate consequence of the following theorem concerning the free energy

$$f(p): = \sum_{n=1}^{\infty} n^{-1} P_n \qquad , \qquad (1)$$

where P_n=Prob(the cluster of the origin contains exactly n sites). How part ii) follows from the proof of part i) will be discussed below.

Theorem 2. For standard nearest neighbor bond percolation on Z^d:

(i)   f is a convex function on [0,1] which is nondifferentiable at p in (0,1) if and only if $Prob_p$ (N > 1) = 1.

(ii) f is differentiable on (0,1).

Before sketching the proofs of Theorem 2 and part ii) of Theorem 1, we discuss the extension of these two theorems to more general models. A translation invariant independent bond percolation model on Z^d is the random graph obtained by starting with vertex set Z^d and edges between every pair of sites and then deleting the $\{x,y\}$ edge with probability $1-p_{y-x}$, independently for every $\{x,y\}$. In analogy with Ising models we express all p's as a function of a single parameter $\beta\varepsilon(0,\infty]$ by setting

$$p_{y-x} = 1 - \exp(-\beta J_{y-x}) \tag{2}$$

with $0 < J_{y-x} < \infty$. The quantities N, $\tau(x,y)$, P_∞, P_n and f are all defined as above but are considered as functions of β. The critical value β_c in $[0,\infty]$ is defined as the obvious generalization of p_c. We call such a model irreducible if $\beta > 0$ $\tau(x,y) > 0$ for all $x,y\varepsilon Z^d$ (i.e. if the additive span of $\{z \varepsilon Z^d : J_z > 0\}$ is all of Z^d).

Theorem 3. For an irreducible translation invariant independent bond percolation model on Z^d:

0) $f(\beta)$ is convex and differentiable on $(0,\infty)$,

i) for any β, either $N=0$ (with prob.1) or $N=1$ (with prob.1),

ii) for any $x_1,\ldots,x_n$ in Z^d, $\tau(x_1,\ldots x_n)$ is a continuous function of β on $(0,\infty]$.

iii) $P_\infty(\beta)$ is continuous except possibly at β_c.

Remarks : a) When $\sum_z J_z = \infty$ and $\beta > 0$, $P_\infty=1$ and part i) of this theorem provides an alternate proof of the previously known result [GKM] that the random graph is connected with probability 1.

b) If d=1 and $z^2 J_z \to 1$ as $z \to \infty$, then $\beta_c\varepsilon (0,\infty)$ [NS2] and $P_\infty(\beta_c) > 0$ [AN]. In these models, the restriction in part iii) of the theorem to points other than the critical point is necessary. Part i) shows that the infinite cluster is unique at the critical point in spite of the discontinuity of P_∞ there.

c) Our results also extend to site percolation, mixed site-bond percolation, and other lattices than (nearest neighbor) Z^d (see [AKN]). In site percolation, f must be modified in order to be convex. However, we have no results for oriented percolation.

d) It is also possible to view the functions $\tau(x_1,\ldots,x_n)$, P_∞ etc. as functions of the infinitely many parameters $\{p_z\}$. It then becomes natural to introduce the L^1 topology. We say that a function $F(\{p_z\})$ is continuous at $\{\bar{p}_z\}$ in this topology if $F(\{p_z\}) \to F(\{\bar{p}_z\})$ whenever $\Sigma |p_z - \bar{p}_z| \to 0$. (Note that $\Sigma \bar{p}_z = \infty$ is allowed in this definition.) We prove that $\tau(x_1,\ldots,x_n)$ as a function of all p_z is continuous in the L^1 topology at each point $\{\bar{p}_z\}$ for which the corresponding model is irreducible. However, P_∞ is in general not continuous in the L^1-topology (at least for d=1).

Sketch of Proof of Part i) of Theorem 2: Let Λ_k denote the cube of side length $2k+1$ centered at 0 in Z^d. We consider two finite random graphs. The first is obtained from the original model by deleting all vertices outside Λ_k; here we denote by M_k^0 the number of clusters and by $\tau_k^0(x,y)$ the two-point connectivity function. The second is obtained by identifying all vertices outside Λ_k as a single point denoted ∞, in this case M_k^+ denotes the number of clusters disconnected from ∞ while $\tau_k^+(x,y)$ denotes the connectivity function. We write $*$ to denote 0 or $+$. Note that for fixed x,y in Z^d,

$$\lim_{k \to \infty} \tau_k^*(x,y) \;=\; \tau^*(x,y) \colon = \begin{cases} \tau(x,y)\,, \quad \text{for} \;\; *=0\,, \\[2em] \tau(x,y) + \text{Prob}(x \;\; \text{and} \;\; y \;\; \text{belong to} \\ \text{distinct infinite clusters), for} \;\; *=+. \quad (3) \end{cases}$$

We define

$$f_k^* \;=\; (2k+1)^{-d}\; E(M_k^*)\,. \qquad (4)$$

By standard arguments, one can show that $f_k^* \to f$ for $* = 0$ or $+$. Since the addition of a previously deleted edge $\{x,y\}$ either leaves M_k^* unchanged if x and y were previously connected, or else reduces M_k^* by exactly one, it follows that

$$\frac{\partial}{\partial p} f_k^* \;=\; -\,(2k+1)^{-d}\; \Sigma' P(B_{xy,k}^*)\,, \qquad (5)$$

where $\sum'$ is the sum over all nearest neighbor pairs $\{x,y\}$ and $B^*_{xy,k}$ is the event that there is no $*$-path from x to y that avoids the edge $\{x,y\}$.

Now f^*_k is convex since each summand in (5) is clearly decreasing in p. f^*_k is also smooth (actually a polynomial). Furthermore, for large k $\text{Prob}(B^*_{xy,k}) = (1-p)^{-1}(1-\tau_k(\overset{*}{x},y))$ is decreasing in k for $*=0$ and increasing for $*=+$. It follows by standard arguments that f is convex, $\tau^0(x,y)$ is left-continuous, $\tau^+(x,y)$ is right-continuous and

$$\frac{\partial f}{\partial p} (p\pm 0) = - (2(1-p))^{-1} \sum_z{}'' (1-\tau^*(0,z)) \tag{6}$$

where $\sum''$ is the sum over the neighbors of the origin, and $*$ equals $+$ for the right derivative and 0 for the left derivative. The proof is completed by use of (3) and the fact that $\text{Prob}(N > 1)=1$ is equivalent to $\text{Prob}(0$ and z belong to distinct infinite clusters$) > 0$ (e.g. by arguments of [NS1]).

Sketch of Proof of Part ii) of Theorem 2: The standard lattice animal expansion for P_n yields the identity

$$\frac{\partial}{\partial p} P_n = \sum_{m,\ell} [\frac{m}{p} - \frac{\ell}{1-p}] a_{nm\ell} p^m (1-p)^\ell , \tag{7}$$

where $a_{nm\ell}$ is the number of bond lattice animals with n connected vertices, m occupied edges and ℓ vacant edges. To show differentiability of f, one must obtain a good bound on the RHS of (7). This is done by decomposing the sum into the two parts for which

$$|p^{-1}m - (1-p)^{-1} \ell| > K (n \ln(n))^{1/2} \quad \text{and}$$

$$|p^{-1}m - (1-p)^{-1}\ell| < K (n \ln(n))^{1/2}, \text{ respectively.}$$

The former is estimable by large deviation methods (see [K2, Lemma 5.1]). The resulting bound, uniform for p in compact subsets of $(0,1)$, is

$$| \frac{\partial}{\partial p} P_n | = 0 (n^{-K'} + (n \ln (n))^{1/2} P_n). \tag{8}$$

K' can be made as large as desired by choosing K large. Combining (8) with (1) and standard arguments completes the proof.

Sketch of Proof of Part ii) of Theorem 1: To see that part (i) of Theorem 1 implies part ii) for the two-point function, simply note the definition (3) of $\tau^*(x,y)$ and the fact, discussed above, that $\tau^0(x,y)$ is left-continuous and $\tau^+(x,y)$ is right-continuous, and finally that $\tau^0(x,y) = \tau^+(x,y)$ by (6) and Theorem 2 ii).

Continuity of $\tau(x_1,\dots,x_n)$ can be obtained by defining obvious analogues $\tau^0(x_1,\dots,x_n)$ and $\tau^+(x_1,\dots,x_n)$ of $\tau^0(x,y)$ and $\tau^+(x,y)$, respectively, and by observing that

$$0 < \tau^+(x_1,\dots,x_n) - \tau^0(x_1,\dots,x_n) < \sum_{i \neq j} \text{Prob}(x_i \text{ and } x_j \text{ belong to distinct}$$

$$\text{infinite clusters}) = \sum_{i \neq j} [\tau^+ (x_i,x_j) - \tau^0(x_i,x_j)] = 0.$$

Acknowledgements

The development of some of the results reported here was aided by the simultaneous presence of the authors at the I.M.A., University of Minnesota, during February, 1986. We thank S. Orey, G. Sell, H. Weinberger and the IMA staff for their hospitality.

References

[ACCFR]   Aizenman, M., Chayes, J.T., Chayes, L., Fröhlich, J. and Russo, L., On a sharp transition from area law to perimeter law in a system of random surfaces, Comm. Math. Phys. 92 (1983), 19-69.

[AKN]   Aizenman, M., Kesten, H. and Newman, C.M., Uniqueness of the infinite cluster and continuity of connectivity functions for short and long range percolation, submitted to Comm. Math. Phys.

[AN]   Aizenman, M. and Newman, C.M., Discontinuity of the percolation density in one-dimensional $1/|x-y|^2$ percolation models, to appear in Comm. Math Phys.

[BH]   Broadbent, S.R. and Hammersley, J.M., Percolation process, Proc. Cambr. Phil. Soc 53 (1957), 629-641 and 642-645.

[BK]   van den Berg, J., and Keane, M., On the continuity of the percolation probability function, Contemporary Mathematics 26 (1984), 61-65.

[F]   Fisher, M.E., Critical probabilities for cluster size and percolation problems, J. Math Phys. 2 (1961), 620-627.

[GKM]   Grimmett, G. R., Keane, M. and Marstrand, J.M., On the connectedness of
        a random graph, Math. Proc. Cambr. Phil. Soc. 96 (1984), 151-166.

[GKR]   Gandolfi, A., Keane, M. and Russo, L., On the uniqueness of the infinite
        occupied cluster in dependent two-dimensional percolation, Delft
        University of Technology,/Rome University preprint, 1986.

[H]   Hammersley. J.M., Percolation processes, lower bounds for the critical
        probability, Ann. Math Statist. 28 (1957), 790-795.

[Ha]   Harris, T.E., A lower bound for the critical probability in a certain
        percolation process, Proc. Cambr. Phil. Soc. 56 (1960), 13-20.

[K1]   Kesten, H., The critical probability of bond percolation on the square
        lattice equals $\frac{1}{2}$, Comm. Math. Phys. 74 (1980), 41-59.

[K2]   Kesten, H., Percolation Theory for Mathematicians, Birkhäuser, 1982.

[NS1]   Newman, C.M. and Schulman, L.S., Infinite clusters in percolation models,
        J. Stat. Phys. 26 (1981), 613-628.

[NS2]   Newman, C.M. and Schulman, L.S., One-dimensional $1/|j-i|^{s}$ percolation
        models: the existence of a transition for s < 2, Comm. Math. Phys. 104
        (1986), 547-571.

[R]   Russo, L., On the critical percolation probabilities, Z. Wahrsch. verw.
        Geb. 56 (1981), 229-237.

Survival of Cyclical Particle Systems

Maury Bramson

School of Mathematics
Vincent Hall
University of Minnesota
Minneapolis, Minnesota 55455

and

David Griffeath

Department of Mathematics
Van Vleck Hall
University of Wisconsin
Madison, Wisconsin 53706

We consider here a continuous time interacting particle system on Z with possible states $0,1,...,N-1$ at each site. We denote by $\xi_t(i)$ the state taken at site i and time t. Our models, which we call <u>cyclical particle systems</u>, are specified by the following transition rates:

$$\xi_t(i) \to \xi_t(i)+1 \quad (\text{mod } N) \quad \text{at rate } \lambda,$$

where

$$\lambda = |\ \{ j=\pm 1: \xi_t(i+j) = \xi_t(i) + 1 \ (\text{mod } N)\}\ |.$$

Note in particular that i cannot change state until at least one of its immediate neighbors has state $\xi_t(i) + 1$. For convenience, one may equip ξ with a space-time percolation substructure. Arrows from i to $i+1$ and $i-1$ are each put down in a Poisson manner with rate 1. A state change is induced at (t,i) by an arrow at time t which enters i from $i+1$ or $i-1$, where the state is $\xi_t(i) + 1$. In diagram 1 below, numbers at the points of arrows indicate state changes in the realization ($N=5$).

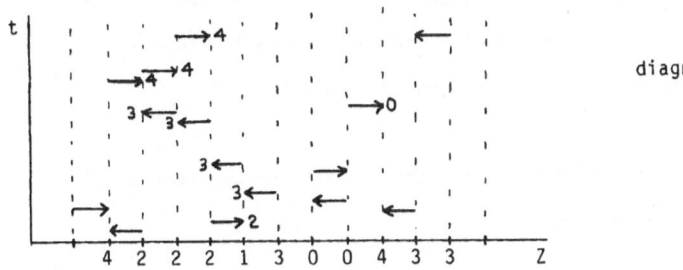

diagram 1

Here, we will always assume that ξ_0 has product measure with

$$P(\xi_0(i) = k) = 1/N \quad \text{for} \quad k=0,1,\ldots,N-1.$$

As a special case of a cyclical particle system, one has N=2, which is the much-studied voter model. (See Clifford and Sudbury [1], Holley and Liggett [2] and Liggett [3].) One can in this case think of 0 and 1 as two different opinions or traits which randomly displace one another as time evolves. For N=3, one can somewhat facetiously consider the 1-state to denote the presence of a rabbit, the 2-state that of a coyote, and the 0-state to be empty. Of course, for N ≥ 4 one can easily conjure up similar models. The point in considering cyclical particle systems is that they represent a first step in the study of the structure for interacting particle systems with more than two states, about which little is known. Cyclical systems are in some sense an extreme case of interacting systems for which all states communicate; the opposite extreme occurring when each state is directly accessible from all other states.

The question we shall consider here is : for which N does the cyclical particle system become <u>stuck</u>? That is,

$$\lim_{t \to \infty} \xi_t(i) \quad \text{exists for all i .}$$

If the system never becomes stuck, we say it <u>survives.</u> Note that being stuck at one site is equivalent to being stuck at all sites. This question is certainly basic for understanding the large time behavior of cyclical systems. In our case, it has by itself presented us with more than enough food for thought. We can show the following.

<u>Theorem 1.</u> Let ξ_t denote the cyclical particle system with N states and product initial measure with $P(\xi_0(i)=k)= 1/N$ for $k=0,1,\ldots,N-1$. Then ξ_t survives for N ≤ 4 and becomes stuck for N ≥ 5 with the probability 1.

As discussed at the end of the talk, the question of survival on Z^d is completely open for d ≥ 2.

Most of the remainder of the talk will be spent summarizing the arguments involved in Theorem 1. The demonstration of survival for N < 4 is simple; we proceed with it first.

Survival for N < 4. Survival for N = 2,3,4 follows in each case from the ergodicity of the system. We consider N=4; the other cases are simpler. First note that if the system were to become stuck, then at t=∞, each site would either take state 0 or 2, or state 1 or 3. Let A_E denote the set of realizations which become stuck and for which $\xi_\infty(i)$ equals 0 or 2 for each i, and let A_0 denote the corresponding realizations with $\xi_\infty(i)$ equal to 1 or 3. By symmetry,

$$(1) \qquad P(A_E) = P(A_0).$$

On the other hand, owing to the ergodicity of the system,

$$(2) \qquad P(A_E) = 0 \text{ or } 1.$$

To see this, employ the percolation substructure described above and set

$$\mathcal{E}_i = \sigma\text{-algebra generated at } i \in \mathbb{Z} \text{ by arrows entering } i$$
$$\text{and by the initial state at } i.$$

Also set $\mathcal{F}_\infty = \sigma(\mathcal{E}_i, -\infty < i < \infty)$ and let $\mathcal{D}$ denote the corresponding invariant σ-algebra. By the independence of the $\mathcal{E}_i$, $\mathcal{D}$ is trivial. Since $A_E \in \mathcal{D}$, (2) follows. From (1) and (2) one therefore concludes that

$$P(A_E) = P(A_0) = 0,$$

and therefore the system survives for N=4.

It is intuitively fairly obvious that for large enough N, the corresponding cyclical particle system must become stuck. For $N=10^6$, for instance, the gaps between states of neighboring sites are typically of the same order of magnitude; in order for one of such neighbors to affect the other, it must undergo many state changes. These state changes must come from somewhere. Other pairs of sites are too busy attempting to bridge their own gaps however to provide any assistance. For N > 8, we provide a fairly detailed argument which is based on these ideas.

For N=6,7 one must work harder; a brief sketch of the required modification is provided. The proof for N=5 employs the same setup, but is much messier and is therefore omitted.

The case $N \geq 8$. We first set

$$c_t(i) = \min\{\pm(\xi_t(i) - \xi_t(i-1)) \bmod N\}$$

and

$$\phi(c) = 0 \qquad \text{if} \quad c=0$$
$$= -(c-2) \quad \text{if} \quad 1 < N/2.$$

The boundary between i-1 and i will be referred to as a live boundary if $c_t(i)=1$ and as a blockade if $c_t(i) > 2$. The purpose of ϕ is to measure the magnitude of a blockade. Important points to keep in mind are (1) live boundaries cannot be created but only moved and (2) at least $\phi(c_t(i))$ live boundaries must be expended to overcome the blockade between i-1 and i if $c_t(i) > 2$. (Note that for c=2, no loss in live boundaries need occur, as illustrated by the transitions $44\widehat{3}55 \rightarrow 44\widehat{4}55 \rightarrow 44\widehat{5}55$, etc.)

The basic argument consists of showing that (A) a state arising from a site far from the site 0 has small chance of ever occupying 0 and (B) if 0 is occupied only by states from a finite number of sites, then 0 eventually becomes stuck. (A) is the main part of the argument, for which we introduce

$$\psi(-n) = \sum_{i=0}^{n-1} \phi(c_0(-i)),$$

$$\psi(n) = \sum_{i=1}^{n} \phi(c_0(i)),$$

for $n \epsilon Z^+$. Roughly, one needs $\psi(-n) > 0$ for a state initially at -n to ever occupy 0 (that is, for offspring from -n to eventually make their way over to 0). Actually, this is not quite true; the statement is however correct for the corresponding process $\tilde{\xi}_t$, where ξ_t is restricted to the nonpositive halfline.

To see this, let

$$T_{-i}^{-n} = \text{first time the state arising at -n occupies -i}$$

and

$$A^{-n} = \{T_0^{-n} < \infty\},$$

$0 < i < n$. It is not difficult to see that by T_{-i}^{-n}, at least $-\phi^-(c_0(-i))$ live boundaries originating in $(-n,0]$ have been expended at $-i$. Since $T_{-i}^{-n} < T_{-j}^{-n}$ for $i > j$, it follows that by T_0^{-n}, at least

(3) $\sum\limits_{i=0}^{n-1} \phi^-(c_0(-i))$ live boundaries originating in $(-n,0]$ have been expended in $(-n,0]$.

Of course,

(4) $\qquad\qquad \sum\limits_{i=0}^{n-1} \phi^+(c_0(-i)) = $ # of live boundaries originating in $(-n,0]$.

One concludes from (3) and (4) that

(5) $\qquad\qquad \psi(-n) = \sum\limits_{i=0}^{n-1} \phi^+(c_0(-i)) - \sum\limits_{i=0}^{n-1} \phi^-(c_0(-i)) > 0$

under A^{-n}. (One can in fact check that strict inequality holds.)

Now, since $\xi_0(-i)$, $i \in Z^+$, are IID with uniform distribution, $\phi(c_0(-i))$ are also IID. On the other hand, one can check that for $N > 8$,

(6) $\qquad\qquad E[\phi(c_0(-i))] < 0.$

(For N=8, $E[\phi(c_0(-i))] = -1/4$.) It is therefore standard that

(7) $\qquad\qquad \psi(-n) = \sum\limits_{i=0}^{n-1} \phi(c_0(-i)) \to -\infty$

w.p.1 as $n \to \infty$. Comparing (5) and (7), one sees that

(8) $\qquad\qquad P(\bigcup\limits_{n=k}^{} A^{-n}) \to 0 \text{ as } k \to \infty.$

(8) is the desired result for (A), but refers to the process $\tilde{\xi}_t$ defined on the negative half-line rather than to ξ_t. One can nonetheless check that for ξ_t,

(9) $\qquad\qquad A^{-n} \subset \{ \psi(-n) + R > 0 \},$

where

$$R = \max_{n} \sum_{i=1}^{n} \phi(c_0(i)).$$

Owing to the obvious analog of (7) over the positive integers, $R < \infty$ w.p.1. Applying (7) a second time, (8) therefore holds for ξ_t as well as for $\tilde{\xi}_t$. By symmetry,

$$(10) \qquad\qquad P(\bigcap_{n=k}^{\infty} A^n) \to 0 \text{ as } k \to \infty$$

also holds.

The limits (8) and (10) show that the probability of 0 ever being occupied by a state originating further than n away goes to 0 as $n \to \infty$. On the other hand, states originating from the sites in $[-n,n]$ cannot without help from the outside indefinitely induce state changes at 0. To see this, let $d(j) = \pm 1$ be the site of the immediate parent responsible for the jth change of state at 0, and let $D(j)$ be the site originally occupied by this state. $S(j)$ will denote the time at which this j^{th} change of state occurs at 0, and $\alpha(k)$ the value at which $d(\alpha(k))$ changes sign for the kth time. The two following lemmas are then easy to show.

Lemma 1. Suppose that the site 0 is never occupied by states originating outside $[-n,n]$. If $S(j) < \infty$ for all j, then $d(j)$ changes sign infinitely often.

Lemma 2. The state arising from $D(\alpha(k)-2)$ no longer exists (anywhere) by time $S(\alpha(k))$.

Together, Lemma 1 and Lemma 2 show that if 0 is never occupied by states originating outside $(-n,n]$ and if $S(j) < \infty$ for all j, then by time $S(\alpha(k))$ the states originating from at least k sites in $(-n,n]$ no longer exist. For $k > 2n + 1$ this clearly gives us a contradiction. Together with (8) and (10), this shows that $S(j) = \infty$ for large enough j. Thus, as $t \to \infty$, the site 0 (and hence all sites) becomes stuck for $N \geq 8$.

The cases N=6,7. One may use the same basic framework here as before, but one must work harder since (6) no longer holds for N=6 or N=7. The basic point is that $\phi(c)$ as constructed before is overly conservative im measuring the ability of a blockade to withstand live boundaries. In particular, live boundaries sometimes build up blockades rather than break them down, as evidenced by the transition

(11)   ...356... → ...366.....   .

To incorporate some of these effects, define $\phi'(c,j)$ so that

$$\phi'(c,j) = \phi(c) \quad \text{if} \quad c \neq 2 \, ,$$
$$\phi'(2,1) = 0, \quad \phi'(2,-1) = -2 \, ,$$

where $0 < c < N/2$ and $j = -1,1$. Also, set

$$S_i = \inf \{s : c_s(i) \neq c_0(i)\},$$
$$e(i) = -1 \quad \text{if} \quad c_{S_i}(i) = 3$$
$$= 1 \quad \text{otherwise.}$$

We may now count live boundaries and blockades as before, but use ϕ' instead of ϕ. $\phi'(c,j) = \phi(c)$ except at $c=2$ and $j = -1$; this difference takes into account the possibility of a live boundary being expended to change $c_0(i) = 2$ into the blockade $c_{S_i}(i) = 3$. By symmetry,

$$P(c_0(i) = 2, e(i) = -1) = P(c_0(i) = 2, e(i) = 1).$$

It is therefore an easy matter to check that

(12) $E[\phi'(c_0(i), e(i))] < 0$

for $N = 6,7$. (For N=6, $E[\phi'(c_0(i), e(i))] = -1/6$.) From (12) one can conclude as from (6) that $\psi(-n) \to -\infty$ as $n \to \infty$. In this case however $\phi'(c_0(i),e(i))$, $i \in Z$, are not independent. One therefore needs to group sites into blocks of size M, M large, which are nearly independent (except near the boundaries). One can then proceed as for $N > 8$ to conclude that

$$\psi(-n) + R \to -\infty \quad \text{as } n \to \infty.$$

The remainder of the proof is the same as for $N > 8$.

The proof that the cyclical particle system becomes stuck for N=5 proceeds along the same basic lines. In this case, however, one needs to construct a considerably more sensitive variant ϕ'' of ϕ' to incorporate the various cancellative behavior of live boundaries relative to one another, from which one obtains $E[\phi''] < 0$. For instance, fortuitous interactions may case transitions $c_t(i) \to c_t(i)+1$ or oscillations of $(\xi_t(i) - \xi_t(i-1))$ mod 5 from 2 to 3. To employ these effects one needs to do considerable bookkeeping, which makes the proof fairly tedious.

Before finishing, we would like to say a few words about the behavior of cyclical particle systems in dimensions $d > 2$, and our frustrations in this regard. Of course there is nothing essentially one dimensional in the concept of cyclical systems. One can just as well specify that for $i \in Z^d$,

$$\xi_t(i) \to \xi_t(i)+1 \text{ (mod N) at rate } \lambda,$$

where λ is the number of neighbors in state $\xi_t(i)+1$ (mod N), and thus extend the setup to any dimension. However, we are embarrassed to admit that we can say nothing here about whether or not the system survives for $N > 5$. For instance, if $d = 2$ and $N = 10^{12}$, does the system become stuck? Does the critical value of N increase to ∞ as $d \to \infty$? In each case, basic intuition gives an obvious yes. Simulation on finite tori with product initial measure indicates, in particular, that for $d = 2$, $N \approx 13$. Our basic analytic techniques break down here, however. Most seriously, our earlier observation that live boundaries cannot be created but only moved no longer holds; branching is quite possible. Our bookkeeping involving ψ is therefore no longer valid.

The computer is also no longer so cooperative if one plugs in different initial data. For instance, we viewed the case $N = 20$ on a 120x120 torus, and considered product initial measure outside a specified 60x60 square, where the state 0 was initially assigned. (The choice of parameters was motivated by a computable modification of the process.) To our discomfort, the "square" - the region consisting of large blobs of colors - kept growing until it had eaten its way around the torus. (Our thanks to Robert Fish for writing and running the program.) Since the probability that a square becomes stuck in a fixed amount of time very likely decreases exponentially fast with the size of the square, this suggests that survival may indeed occur under product initial measure with N=20 and $d = 2$. One may just require an unreasonably large torus to show this. Maybe the critical value of N for d=2 occurs at some larger value, say N ≈ 80. One could also hazard the quite unnatural guess that the system survives for all N. Suggestions are welcome.

References

[1] Clifford, P. and Sudbury, A. (1973) A model for spatial conflict. Biometrika 60, 581-588.

[2] Holley, R. and Liggett, T.M. (1975) Ergodic theorems for weakly interacting systems and the voter model. Ann Probab. 3, 643-663.

[3] Liggett, T.M. (1985) Interacting Particle Systems. Springer-Verlag, New York.

EXPANSIONS IN STATISTICAL MECHANICS AS PART OF THE THEORY OF PARTIAL DIFFERENTIAL EQUATIONS

D.C Brydges
Department of Mathematics
Mathematics and Astronomy Bldg
University of Virginia
Charlottesville, Va 22903

Abstract

We study perturbations of gaussian processes using a partial differential equation in (infinitely) many variables which describes what infinitesimal change in the perturbation compensates an infinitesimal change in the covariance. We derive a series representation for the solution by iterating an integral equation form of the flow equation and show that the series is majorised for short times by a corresponding solution of a Hamilton-Jacobi equation when the initial data is bounded and analytic. The resulting series solutions are generalizations of the Mayer expansion in statistical mechanics. This approach gives a remarkable identity for "connected parts" and accurate estimates which include criteria for convergence of iterated Mayer expansions.

Introduction

In the theory of critical phenomena in equilibrium statistical mechanics and the related problem of constructing euclidean quantum field theories the mathematical problem is the study of integrals of the form

$$Z(\varphi) = \int d\mu(\overline{\varphi}) e^{-V(\overline{\varphi} + \varphi)} \equiv (\mu * e^{-V})(\varphi)$$

where $\varphi = (\varphi_x)$ is a gaussian process with joint probability distribution $d\mu$ with mean zero and covariance C_{xy}. To stay away from technicalities I shall assume that φ is indexed by a countable set Λ_∞ but it is possible to do much of the following analysis for random gaussian fields on R^d. The perturbation V depends on a finite subset Λ of Λ_∞ and is an approximately *additive* functional (*local* in physics terminology). This means V has the form

$$V \simeq \sum_{x \in \Lambda} v_x(\varphi_x)$$

where the precise meaning of $\simeq$ will be discussed later.

The object in analysing Z is to determine it's behaviour as Λ increases to Λ_∞ so that the perturbation becomes dependent on infinitely many variables. An easy case of this problem is when V is exactly additive and the covariance matrix vanishes off the diagonal, because in this case Z factors into a product of one dimensional integrals. We see in this example that unless the values of the one dimensional integrals happen to cluster at one Z will be either zero or infinite in this limit but that derivatives with respect to φ variables of $\log Z$ have respectable limits. The expansion techniques of statistical mechanics are designed to discover the behaviour of these quantities when the integral approximately factors.

In this lecture I will talk about some work with Tom Kennedy [1] in which we rederive results usually phrased in terms of expansions by studying a differential equation for $\partial/\partial\varphi_x \log Z$ which expresses how it changes when the covariance C is infinitesmally deformed. The results I discuss are not in important respects new but the proofs are cleaner and we have found an attractive "tree graph identity" which is an improvement on previous formulas of this kind, [2].

The Flow Equations

We define $V(t) \equiv V(t,\varphi)$ by

$$V(t) = -\log \mu_t * e^{-V(0)} \qquad (1)$$

where $V(0) \equiv V(t = 0, \varphi) \equiv V(\varphi)$ and μ_t has covariance $C_{xy}(t)$ where

t parametrizes a deformation, thus we assume that C(t) is
differentiable in t, its deivative Ċ is positive as a form on 1^2 and
as a form on 1^2

$$\lim_{t \to 0} C(t) = 0$$

so that the notation V(0) for V is consistent, since the weak limit
of a gaussian measure as its covariance tends to zero is δ(φ) the
point mass at zero.

By an easy calculation we find that V(t) obeys the partial
differential equation

$$\frac{\partial V}{\partial t} = \frac{1}{2} (V_{\varphi\varphi} - V_\varphi^2) \tag{2}$$

where the φφ subscript denotes a Laplacian in all the φ's, namely

$$V_{\varphi\varphi} \equiv \sum_{x,y} \dot{C}_{xy}(t) \frac{\partial^2 V}{\partial\varphi_x \partial\varphi_y}$$

$$V_\varphi^2 \equiv \sum_{x,y} \dot{C}_{xy}(t) \frac{\partial V}{\partial\varphi_x} \frac{\partial V}{\partial\varphi_y} \ . \tag{3}$$

Estimates on the Solution for Short Time

In outline our argument is that the maximum principle, i.e.,
positivity of the fundamental solution of the heat equation along with
a Cauchy-Kowaleska type of existence proof shows that V and its
derivatives may be estimated in terms of the solution of a similar
equation with the Laplacian omitted and a sign changed, i.e.,

$$\frac{\partial V}{\partial t} = \frac{1}{2} V_\varphi^2 \ . \tag{4}$$

This is a Hamilton-Jacobi equation. It is even possible to majorise
norms of the original V (which   measure how "non local" V is) in
terms of the solution of an equation of this type with only one φ
variable. Thus define:

$$V_{x_1, \ldots, x_M} \equiv \frac{\partial^M V}{\partial\varphi_{x_1} \cdots \partial\varphi_{x_m}}$$

$$V_M \equiv \sup_{\varphi \text{ real}, x} \frac{1}{M!} \sum_{x_2, \ldots, x_M} | V_{x, x_2, \ldots, x_M}(t, \varphi)| \tag{5}$$

$$|\dot{C}(t)| \equiv \sup_{x} \sum_{y} |\dot{C}_{xy}(t)| \ .$$

In (5) the sum is omitted if M = 1.

Theorem 1: <u>Suppose that the power series</u> *in one variable* φ

$$v^{(0)}(\varphi) \equiv \sum_{M=1}^{\infty} V_M(0) \ \varphi^M$$

<u>has a non zero radius of convergence, then</u>

$$\frac{\partial v}{\partial t} = \frac{1}{2} \ |\dot{C}(t)| \ \left[\frac{\partial v}{\partial \varphi}\right]^2 \ , \qquad v(0,\varphi) = v^{(0)}(\varphi)$$

<u>defines for</u> t <u>small, a function</u> v(t,φ) <u>which is analytic near</u> $\varphi = 0$. <u>For all</u> t <u>for which</u> v <u>exists the flow equation (2) has a unique solution which is analytic in the initial data and bounded according to</u>

$$V_M(t) \le v_M(t), \qquad M = 1,2,\dots \ ,$$

<u>where</u> $v_M(t)$ <u>is the</u> Mth <u>coeficient of the power series for</u> v_M:

$$v(t,\varphi) = \sum_{M} v_M(t) \ \varphi^M$$

<u>In particular these bounds hold if</u> t <u>is sufficiently small that</u>

$$(\int_0^t |\dot{C}(t)| \ ds \ ) \sup_{M \ge 0} [M \ v_M(0)]^{2/M} < \frac{1}{4} \ .$$

Burger's equation is the derivative of the flow equation with respect to φ. Burger's equation and a one variable bound like ours has appeared for somewhat similar reasons in work on the Ising model [3]. Before proving this theorem I would like to make some comments.

(A) The phrase "analytic in the initial data" means that if V(0) is replaced by zV(0) then V(t) is analytic in z. In fact the proof will show that V(t) has a power series in z convergent for $|z| \le 1$.

(B) Usually in discussing Burger's equation one uses the formula (1) to analyse the equation. For us the situation is the other way around because the explicit formula for the solution is not very informative about the $\Lambda \to \Lambda_\infty$ limit, whereas the equations can be. In particular notice that this theorem makes no reference to the cardinality of Λ and provides bounds which are uniform in Λ provided the hypotheses on $v^{(0)}$ hold uniformly. It prepares the way for the limit $\Lambda \to \Lambda_\infty$.

(C) The connection with the Mayer expansion in statistical mechanics comes when we specialise the initial conditions to be a local trigonometric polynomial such as

$$V(0,\varphi) = z \sum_{x \in \Lambda} \exp(i \varphi_x) \; .$$

There is the following well known identity, the *sine gordon transformation*, which connects the gaussian integrals we have just been discussing to statistical mechanics.

$$\mu * \exp(- V(0)) = \sum_N \frac{z^N}{N!} \sum_{x_1,\ldots,x_N \in \Lambda} \exp(- \tfrac{1}{2} \sum_{i,j} C_{x_i x_j}) \cdot$$

$$\cdot \exp[i \sum_j \varphi_{x_j}] \; .$$

Proof: Expand the exponential on the right hand side, interchange the sum over N and the convolution and then explicitly evaluate the $d\mu$ integral. End of proof.

The right hand side is the partition function for a Grand Canonical Ensemble of N particles in states $x_1,\ldots,x_N \in \Lambda$ with activities $z \exp[i \varphi]$ and two body interactions C_{xy}. The Mayer expansion is the formal power series for the logarithm:

$$\log \left[\mu * \exp(- V(0)) \right] = \sum_M \frac{z^M}{M!} \sum_{x_1,\ldots,x_M} m(x_1,\ldots,x_M) \cdot$$

$$\cdot \exp[i \sum_j \varphi_{x_j}] \; .$$

Theorem 1 implies that this series is convergent if $C = C(t)$ with t small as in the last line of the theorem.

(D) *The Renormalisation Group* is a special type of deformation of C, namely a change of length scale. Wilson long ago obtained the flow equations in this case [4] and Gallavotti et al. have studied a discrete time version of the flow equations by the same iterative scheme as we are using [5]. Let us be heuristic for a moment and consider the (generalized) gaussian process on R^d with covariance

$$C(t,x,y) \equiv (e^{-2t} - \Delta)^{-1}(x - y)$$
$$\equiv (2\pi)^{-d} \int (e^{-2t} + k^2)^{-1} e^{-ik.(x - y)} dk .\qquad (6)$$

This covariance is scale covariant in the sense that if $\psi(x)$ has covariance $C(t = 0)$ then $\varphi(x) \equiv e^{\beta t} \psi(e^{-t}x)$ has covariance $C(t)$ where $\beta \equiv (d - 2)/2$. If we were to study the flow equation for this problem it would be possible to change $\dot{C}(t)$ to $\dot{C}(0)$ in (2) by changing variables to ψ. We are of course discussing a highly formal object - a partial differential equation in a continuum of variables - but the integral equation version of these equations is better defined. In the spirit of theorem 1 let us consider a one ψ version of the problem. It is

$$\frac{\partial}{\partial t}\bar{V} = \frac{1}{2} \bar{V}_{\psi\psi} - \beta \psi \bar{V}_{\psi} + d \bar{V} - \bar{V}_{\psi} \bar{V}_{\psi}$$
$$\equiv (-H + d) \bar{V} - \bar{V}_{\psi} \bar{V}_{\psi} .$$

where H is the Hermite operator. The $\beta \psi \partial/\partial\psi$ comes from the scaling of φ. The Laplacian is now time independent. The $d \bar{V}$ comes from an exponential scaling of the V by e^{td} which is necessary because the sums over x and y in the first term in (2) become integrals and it scales as a volume whilst the second term scales as a volume squared. In terms of this equation we can explain some of Wilson's view of the renormalization group by noting that if we linearise around the fixed point $\bar{V} = 0$ by neglecting the second term and solving the remaining equation by inserting the Hermite expansion

$$\bar{V}(t,\psi) = \Sigma c_n(t) h_n(\psi) ,$$

where h_n is the nth Hermite polynomial, then the equation becomes

$$c_n(t) = c_n(0) e^{(d - n\beta) t} .$$

Thus the linearised equation has for $d > 2$ (so that $\beta > 0$) a finite dimensional unstable manifold with coordinates $c_0, \ldots, c_r$ where r is the largest integer such that $d - r\beta \geq 0$ and an infinite dimensional stable manifold coordinatized by the remaining c_i's. Is it possible to prove that there is a local diffeomorphism defined near zero mapping the linearised stable and unstable manifolds onto corresponding stable and unstable manifolds for the above flow equation? For finite systems of ODE's this is a well studied type of problem [6] but not so well understood im the partial differential equation context. A result of this sort is Wilson's perspective on the problem of constructing euclidean quantum field theories: we consider a local perturbation of a gaussian process on a lattice $(\varepsilon Z)^d \subset R^d$ with the covariance $C(t)$ given in (6) with Δ for R^d replaced by the Laplacian for the lattice. The effect of deforming the covariance by changing t is to alter the ratio between ε and the length scale in $C(t)$ by e^{-t}. Rescaling by casting the flow in terms of the Ψ variable changes the t in $C(t)$ back again. Thus the net result is that the ε is deformed to zero in the limit $t \to \infty$. The result should be a continuum process. A procedure such as this which involves taking a limit outside the $d\mu$ integrals is necessary because if $d \geq 2$ the covariance given by (6) is singular at $x = y$. This means that almost surely φ is not a function so that local perturbations may be impossible to define inside the $d\mu$ integral. Our discussion of the flow in the neighborhood of the fixed point $\overline{V} = 0$ suggests the following class of $t \to \infty$ limits: for each T in a sequence which tends to infinity impose initial data $\overline{V}(0)$ close to the fixed point *depending on* T so that $c_1(T), \ldots, c_r(T)$ have specified values $c_1, \ldots, c_r$. Clearly the result in the $T \to \infty$ limit must be a point on the unstable manifold so all these continuum limits lie on the unstable manifold and are parametrized via the diffeomorphism from the linearized to nonlinearized flows by $c_1, \ldots, c_r$. We also see that information is lost in taking the limit because an infinite dimensional manifold of initial data leads only to a finite dimensional manifold of limits. The necessity of adjusting the initial data as the limit is taken is the well known phenomenon of renormalization.

Proof of Theorem 1

The flow equation for V is

$$\frac{\partial}{\partial t} V = \frac{1}{2} \sum_{y,z} \dot{C}_{yz}(t) \left[\frac{\partial^2}{\partial \varphi_y \partial \varphi_z} V - \frac{\partial V}{\partial \varphi_y} \frac{\partial V}{\partial \varphi_z} \right], \quad y, z \in \Lambda .$$

Since we are only considering bounded solutions of these equations we can rewrite them as integral equations

$$V(t) = \mu_t * V(0) - \frac{1}{2} \int_0^t ds \sum_{y,z} \dot{C}_{yz}(s) \mu_{[t,s]} * \left[V_y(s) V_z(s) \right] .$$

$\mu_{[t,s]}$ has covariance $C(t) - C(s)$. We differentiate this with respect to $\varphi_{x_1}, \ldots, \varphi_{x_M}$ and obtain

$$V_I = \mu_t * V_I(0) - \frac{1}{2} \sum_{J \subset I} \int_0^t ds \sum_{y,z} \dot{C}_{yz}(s) \mu_{[t,s]} * \left[V_{y,J} V_{z,I \backslash J} \right] ,$$

$$\tag{7}$$

where $I \equiv \{1, \ldots, M\}$ and we have used the set subscripts I, J and $I \backslash J$ to denote multiple derivatives with respect to φ's at x_i for $i \in J$, etc. We can insert these equations into themselves and thereby generate a series for any given derivative with coeficients involving only derivatives of the initial conditions. This becomes a power series in z if we replace $V(0)$ by $z V(0)$. We also find by taking absolute values and supremums over φ that

$$V_M(t) \leq V_M(0) + \frac{1}{2} \sum_{p=0}^M \int_0^t ds \, |\dot{C}(s)| \, (p+1) \, V_p(s).$$

$$\cdot (M - p + 1) V_{M-p+1}(s), \quad M \geq 1. \tag{8}$$

The main point in deriving this estimate is that μ's disappear on taking supremums because they are probability measures. The iteration of the corresponding equality produces a series that majorises the formal series for V obtained by iterating (7). Furthermore if we define $v(t, \varphi)$ by

$$v(t,\varphi) \equiv \sum_{M \geq 1} V_M(t) \; \varphi^M,$$

then (8) with equality instead of inequality becomes

$$v(t,\varphi) = v(0,\varphi) + \frac{1}{2} \int_0^t ds \; |\dot{C}(s)| \; \left[\frac{\partial v(s,\varphi)}{\partial \varphi}\right]^2 \qquad (9)$$

which is the integral equation corresponding to

$$\frac{\partial v}{\partial t} = \frac{1}{2} |\dot{C}| \; \left[\frac{\partial v}{\partial \varphi}\right]^2, \qquad v(t = 0) = \sum_{M \geq 1} V_M(0) \; \varphi^M . \qquad (10)$$

We will now prove that the iteration of (7) converges by replacing
$v(0)$ by $z \, v(0)$ in (10) and showing that this equation has a
solution analytic in z in a neighborhood of $\{z: |z| \leq 1\}$ and in φ
near $\varphi = 0$ for t satisfying the bound in theorem 1. It is enough
to replace the initial condition on v by the majorising series

$$v(0.\varphi) = - z \sum_{M \geq 1} \frac{K^M}{M} \; \varphi^M = - z \log (1 - K \, \varphi)$$

where $K \equiv \sup_M [M V_M(0)]^{1/M}$. By changes of variable

$$\tau = K^2 \int_0^t |\dot{C}(s)| \; ds, \qquad \Psi = K \, \varphi$$

the equations are transformed to

$$\frac{\partial v}{\partial \tau} = \frac{1}{2} \left[\frac{\partial v}{\partial \Psi}\right]^2, \qquad v(\tau = 0) = - z \log (1 - \Psi) .$$

This equation may be integrated explicitly by the method of
characteristics (Courant and Hilbert, Vol 2) and the reader can check
that the solution is

$$v = - z \log (1 - \Psi_0) - \frac{1}{2\tau} (\Psi - \Psi_0)^2$$

where Ψ_0 is the solution of

$$\frac{z}{1 - \Psi_0} + \frac{1}{\tau} (\Psi - \Psi_0) = 0$$

which satisfies $\Psi_0 = \Psi$ when $\tau = 0$. v is clearly analytic near $\Psi =$
0 for $0 \leq |z \, \tau| < 1/4$ which by our change of variable above is the
same as the bound in the theorem since $|z| \leq 1$. We conclude from
these arguments that the series generated by iterating (7) is
convergent for small time in the V_M norms defined above. This
implies that the series are uniformly convergent series of analytic
functions so that for example the series for V solves the integral

equation and therefore the flow equation. *End of Proof.*

Trigonometric Initial Data

Theorem 1 does not take advantage of the smoothing properties of the Δ. In the special case of the trigonometric initial data discussed in (C). It is easy to do better: Fix $T \geqslant 0$ and consider for example the initial data

$$V(0) = z \sum_{x \in \Lambda} \; : \; \exp[\; i \; \varphi_x \;] \; :_T$$

where $: \quad :_T$ *(normal ordering with respect to the covariance at* $t = T$) is defined by

$$: \exp[\; i \; \varphi_x \;] \; :_T \; \equiv \; \exp[\; \tfrac{1}{2} \; C_{xx}(T) \;] \; \exp[\; i \; \varphi_x \;] \; .$$

We assume that $\dot{C}_{xx}$ is independent of x and set $c(s) \equiv \dot{C}_{xx}(s)$

Theorem 2: **For T sufficiently small that**

$$|z| \int_0^T |\dot{c}(s)| \; \exp\{ \int_s^T c(\tau) \; d\tau \} \; ds < e^{-1}$$

The *one variable* equation

$$\frac{\partial \overline{v}}{\partial t} = \tfrac{1}{2} \; \exp\left[\int_t^T c(\tau) \; d\tau \right] \; |\dot{c}(t)| \; \left[\frac{\partial \overline{v}}{\partial \varphi} \right]^2$$

$$\overline{v}(t = 0) = z \; e^{\varphi}$$

has a solution at least up to time T and

$$V_M(T) \leq v_M(T) \; .$$

If Theorem 1 were applied we would achieve essentially the same result but without the normal ordering and with $c(s)$ set to zero.

The proof of this theorem is in [1]. The idea is very simple. In the integral equation version of the flow equation used in the proof of theorem 1 the initial data is convoluted by μ_t. Since the action of this convolution is to multiply $V(0)$ by an exponentially small factor we can track this factor through the proof of theorem 1 and the equation which dominates the norms of V now turns out to have an extra linear term in the right hand side. The exponential damping produced by the extra linear term in the right hand side permits us to consider initial data which is exponentially larger, i.e., normal ordered.

Tree Graphs and the Hamilton-Jacobi Equation

The theme has been to find relationships between the second order equations (2) and first order equations such as (4). I would like to interpret this in terms of Feynman graphs. It is a standard fact that the logarithm of the partition function admits a formal power series expansion in powers of z whose terms are labelled by connected graphs. The least connected type of connected graph is a tree graph and it is in fact these graphs which are generated by the first order equations we obtain by omitting the Laplacian terms. The terms labelled by these graphs are the "least local" parts of V .

Notice that if the Laplacian part of (1) is omitted we get a Hamilton-Jacobi equation.

$$\frac{\partial V}{\partial t} = -\frac{1}{2} V_\varphi^2 .$$

By the action principle of classical mechanics the solution to this equation is, at least locally,

$$V(t,\varphi) = V(0,\varphi(0)) + \int_0^t L(s,\dot{\varphi}(s)) \, ds$$

where the integral is taken along a path $t \to \varphi(t)$ which is an extremum for the right hand side and L is the Lagrangian corresponding to (4) which is

$$L(s,u) = \frac{1}{2} \sum u_x \, (\dot{C}^{-1})_{xy}(s) \, u_y .$$

By using the Euler Lagrange equation to find the path $\varphi(t)$ which extremises the ds integral subject to fixed endpoints we can rewrite the solution as

$$V(t,\varphi) = V(0,\varphi + \Psi) + \frac{1}{2} \Sigma \, \Psi_x \, (C^{-1})_{xy}(s) \, \Psi_y \,)$$

where Ψ is an extremum for the right hand side. (Ψ is what used to be $\varphi(0)$). If $V(0)$ is replaced by $z\,V(0)$ then it is a standard result in Feynman graphology that $V(t)$ has a formal power series in z which for simplicity I state only for local initial data, i.e., $V(0,\varphi) = \Sigma \, v(\varphi_x)$

$$V(t,\varphi) = \sum_N \frac{(-z)^N}{N!} \sum_{x_1,\ldots,x_n} \sum_T \prod_{ij \in T} C_{x_i x_j}(t) \prod_i v^{(n_i)}(\varphi_{x_i}) \quad (11)$$

where T is summed over all tree graphs on N labelled vertices $\{1,\ldots,N\}$, n_i is the number of lines in T incident to i and $v^{(n)}$ denotes the nth derivative of v. The equation (4) that we have been using generates tree graphs with only positive signs because we are forced to have a positive sign in front of the right hand side of the Hamilton-Jacobi equation when we throw out the Laplacian. The following theorem results from the same methods as Theorem 1.

Theorem 3 <u>Given $V(0,\varphi)$ define</u>

$$\bar{V}(0,\Psi) \equiv \sum_N \frac{1}{N!} \sum_{x_1,\ldots,x_N} \sup_\varphi |\, V_{x_1,\ldots,x_N}(0,\varphi)\,| \, \Psi_{x_1}\cdots\Psi_{x_N} \, .$$

<u>Assuming this series has a non zero radius of convergence we may extend</u> <u>$\bar{V}$ to be a function of t which is analytic in Ψ near zero by</u> <u>solving the Hamilton-Jacobi equation</u>

$$\frac{\partial}{\partial t} \bar{V} = \frac{1}{2} \sum_{x,y} |\, \dot{C}_{xy}(t)\,| \, \frac{\partial \bar{V}}{\partial \Psi_x} \frac{\partial \bar{V}}{\partial \Psi_y}$$

<u>with the prescribed inital data. This equation will have such a</u> <u>solution at least for t small as in theorem 1. Then for all</u> <u>$x_1,\ldots,x_N \in \Lambda$</u>

$$|\, V_{x_1,\ldots,x_N}(t,\varphi)\,| \le \bar{V}_{x_1,\ldots,x_N}(t)$$

<u>where $\bar{V}_{x_1,\ldots,x_N}(t)$ are the coefficients in the expansion of $\bar{V}$:</u>

$$\bar{v}(t,\Psi) \equiv \sum_N \frac{1}{N!} \sum_{x_1,\ldots,x_N} \bar{v}_{x_1,\ldots,x_N}(t) \ \Psi_{x_1}\ldots\Psi_{x_N} \ .$$

It is now a natural question to ask if we can find a more accurate relation between the first order and second order equations instead of these bounds. This is done in [1]. We find there is a formal power series for the solution of (2) in which the right hand side is the same as (11) except that the product over i is preceded by a convolution by a gaussian measure whose covariance depends on T. In the special case where $v(\varphi) = \exp(i\varphi)$ this result is especially simple because the convolution can be performed explicitly and the result is

Theorem 4: <u>The coefficients in the formal power series expansion for</u>
<u>V:</u>

$$V(t,\varphi) = - \sum_N \frac{z^N}{N!} \sum_{x_1,\ldots,x_N} m(t,x_1,\ldots,x_N) \ \exp[i \sum_j \varphi_{x_j}] \ ,$$

<u>are given by</u>

$$m(t,x_1,\ldots,x_N) = (-1)^{N-1} \sum_T \prod_{ij \in T} \left[\int_0^t ds_{ij} \ \dot{C}_{ij}(s_{ij}) \right].$$

$$\cdot \exp\left[- \frac{1}{2} \sum_{k,l} \int_{s(k,l)}^t ds \ \dot{C}_{kl}(s) \right]$$

<u>where T is summed over all connected tree graphs on the labeled</u>
<u>vertices</u> $\{1,\ldots,N\}$, $s(k,l)$ <u>is defined by</u>

$$s(k,l) \equiv \sup \{s_{ij}: ij \in \text{unique path in } T \text{ joining } k \text{ and } l\} \ .$$

This identity is obtained by proving that $V(t,\varphi)$ as given by this formula formally solves the flow equation (2). The derivative with respect to t can act either in the exponent in the formula for m or on one of the upper limits in the ds_{ij} integrals. It is easy to verify that the first possibility corresponds to the $V_{\varphi\varphi}$ in (2) and the second corresponds to the $V_\varphi{}^2$ term in (2). More details can be found in [1]. This is an improved version of the Tree Graph Identities found by Battle and Federbush [7]. An important feature of this identity is that the positivity of $\dot{C}$ implies that

$$\sum_{kl:\ s(k,l)\ \leq\ s} \dot{C}_{kl}(s) \geq 0$$

which implies:

$$| m(t,x_1,\ldots,x_N) | \leq \sum_T \prod_{b \in T} \int_0^t ds_b\ |\dot{C}_b(s_b)|$$

whivh is a result closely related to theorem 3.

Conclusions

How good are these theorems and how should they be inproved? We would like to use them to study critical phenomena and the related problem of constructing perturbations of generalised fields which is the basic problem of constructive field theory. These are global existence problems. In the theory of ODE's such problems are tackled by using a short time existence theorem repeatedly to maximally extend a trajectory until some quantity becomes infinite. In a series of profound papers [8] Gawedski and Kupiainen have accomplished the equivalent step in statistical mechanics by discovering how to apply a cluster expansion repeatedly to a functional integral. A cluster expansion is a short time existence theorem for our PDE. The theorems we have presented cannot be used directly for the same purpose because they are restricted to bounded initial data. The one φ equation discussed under the heading of the renormalisation group was included to show that in the neighborhood of a fixed point a global existence theorem is feasible but one must be prepared to consider solutions with polynomial growth in φ e.g., Hermite polynomials.

Aside from this, theorem 2 in particular, is surprisingly accurate considering its rather trivial proof. We have not formulated a version that applies to continuum gaussian fields directly but we can for example take C_{xy} to be the restriction to $x,y \in (\varepsilon Z)^d$ of a covariance for a continuous gaussian field $\varphi(x)$ defined on R^d and take the initial data to have the form

$$V^{(\varepsilon)}(0,\varphi) \equiv z \sum_{x \in \Lambda} \varepsilon^2 : \cos \varphi(x) :_T$$

where $\Lambda \equiv (\varepsilon Z)^2 \cap \Omega$ where Ω is a bounded subset of R^2, ε is a lattice spacing so V is a Riemann approximation to an integral. The Riemann sums propagate through the theorems in a natural manner and although the flow equatons are as yet not properly defined in the $\varepsilon \to 0$ limit the corresponding integral equations are well defined in the limit and the term by term limit of the series which they generate for $V^{(\varepsilon)}(t)$ exists and is the power series for V given by

$$V(t,\varphi) \equiv - \log \int d\mu_t(\overline{\varphi}) \exp(z \int_\Omega : \cos [\overline{\varphi}(x) + \varphi(x)] :_T dx) .$$

The condition in theorem 2 is, in the limit as $\varepsilon \to 0$,

$$z \int_0^T |\dot{C}(s)| \exp\{ \int_s^T c(r) dr\} ds < e^{-1}$$

where

$$| \dot{C}(s) | \equiv \sup_x \int dy | \dot{C}(x,y) | .$$

In [1,11] we use this criterion to prove the existence of the following limit:

$$Z(\varphi) = \lim_{T \to \infty} \int d\mu_T(\overline{\varphi}) \exp (z \int_\Omega dx :\cos \beta^{1/2}[\varphi + \overline{\varphi}]:_T)$$

where the covariance of $d\mu_T$ is

$$[1 - \Delta]^{-1} - [e^T - \Delta]^{-1}$$

where β and z are required to satisfy

$$\beta < 4\pi \quad \text{and} \quad 2 |z| \beta (1 - \frac{\beta}{4\pi})^{-1} < e^{-1} .$$

Ω is a bounded subset of R^2 with smooth boundary. This is the functional integral associated with the euclidean cosine quantum field theory in two dimensions. It is also by a sine gordon transformation a continuum Yukawa gas in two dimensions. It can be shown to be a perturbation of the generalized gaussian field with covariance $(1 - \Delta)^{-1}$. The expansion given in the proof of theorem 2 and also in theorem 4 is the Mayer expansion of statistical mechanics. The convergence of this expansion for this quite singular model is not a new result [9,10,11] but the previous proofs were much longer. Furthermore these results generalise the work by Göpfert and Mack on the iterated Mayer Expansion [12].

Acknowledgments

This is a summary of results found in collaboration with Thomas Kennedy. In addition, I have greatly benefited from discussions with Konrad Osterwalder and Erhard Seiler on these topics. I thank Hans Weinberger for many useful discussions and the hospitality of the Institute, and I thank Harry Kesten for inviting me to this conference. This research was partially supported by the Sloan Foundation and by National Science Foundation Grant DMS8500516.

REFERENCES

1. Brydges, D.C.: Mayer Expansions and Burger's Equation. University of Virginia Preprint.

2. Battle, G., Federbush, P.: Lett. Math. Phys. 8, (1984), 55.

3. Newman, C.: Unpublished work. See Journal of Statistical Physics, 27, No. 4, 1982, p. 836.

4. Wilson, K. G.: Phys. Rev. B4 (1971), 3174. Kogut, J. G., and Wilson, K. G., Phys. Reports 12 (1974), 263.

5. Gallavotti, G., Nicolo, F., Renormalization Theory in Four Dimensional Scalar Fields (I). Comm. Math. Physics 100, 545, (1985).

6. Carr, J.: Applications of Center Manifold Theory. Applied
    Mathematical Sciences 35, Springer Verlag New York 1981.

7. See reference 2 and also Theorem 3.1 in Brydges, D. C.: A short
    course on cluster expansions: in critical phenomena, random
    systems, gauge theories, Les Houches, Session LIII, 1984, eds.
    K. Osterwalder, R. Stora, Elsevier science publishers, 1986.

8. Gawedski K., Kupiainen, A.,: in Scaling and self similarity in
    physics - Renormalization in statistical mechanics and dynamics.
    Fröhlich, J. (ed.): Boston, Basel, Stuttgart: Birkhäuser 1983.

9. Fröhlich, J., Seiler, E.: Helv. Phys. Acta 49, 889 (1976).
    Fröhlich, J. Comm. Math. Phys. 47, 233 (1976).

10. Benfatto, G.: Journal of Statistical Physics 41, Nos. 3/4, 671,
    (1985).

11. Brydges, D. C.: Journal of Statistical Physics 42, Nos 3/4, 245,
    (1986).

12. Göpfert, M., Mack, G.: Comm. Math. Phys. 81, 97 (1981) and 82,
    545 (1982.

The Mean Field Bound for the Order Parameter
of Bernoulli Percolation

J. T. Chayes and L. Chayes
Laboratory of Atomic and Solid State Physics
Cornell University
Ithaca, New York 14853

Abstract

We consider a general, translation invariant bond percolation model on $\mathbb{Z}^d$ with bonds characterized by couplings $\{J_x | x \in \mathbb{Z}^d\}$ and an inverse temperature parameter ϑ, with nontrivial critical value ϑ_c. We prove several inequalities including: (1) a differential inequality for the infinite cluster density, $P_\infty(\vartheta)$; and (2) an inequality relating the backbone density, $Q_\infty(\vartheta)$, to $P_\infty(\vartheta)$ and the expected size of finite clusters, $\chi'(\vartheta)$. If the above quantities exhibit critical scaling with exponents "defined" by $P_\infty(\vartheta) \sim |\vartheta - \vartheta_c|^\beta$, $Q_\infty(\vartheta) \sim |\vartheta - \vartheta_c|^{\beta_Q}$ and $\chi'(\vartheta) \sim |\vartheta - \vartheta_c|^{-\gamma'}$ as $\vartheta \downarrow \vartheta_c$, these inequalities imply the mean field bounds: $\beta \leq 1$ and $2\beta \leq \beta_Q \leq \beta + \gamma'$. Furthermore, a magnetic backbone exponent, δ_Q, is defined analogously to the standard magnetic backbone exponent, δ. Again assuming critical scaling, our inequalities also imply the mean field bounds $\delta \geq 2\delta_Q$ and $\delta_Q \geq 1$.

1. Introduction

In these notes, we present a proof of the mean field inequality for the infinite cluster density in Bernoulli percolation. We also prove an inequality relating the "backbone" density, Q_∞, to the infinite cluster density, P_∞, and the expected size of finite clusters, χ'. These results first appeared in [1] and were discussed by one of us (JTC) at the IMA conference on Percolation Theory and Ergodic Theory of Infinite Particle Systems.

There are three differences between the results presented here and the version in [1].

(I) While in [1] only nearest neighbor site models were explicitly discussed, here we consider a rather general class of independent percolation models, namely all nontrivial translation invariant independent bond percolation models on the hypercubic lattice $\mathbb{Z}^d$. (Note, however, that even this does not include all possible independent percolation models.) Our reason for treating the general cases stems from the recent surge of interest in percolation problems involving long-range couplings [2-7]. The class of models we consider is described in some detail in Section 2; here we need only note that these models depend on a single (temperature-like) parameter which we denote by θ. In terms of the parameter θ, our basic inequalities are

$$P_\infty(\theta) \leq P_\infty^2(\theta) + c_1(\theta) P_\infty(\theta) dP_\infty(\theta)/d\theta \qquad (1)$$

and

$$P_\infty(\theta) \leq c_2(\theta) Q_\infty(\theta) \chi'(\theta) , \qquad (2)$$

where $c_1(\theta)$ and $c_2(\theta)$ are nonsingular functions. It is easy to see, as is shown below, that if $P_\infty(\theta)$, $Q_\infty(\theta)$ and $\chi'(\theta)$ exhibit "scaling" as θ tends to its critical value θ_c, with critical exponents "defined" by $P_\infty(\theta) \sim |\theta - \theta_c|^\beta$, $Q_\infty(\theta) \sim |\theta - \theta_c|^{\beta_Q}$ and $\chi'(\theta) \sim |\theta - \theta_c|^{-\gamma'}$, then (1) and (2) imply the exponent bounds

$$\beta \leq 1 \qquad (1')$$

and

$$\beta + \gamma' \geq \beta_Q . \qquad (2')$$

As will be explained below, the inequality (1') holds in a rather strong sense.

(II) In addition to exercising the inalienable right of IMA authors to introduce unnecessary complications, another purpose of these notes is to present a strengthened version of the proof that appeared previously. In [1], inequalities (1) and (2) were proved under an assumption concerning the structure of infinite clusters: namely, with probability one, every infinite cluster has a backbone. Here we will show that if this structural assumption fails at any parameter value θ_0, then θ_0 is a point of discontinuity of $P_\infty(\theta)$. Since (1) is a lower bound on $dP_\infty(\theta)/d\theta$, this shows that (1) always holds. Coincidently, Aizenman, Kesten and Newman [8] have recently announced a proof of continuity of $P_\infty(\theta)$ for all $\theta \neq \theta_c$. Thus we have that for $\theta > \theta_c$, with probability one, every infinite cluster has a backbone. Evidently, the inequality (2) also holds for all $\theta \neq \theta_c$.

(III) Finally, we generalize our second inequality (2) to the case of a nonzero external field h. As will be explained below, if the critical behavior as h↓0 of the magnetic backbone is characterized by an exponent δ_Q, this in-field inequality implies

$$\delta_Q \geq 1 \tag{3'}$$

The exponent inequalities (1'), (2') and (3') are known as mean field bounds since they hold as equalities on the Bethe lattice and are expected to be equalities in sufficiently high dimension.

The organization of these notes is as follows. In Section 2, we precisely define our models and discuss the principal "ingredients" in the proof of the above inequalities. Two of these are relatively well-known: the Russo formula [9] and the van den Berg – Fiebig – Kesten inequality [10-11]. The third is the notion of a *backbone* of an infinite cluster, which enters as a (crucial) intermediate step in the derivation of the first inequality and is explicit in the others. In Section 3, we derive the differential inequality (1) under our structural assumption, and then remove the assumption by demonstrating that it implies a discontinuity in $P_{oo}(\theta)$. Section 4 is devoted to the backbone. There we show how minor modifications of the derivation of (1) lead to the inequality (2) and hence a bound on β_Q. Finally, we generalize (2) to obtain a bound on δ_Q.

2. Preliminaries

Definition of the Models. The percolation models we consider are defined as follows: We start with the hypercubic site lattice $\mathbf{Z}^d$ and take a subset $\mathbb{K} \subset \mathbf{Z}^d \backslash \{0\}$ (which need not be finite). For every $x \in \mathbf{Z}^d$, edges are drawn between x and the $x + \varkappa$, $\varkappa \in \mathbb{K}$. It is assumed, for obvious reasons, that when this procedure has been completed, we are left with a single connected set, i.e. any given point in $\mathbf{Z}^d$ can be reached from the origin by walking along a finite number of edges. The set of all edges will be denoted by $\mathbb{B}$.

Next, we define the bond occupation probabilities. For each $\varkappa \in \mathbb{K}$, we choose a number $p_\varkappa \in (0,1)$ parameterized by

$$p_\varkappa = 1 - e^{-\theta J_\varkappa} . \tag{4}$$

Each edge is assigned the probability $p_\varkappa$ -- depending on the $\varkappa \in \mathbb{K}$ which is the vector difference between its endpoints -- and declared to be "occupied" with probability $p_\varkappa$ and "vacant" otherwise. In the models we consider here, these occupation events are independent. A set of occupied bonds will sometimes be identified with a subset of $\mathbb{B}$.

It should be pointed out that, in these notes, the only variable parameter that will be considered is ϑ; thus the $\{J_x\}$ are first chosen (for once and all) and then we allow ϑ to run through the positive reals. Observe that when $\vartheta = 0$ or $\vartheta = \infty$, the model has all sites isolated or all sites tangled together in one component. We will say that the model has nontrivial content if these situations can (qualitatively) persist at non-extreme values of temperature.

In particular, let ω denote a configuration of occupied and vacant bonds on $\mathbb{Z}^d$ and denote by $C_0(\omega)$ the sites:

$$C_0(\omega) = \{x \in \mathbb{Z}^d | x \text{ connected to 0 by a path of occupied bonds in } \omega\}, \qquad (5)$$

and by $|C_0|$ the number of sites in C_0. The models we consider have the property that for ϑ is sufficiently small, $|C_0|$ is finite with probability one, while for ϑ large enough, one finds there is an infinite component, i.e. $|C_0| = \infty$ with nonzero probability. The supremum of all values of ϑ for which there are no infinite clusters will be called ϑ_c. We further define

$$P_\infty(\vartheta) \equiv \text{Prob}_\vartheta[0 \text{ is part of an infinite cluster}] \qquad (6)$$

so that for $\vartheta > \vartheta_c$, $P_\infty(\vartheta) > 0$.

Remark. It is worth examining the conditions under which a model is nontrivial. If the $\{J_x\}$ satisfy $\Sigma_x J_x = \infty$, then for all nonzero ϑ, $P_\infty(\vartheta) = 1$. This is easily seen by observing that, under such circumstances, any fixed site is connected to infinitely many sites with probability one. Hence, it cannot be the case that the origin is in a finite cluster. (See [2] and [8] for a more detailed discussion of the properties of these models.) On the other hand, should $\Sigma_x J_x$ be finite, then for ϑ sufficiently small, the expected number of sites attached to the origin by single bonds is smaller than one. An elementary branching process argument can then be employed to show that $E_\vartheta[|C_0|]$ is finite, and hence $P_\infty(\vartheta) = 0$. A sensible requirement is therefore to assume that $\Sigma_x J_x$ converges; we define

$$T \equiv \Sigma_x J_x < \infty \qquad (7)$$

and

$$J_{MAX} \equiv \max_x J_x . \tag{8}$$

In dimension larger than one, all models satisfying the requirement (7) (and the assumption on $\mathbb{K}$) are nontrivial. This is easily shown by a contour argument which dates back to Peierls [12] and was first used in the context of percolation by Broadbent and Hammersley [13]. The interesting borderline cases therefore occur in one dimension. It has been established [4] that if $J_x \sim |x|^{-s}$ with $1 < s \leq 2$, then the model is nontrivial. Earlier, it was shown [3] that if $s > 2$, the model cannot support an infinite cluster; a stronger statement was then proved [5], namely that if

$$|x|^2 J_x \longrightarrow 0 , \tag{9}$$

the same result holds. Many of the above statements (with the notable exception of (9)) can also be derived by using a comparison method developed in [6] and previously established analogous results for Ising systems.

Before deriving the differential inequality (1), let us examine its consequences for the models under consideration. In the $(\vartheta, \{J_x\})$ language, the inequality is

$$P_\infty(\vartheta) \leq P^2_\infty(\vartheta) + \vartheta \exp[\vartheta J_{MAX}] P_\infty(\vartheta) dP_\infty(\vartheta)/d\vartheta . \tag{10}$$

For $\vartheta > \vartheta' > \vartheta_c$, we can divide both sides of (10), evaluated at ϑ', by $P_\infty(\vartheta')$ to obtain

$$1 \leq P_\infty(\vartheta') + \vartheta' \exp[\vartheta J_{MAX}] dP_\infty(\vartheta')/d\vartheta$$

$$\leq \exp[\vartheta J_{MAX}] d/d\vartheta'[\vartheta' P_\infty(\vartheta')] . \tag{11}$$

If we integrate equation (11) up from ϑ_c (and discard possible contributions from singular pieces), we get

$$P_\infty(\vartheta) \geq [\vartheta_c/\vartheta] P_\infty(\vartheta_c) + (\exp[-\vartheta J_{MAX}]/\vartheta)(\vartheta - \vartheta_c)$$

$$= P_\infty(\vartheta_c) + \text{const.}(\vartheta - \vartheta_c) + O(\vartheta - \vartheta_c)^2 , \tag{12}$$

which is our principal result. Obviously, (12) implies that $P_\infty(\vartheta)$ rises from its limiting value faster than linearly. I.e., if it is the case that $P_\infty(\vartheta)$ tends to zero at ϑ_c with critical index β "defined" by

$$P_\infty(\vartheta) \sim |\vartheta - \vartheta_c|^\beta , \tag{13}$$

we have

$$\beta \leq 1 . \tag{14}$$

Observe that if one interprets the "definition" (13) of β as $\beta = \lim_{\theta \downarrow \theta_c} \log P_\infty(\theta)/\log|\theta - \theta_c|$

then (12) is in fact a stronger statement than (14); indeed, (12) provides an explicit lower bound on the percolation probability.

It is worth noting that there is an example [5], namely the one-dimensional problem with $x^2 J_x \to 1$, where (13) is violated in the sense that $P_\infty(\theta_c)$ is positive. In this instance, equation (12) provides a bound on the "cusp singularity" (if any) as $\theta \downarrow \theta_c$.

Ingredients. The proof of the inequality (1) is quite straightforward; it involves only three ingredients. Two of these, a and b below, are in fact well-known. The third, the backbone cluster, although popular in the physics literature, has been largely overlooked by workers interested in rigorous percolation theory.

a) *Russo's Formula*

Russo's formula [9] was first derived in the context of nearest neighbor models, although it extends easily to the models considered here. Our use of the Russo formula is that it allows us to identify the major class of graphs which contribute to the infinite cluster density.

We must start with the notion of a positive event.

Definition. The collection of configurations of occupied bonds comes equipped with a natural partial order. In particular, if ω and ω' are bond configurations, we say that $\omega' \ll \omega$ if $\omega' \subset \omega$. Functions which are non-decreasing with respect to this partial order are called *positive;* events are positive when their indicators are positive functions. *Negative* functions (and events) are defined by replacing the word "non-decreasing" with "non-increasing" in the above statement.

We also need the concept of pivotal (or articulation) bonds, as introduced by Russo [9,14]. This concept is most useful in the context of positive events.

Definition. Let A be an event, b a bond and ω a configuration of occupied bonds. The bond b is said to be an *articulation bond* for the event A in configuration ω if the occurrence of A in ω depends crucially on the status of the bond b. To be specific, define ω_b^+ to be the configuration which agrees with ω on the complement of b and has the bond b occupied, and ω_b^- to be the corresponding configuration with b vacant. The bond b is an articulation bond for the event A in ω if $|\mathbf{1}_A(\omega_b+) - \mathbf{1}_A(\omega_b-)| = 1$, i.e. A happens in ω_b+ but not ω_b- or the other way around. The set of articulation bonds for A in ω will be denoted by $\delta A(\omega)$, so the event that b is an articulation bond may be expressed as $b \in \delta A(\omega)$, which we occasionally denote by $\delta_b A$. As usual, we denote by $|\delta A(\omega)|$ the size (number of bonds) in the articulation set.

The formula of Russo can be expressed as follows:

Proposition 1 (Russo's formula, [9]). *Consider the nearest neighbor independent percolation model with the single parameter* p. *Let* A *be a positive event, defined on a finite collection,* Λ, *of bonds. Then*

$$d\text{Prob}[A]/dp = \sum_{b \in \Lambda} \text{Prob}[b \in \delta A(\omega)] = E[|\delta A|]$$

Proof. Imagine that all the bonds of Λ are assigned different probabilities, p_b. The derivative can then be computed by summing the partial derivatives, $\partial \text{Prob}[A]/\partial p_b$, evaluated at $p_b = p$. Picking any $b \in \Lambda$ and summing over all configurations $\omega \subset \Lambda \backslash b$, we may write the identity

$$\text{Prob}(A) = \sum_{\omega \subset \Lambda \backslash b} \text{Prob}(\omega)[p_b \mathbf{1}_A(\omega_b+) + (1 - p_b)\mathbf{1}_A(\omega_b-)] , \tag{15}$$

which, when differentiated, yields

$$\partial \text{Prob}(A)/\partial p_b = \sum_{\omega \subset \Lambda \backslash b} \text{Prob}(\omega)[\mathbf{1}_A(\omega_b+) - \mathbf{1}_A(\omega_b-)] . \tag{16}$$

Since the event A was stipulated to be positive, it is seen that the term in the sum vanishes unless $b \in \delta A(\omega)$. Summing over all bonds in Λ gives the desired formula. ▨

A few remarks are in order.

I) The above formula is only strictly valid for events which are defined in finite volumes. For certain infinite volume events, it must surely fail. Indeed, Russo [14] pointed out that a tail event can be defined by the requirement that it have *no* articulation bonds with probability one. Evidently, if a tail event is also positive, the formula cannot hold. Nonetheless, it is easy to see that the Russo formula always holds as an inequality:

$$d\text{Prob}[A]/dp \geq \sum_{b \in \mathbb{B}} \text{Prob}[b \in \delta A(\omega)] , \tag{17}$$

(where, if necessary, we interpret the l.h.s. in terms of the appropriate Dini derivative.) For the event which is of principal interest here, namely that the origin is part of an infinite cluster, it has been explicitly shown [15], that Russo's formula is valid as an equality; this is actually unimportant for us because we can get away with using the weaker form written above.

II) For models parameterized according to an inverse temperature ϑ, a massive derivation following the above lines yields the *amalgamated Russo formula* (ARF):

$$d\text{Prob}[A]/d\vartheta = \sum_{b \in \mathbf{B}} \text{Prob}[b \in \delta A(\omega)]J_b \, e^{-\vartheta J_b}, \tag{18}$$

which again (in the worst cases) holds as an inequality.

III). For a given $b \in \mathbf{B}$, it is worthwhile to visualize the event that b is an articulation bond for the connection of the origin to infinity. Denoting the event of connection of the origin to infinity by $\mathbf{P}_\infty(0)$ (so that $P_\infty(\vartheta) = \text{Prob}_\vartheta[\mathbf{P}_\infty(0)]$), the articulation event is $\delta_b \mathbf{P}_\infty(0)$. By definition, this means that the removal of b disconnects the origin from infinity, i.e. all infinite paths which start from the origin must pass through the bond b. Alternatively, self-avoiding paths starting from 0 which do not employ b must stop somewhere; thus we may envisage a "cutting surface" passing through the bond b and separating the origin from infinity. The presence or absense of b determines whether or not this surface is punctured. Graphically, the above paragraph may be expressed as

$$\delta_b(\mathbf{P}_\infty(0)) =$$

(19)

b) *The v-BKF Inequalities*

Here, we will describe a relatively recent class of inequalities which were introduced by van den Berg and Kesten [10] and extended by van den Berg and Fiebig [11]. Although primitive forms of these inequalities can be derived case by case -- usually on the basis of geometric considerations -- in the "all bonds possible" formulation presented here, we are more than happy to use the general results of [10] and [11].

We first need the notion of events which occur disjointly.

Definition. Let A be an event and let $\Lambda \subset \mathbb{B}$. The event A may "occur on Λ" in the sense that just by knowing the configuration restricted to Λ, it can be determined that the event A has occurred. In particular, let us define $A_{|\Lambda} \subset A$ to be the event

$$A_{|\Lambda} = \{\omega \in A | \, \omega' \in A \text{ for all } \omega' \text{ such that } \omega' = \omega \text{ on each bond in } \Lambda\} . \quad (20)$$

For example, if A is a positive event, then $A_{|\Lambda}$ are the configurations of Λ for which A occurs even if all bonds on $\mathbb{B}\backslash\Lambda$ are vacant.

Two events, A and B, are said to *occur disjointly* if there are disjoint sets, Λ, Ξ on which A, B occur; i.e. A and B do not interfere. The notation for this is A∘B:

$$A \circ B = \{\omega \in A \cap B| \, \exists \Lambda, \Xi \subset \mathbb{B}, \Lambda \cap \Xi = \emptyset , \omega \in A_{|\Lambda} \cap B_{|\Xi}\} . \quad (21)$$

It should be remarked that the above defines *bond* disjoint events. There is a stronger notion of *vertex* disjointness which one can define; however this does not yet seem to have useful applications.

Proposition 2. *If* A *and* B *are both positive (or both negative) events, then* [10]
$$\text{Prob}[A \circ B] \leq \text{Prob}[A] \, \text{Prob}[B].$$

Furthermore, the above inequality also holds if A *and* B *are intersections of positive and negative events* [11].

c) *The Infinite Backbone Cluster.* In problems concerning *bulk* properties of percolation-like systems, e.g. random resister networks, the mere existence of an infinite cluster is insufficient to ensure that the medium has nontrivial transport characteristics. As was first pointed out (experimentally!) by Last and Thouless [16], only a vanishingly small percentage of the infinite cluster is relevant for bulk transport in the critical regime.

Bulk properties are often determined by how well opposing faces of (large) cubes are connected; i.e. by the number of disjoint paths of occupied bonds. Thus, in order for a site in the interior of a sample to contribute to the overall transport properties, it must be connected by at least two separate routes to the boundary of the sample. In the infinite volume limit, this amounts to the statement that those sites relevant for transport belong to two disjoint infinite paths. This motivates the definition of the backbone of an infinite cluster.

Definition. A site $j \in \mathbf{Z}^d$ is said to belong to an infinite *backbone* if it is connected to infinity by two disjoint paths of occupied bonds. We denote by $\mathbf{Q}_\infty(0)$ the event

$$\mathbf{Q}_\infty(0) = \{\omega | \text{ the origin belongs to a backbone}\}, \qquad (22)$$

or, more succinctly,

$$\mathbf{Q}_\infty(0) = \mathbf{P}_\infty(0) \circ \mathbf{P}_\infty(0), \qquad (23)$$

and define $Q_\infty(\theta) \equiv \text{Prob}_\theta(\mathbf{Q}_\infty(0))$.

3. A Mean Field Bound for the Order Parameter

From the preceding discussion, it is clear that the properties of the backbone are both natural and crucial for the understanding of transport phenomena. Here we show that the backbone is also central to an understanding of the order parameter, P_∞, in "ordinary" percolation -- at least in the context of mean field bounds. For us, the backbone allows a decomposition of the percolation event into manageable pieces. These pieces then can be attacked with the tools (a) and (b) discussed above. In order to implement this reasoning, we require:

A structural assumption. The assumption we make is: *Every infinite cluster has a backbone.*

It is possible to imagine infinite clusters without backbones. We will call these *spineless* clusters. The possibility of a spineless infinite cluster should not be immediately dismissed; it may well be the case that such objects describe the high density phase of polymer-like theories other than percolation. However (see remarks at the end of this section), one of the outcomes of these notes will be a proof that above θ_c, the possibility of a spineless infinite cluster is of measure zero.

We will temporarily assign the notation $\mathbf{S}_\infty(0)$ to the event that the origin belongs to a spineless infinite cluster, and $S_\infty(\theta)$ to the probability of $\mathbf{S}_\infty(0)$. Thus our assumption is $S_\infty(\theta) = 0$.

The differential inequality. In this subsection, we will take as given our structural assumption, so that in any configuration contributing to $\mathbf{P}_{\infty}(0)$, the origin is w.p.1 either in a backbone or in a "dangling end" of an infinite cluster. It is worth observing that the statement that the origin belongs to a backbone infinite cluster is another way of saying that there are no articulation bonds for the event $\mathbf{P}_{\infty}(0)$. The converse is also true: namely if the origin is in a dangling end, we can find an articulation bond. This is the key to the derivation.

Assuming then that $S_{\infty}(\theta) = 0$, we may write

$$\mathbf{P}_{\infty}(0) = \mathbf{Q}_{\infty}(0) \cup \mathbf{P}_{\infty}(0) \backslash \mathbf{Q}_{\infty}(0), \qquad (24)$$

the second term representing the contribution due to the dangling ends. This gives us

$$P_{\infty}(\theta) = Q_{\infty}(\theta) + \text{Prob}_{\theta}[\mathbf{P}_{\infty}(0) \backslash \mathbf{Q}_{\infty}(0)] . \qquad (25)$$

Using the van den Berg - Kesten inequality, we can bound the first term in (25) above by $P_{\infty}^2(\theta)$. In order to make progress on the dangling end term, we must establish an elementary result.

Proposition 3. *If the origin is in a dangling end of an infinite cluster, there is a unique bond b satisfying*:

> i) b *has one endpoint in the backbone of the infinite cluster*
>
> ii) b *is an articulation bond for the event* $\mathbf{P}_{\infty}(0)$.

Remark. The above bond is, of course, just the one which attaches to the backbone the dangling end containing the origin. Although morally this constitutes a proof, in the spirit of the introductory remarks, we will proceed with a formal restatement of the above sentence.

We first need a lemma which constitutes an alternative description of of the backbone cluster.

Lemma 4. *Suppose that* j *and* j* *are (distinct) sites and both part of a backbone infinite cluster. Then there are two (bond) disjoint infinite paths,* $_{\mathcal{P}}$ *and* $_{\mathcal{P}}$* *such that* $_{\mathcal{P}}$ *passes through* j *and* $_{\mathcal{P}}$* *passes through* j*.

Proof. If j and j* are in *separate* backbones, this is trivial. Let us assume otherwise and construct a self-avoiding path $W_{j*,j}$ between j* and j using only bonds in the backbone. Since j is a backbone site, we can construct two disjoint infinite paths emanating from j; call these p_1 and p_2. If either or both p_1 and p_2 pass through j*, we are done and there is no need for $W_{j*,j}$. If it is not the case that both p_1 and p_2 use some bonds of $W_{j*,j}$, then, assuming that p_1 is not the culprit, we can declare p_1 to be the path p, and find $p*$ by joining $W_{j*,j}$ to p_2 and deleting any unnecessary bonds. Finally, if both p_1 and p_2 used bonds in $W_{j*,j}$, one of them -- say p_2 -- used an earlier bond along the walk $W_{j*,j}$ than any bond in $W_{j*,j}$ used by p_1. We can again declare p_1 to be p. To find $p*$ in this case, add all the even earlier bonds along $W_{j*,j}$ (not used by p_2) to p_2 and delete unnecessary bonds.　▨

Proof of Proposition 3. Let $D_E(0)$ be the set of sites in the dangling end of the origin. That is to say, $D_E(0)$ includes all sites which are not in the backbone and are connected to the origin by a path of occupied bonds whose endpoints are not in the backbone.

We claim that at least one site $i \in D_E(0)$ is attached, by a single bond b, to the backbone. Should no such i exist, the cluster of the origin would be finite (or, worse yet, spineless).

Let us show any given $i \in D_E(0)$ cannot be attached by *two* bonds to the infinite backbone. Indeed, let us suppose that b and b* have endpoints {i,j} and {i,j*}, respectively, with both j and j* in a backbone. Using Lemma 4, we can construct two disjoint infinite paths emanating from i, namely $p \cup b$ and $p* \cup b*$, contradicting the fact that i is in a dangling end.

Having found one such $i \in D_E(0)$ -- together with its unique bond b -- we claim that two sites cannot have this property. Indeed, letting i,j and b have their previous meanings, and allowing i', j' and b' to be another candidate, a similar contradiction can be reached by constructing a walk between i and i' using only bonds with both endpoints in $D_E(0)$.

Thus all paths starting in $D_E(0)$ must pass through the unique site i and use the unique bond b. Removal of the bond b therefore disconnects $D_E(0)$, and in particular the origin, from infinity.　▨

Using Proposition 3, it is seen that the dangling end term can be written as the disjoint partition of the events, $\mathbf{D}_b(\mathbf{P}_\infty(0))$, that $b \in \mathbb{B}$ connects the origin to the infinite backbone. Graphically, these events may be expressed as

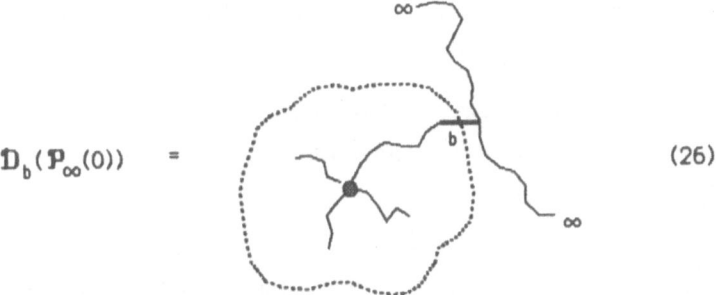

$$\mathfrak{D}_b(\mathbb{P}_\infty(0)) \quad = \qquad\qquad\qquad\qquad\qquad\qquad\qquad (26)$$

From this and equation (19) it is obvious that

$$\mathfrak{D}_b(\mathbb{P}_\infty(0)) = \delta_b\mathbb{P}_\infty(0) \circ \mathbf{b} \circ \mathbb{P}_\infty(\partial b) , \qquad (27)$$

where $\mathbf{b}$ is the event that the bond b is occupied, and $\mathbb{P}_\infty(\partial b)$ is the event that an endpoint of b is connected to infinity. The desired inequality will now be an immediate consequence of equation (27).

Theorem 5. *For any value of ϑ for which the structural assumption holds,*

$$P_\infty(\vartheta) \leq P_\infty^2(\vartheta) + \vartheta e^{+\vartheta J_{MAX}} P_\infty(\vartheta) \, dP_\infty(\vartheta)/d\vartheta .$$

Proof. We have already degraded the $Q_\infty(\vartheta)$ term in (25) to $P_\infty^2(\vartheta)$. Under the structural assumption, the dangling end term may be written as

$$\text{Prob}_\vartheta[\mathbb{P}_\infty(0)\backslash Q_\infty(0)] = \sum_{b\in\mathbf{B}} \text{Prob}_\vartheta[\mathfrak{D}_b(\mathbb{P}_\infty(0))] . \qquad (28)$$

Then by (27), the van den Berg – Fiebig inequality and the Russo formula, we have

$$\text{Prob}_\vartheta[\mathbb{P}_\infty(0)\backslash Q_\infty(0)] \leq P_\infty(\vartheta) \sum_{b\in\mathbf{B}}\text{Prob}_\vartheta[\delta_b\mathbb{P}_\infty(0)](1-e^{-\vartheta J_b})$$

$$\leq P_\infty(\vartheta)\,\exp[\vartheta J_{MAX}] \sum_{b\in\mathbf{B}}\text{Prob}_\vartheta[\delta_b\mathbb{P}_\infty(0)](\vartheta J_b)\exp[-\vartheta J_{MAX}]$$

$$\leq \vartheta\exp[\vartheta J_{MAX}]P_\infty(\vartheta) \sum_{b\in\mathbf{B}}\text{Prob}_\vartheta[\delta_b\mathbb{P}_\infty(0)]J_b e^{-\vartheta J_b}$$

$$= \vartheta\exp[\vartheta J_{MAX}]P_\infty(\vartheta)d/d\vartheta[P_\infty(\vartheta)] . \qquad (29)$$

This completes the proof. ▨

Removal of the structural assumption. A spineless infinite cluster, in an independent percolation model, is a peculiar object. The reason for this is that, analogous to the definition of a backbone in terms of the *absence* of articulation bonds, a spineless infinite cluster can be characterized by the presence of *infinitely many* articulation bonds. This implies that, under small perturbations, such objects would break up, which in turn would force a discontinuity in the percolation probability. This is the content of our next result which, as it turns out, follows closely the spirit of [17].

Proposition 6. *Consider a bond percolation model of the type described here and suppose that $S_\infty(\theta_0) \neq 0$ for some value θ_0. Then θ_0 is a point of discontinuity for $P_\infty(\theta)$. In particular,*

$$P_\infty(\theta_0) - P_\infty(\theta_0^-) \geq S_\infty(\theta_0) .$$

Proof. Pick $\varepsilon \in (0, \theta_0)$, and let us discuss the percolation model at parameter value $\theta_0 - \varepsilon$. One way of doing this is to first occupy bonds at the densities $(1 - e^{-\theta_0 J_b})$, and then to drop "antibonds" with probabilities $R_{\varepsilon, \theta_0}(J_b)$ (to be computed below) with the following rule: if an antibond hits a vacant bond, nothing happens, but should it land on an occupied bond, this bond gets removed. The overall probability of retaining a bond is

$$(1 - e^{-\theta_0 J_b})(1 - R_{\varepsilon, \theta_0}) = 1 - e^{-(\theta_0 - \varepsilon)J_b} , \tag{30}$$

provided that

$$R_{\varepsilon, \theta_0}(J_b) = (e^{+\varepsilon J_b} - 1)e^{-\theta_0 J_b}/(1 - e^{-\theta_0 J_b}) . \tag{31}$$

Note that as $J \downarrow 0$, $R_{\varepsilon, \theta_0}(J_b) \rightarrow \varepsilon/\theta_0$; it is also easy to check that the "worst case" is that of maximum J:

$$R_{\varepsilon, \theta_0}(J_b) \geq R_{\varepsilon, \theta_0}(J_{MAX}) > 0 . \tag{32}$$

The uniform bound (32), together with the abundance of articulation bonds, in essence completes the proof. Indeed, doing a percolation problem at density $\theta_0 - \varepsilon$ in this fashion, we can easily estimate from below the probability that the origin is disconnected from infinity. Stopping temporarily at parameter value θ_0, there are two possibilities: either the origin is already in a finite cluster, in which case it remains in this state after the antibonds have been dropped, or it is in an infinite cluster. In the latter case, there are two further possibilities: the infinite cluster could be spineless or regular. We have

$$1 - P_\infty(\theta_0 - \varepsilon) = 1 - P_\infty(\theta_0) + S_\infty(\theta_0)\text{Prob}_{\varepsilon, \theta}[\mathcal{A}_\infty(0)|\mathcal{S}_\infty(0)] +$$
$$+ [P_\infty(\theta_0) - S_\infty(\theta_0)] \, \text{Prob}_{\varepsilon, \theta}[\mathcal{A}_\infty(0)|\mathcal{P}_\infty(0)\backslash\mathcal{S}_\infty(0)] , \tag{33}$$

where $\mathcal{A}_\infty(0)$ denotes the event that the origin gets disconnected by the antibonds. In equation (33), we have tacitly assumed that $\mathcal{P}_\infty(0)\backslash\mathcal{S}_\infty(0)$ is not of measure zero; otherwise, the final term should be ignored. It is easily seen that

$$\text{Prob}_{\varepsilon, \sigma}[\mathcal{A}_\infty(0)|\mathcal{S}_\infty(0)] = 1 , \tag{34}$$

since there are infinitely many articulation bonds each having a uniform probability of being struck by an antibond. Pitching out the final term in (33), we have

$$P_\infty(\sigma_0) \geq P_\infty(\sigma_0 - \varepsilon) + S_\infty(\sigma_0) . \tag{35}$$

The desired result is established by taking $\varepsilon \downarrow 0$. ▨

Remark. Recently, Aizenman, Kesten and Newman [8] have announced a proof that $P_\infty(\sigma)$ is continuous, except perhaps at σ_c. Thus Proposition 6 shows that above the critical point, the infinite cluster always has a backbone.

Even in the absence of a continuity result, Theorem 5 and Proposition 6 immediately imply our principal result.

Corollary. *For the percolation models of the type described here, for all σ,*

$$P_\infty(\sigma) \leq P_\infty^2(\sigma) + \sigma \exp[\sigma J_{MAX}] P_\infty(\sigma) \, dP_\infty(\sigma)/d\sigma .$$

In particular, for $\sigma \geq \sigma_c$,

$$P_\infty(\sigma) \geq P_\infty(\sigma_c) + \text{Const.}(\sigma - \sigma_c) + O(\sigma - \sigma_c)^2 .$$

4. Mean Field Bounds for the Backbone

The backbone density. Previously, we alluded to the fact that backbone of the infinite cluster may be of independent interest, in particular with regards to its critical behavior (if any). We may thus "define" a backbone density exponent, β_Q, according to

$$Q_\infty(\sigma) \sim |\sigma - \sigma_c|^{\beta_Q}, \tag{36}$$

where $\sim$ is to be understood in the usual sense, e.g. $\beta_Q = \lim_{\sigma \downarrow \sigma_c} \log Q_\infty(\sigma)/\log|\sigma - \sigma_c|$, should this limit exist.

An easy calculation performed on the Bethe lattice yields the result

$$\beta_Q[\text{Bethe lattice}] = 2 , \tag{37}$$

which we will identify as the mean field value. In this first subsection, we will obtain inequalities relating β_Q to other exponents. These will be mean field bounds in the sense that they hold as *identities* in mean field theory.

The first bound is straightforward; in fact, it follows directly from equation (23) and the van den Berg – Kesten inequality:

$$P_{\infty}^2(\theta) \geq Q_{\infty}(\theta) , \tag{38}$$

which, under the usual assumptions, implies

$$\beta_Q \geq 2\beta . \tag{39}$$

Although not particularly difficult to derive, (39) already has interesting consequences in the context of the random resistor and flow problems [18].

We will now show that the decomposition (24) permits an *upper* bound on β_Q in terms of β and a second exponent.

For $\theta > \theta_c$, the average size of the cluster of the origin is, of course, infinite. It may, however, be the case that the average size of the finite cluster of the origin remains bounded, diverging only as $\theta \downarrow \theta_c$. It is anticipated that this situation is typical -- for example, in the nearest neighbor model this can be rigorously established for $d = 2$ [19], and a similar statement can be made for $d = 3$ if θ_c is replaced by a limit of "slab thresholds" [20]; however, it *fails* for the one-dimensional $1/r^2$ model [7]. Nonetheless, we define

$$\chi'(\theta) = E_\theta(|C(0)| ; |C(0)| < \infty) , \tag{40}$$

with the hope that $\chi'(\theta)$ diverges with the power γ':

$$\chi'(\theta) \sim |\theta - \theta_c|^{-\gamma'} \tag{41}$$

as $\theta \downarrow \theta_c$. For those values of θ for which $\chi'(\theta) < \infty$, it is worthwhile to write

$$\chi'(\theta) = \sum_{x \in Z^d} \tau'_{ox}(\theta) , \tag{42}$$

where

$$\tau'_{ox} = \text{Prob}(0 \text{ and } x \text{ are in the same finite cluster}) \equiv \text{Prob}(T'_{ox}) , \tag{43}$$

and T'_{ox} admits the graphical definition:

$$T'_{ox} = \tag{44}$$

From equations (26) and (44), our next theorem follows in a straightforward fashion.

Theorem 7. *For all $\theta > \theta_c$,*

$$P_{\infty}(\theta) \leq C(\theta)[Q_{\infty}(\theta)\chi'(\theta)] ,$$

where $C(\theta)$ is a nonsingular function.

Proof. Using the results of [8] and Proposition 6, we can safely ignore all possible spineless contributions. Let us reevaluate the "dangling end" term in equation (25). As before, we write

$$\text{Prob}_\sigma[\mathbf{P}_\infty(0) \backslash \mathbf{Q}_\infty(0)] = \sum_{b \in \mathbf{B}} \text{Prob}_\sigma[\mathbf{D}_b(\mathbf{P}_\infty(0))] . \tag{45}$$

First, we define $\mathbf{D}'_b(\mathbf{P}_\infty(0))$ to be the event which is identical to $\mathbf{D}_b(\mathbf{P}_\infty(0))$ except that, here, the privileged edge b is vacant, not occupied. Obviously,

$$\text{Prob}_\sigma[\mathbf{D}_b(\mathbf{P}_\infty(0))] = \text{Prob}_\sigma[\mathbf{D}'_b(\mathbf{P}_\infty(0))][1 - e^{-\sigma J_b}]e^{+\sigma J_b} . \tag{46}$$

Focusing attention on a single bond b, with endpoints ∂b_1 and ∂b_2, it is seen that the event $\mathbf{D}'_b(\mathbf{P}_\infty(0))$ can be achieved in one of two ways: either the origin and ∂b_1 are in the same finite cluster and ∂b_2 is in the backbone or the other way around, i.e.

$$\mathbf{D}'_b(\mathbf{P}_\infty(0)) = T'_{0 \partial b_1} \circ \mathbf{Q}_\infty(\partial b_2) \cup T'_{0 \partial b_2} \circ \mathbf{Q}_\infty(\partial b_1) . \tag{47}$$

The sum over all bonds in $\mathbf{B}$ can be performed by first summing over the bonds emanating from a given site x, allowing x to play the role of the "inside site," and then summing over x:

$$\sum_{b \in \mathbf{B}} \text{Prob}_\sigma[\mathbf{D}_b(\mathbf{P}_\infty(0))] = \sum_{b \in \mathbf{B}} \text{Prob}_\sigma[\mathbf{D}'_b(\mathbf{P}_\infty(0))][1 - e^{-\sigma J_b}]e^{+\sigma J_b}$$

$$= \sum_{x \in \mathbf{Z}} \sum_{\varkappa \in \mathbf{K}} \text{Prob}_\sigma[T'_{0x} \circ \mathbf{Q}_\infty(x + \varkappa)][1 - e^{-\sigma J_x}]e^{+\sigma J_x} . \tag{48}$$

Using the van den Berg – Kesten inequality, the right hand side of (48) may be bounded above by

$$\mathbf{Q}_\infty(\sigma) \sum_{x \in \mathbf{Z}} \tau'_{0x} \sum_{\varkappa \in \mathbf{K}}[1 - e^{-\sigma J_x}]e^{+\sigma J_x} \leq \mathbf{Q}_\infty(\sigma)X'(\sigma)[T\sigma \exp(\sigma J_{MAX})] , \tag{49}$$

where T was defined in equation (7). Since $T\sigma \exp(\sigma J_{MAX})$ is a harmless function of σ, and since the contribution to $P_\infty(\sigma)$ due to the origin being in the backbone is just $1 \times Q_\infty(\sigma)$, the theorem has been proved. Furthermore, in any given neighborhood of σ_c, one can find a constant C such that

$$P_\infty(\sigma) \leq C Q_\infty(\sigma)X'(\sigma) . \tag{50}$$

Corollary. *If critical exponents* β, γ' *and* β_Q *exist, in the usual sense, then*

$$\beta_Q \leq \gamma' + \beta . \tag{51}$$

Remark. It is known that a lower bound of one can be obtained on $\aleph'$, and that this also saturates for the Cayley tree. Indeed, the former statement is derived (see [15]) by noting that

$$dP_{\infty}/d\theta = \sum_{b \in B} \text{Prob}[\delta_b(\mathbf{P}_{\infty}(0))] J_b e^{-\theta J_b}. \tag{52}$$

Hence

$$dP_{\infty}/d\theta = \sum_{x \in Z} \sum_{x \in K} \text{Prob}_\theta[T'_{ox} \circ \mathbf{P}_{\infty}(x + x)] J_x$$

$$\leq T P_{\infty}(\theta) \chi'(\theta), \tag{53}$$

implying that "$\aleph' \geq 1$." The calculation $\aleph' = 1$ on the tree is easy to perform; thus the inequality (51) is also of the mean field type.

Magnetic exponent for the backbone. In percolation theory, there is a well-known device for introducing an external field into the problem. One examines the usual configurations of occupied and vacant bonds; however, "infinity" is brought in so as to be close at hand: coming "up" from each site is a "ghost bond" directly connecting the given site to a huge "ghost site" -- which is identified as infinity. The probability that a ghost bond becomes occupied is given by

$$\text{Prob[occupied ghost bond]} = 1 - e^{-h}, \tag{54}$$

where h may be identified as a magnetic field strength.

Observe that if the origin is in an infinite cluster, it is automatically connected to the ghost site (i.e. with probability one for nonzero h). There are, of course, other mechanisims now available to get to infinity.

We denote by $P_n(\theta)$ the probability that the origin belongs to a cluster consisting of exactly n (earthly) sites. If this should happen, the probability that the origin gets hooked up to the ghost is exactly $1 - e^{-nh}$. Thus, the ghost percolation probability, which we denote by $P_*(\theta,h)$, is given by

$$P_*(\theta,h) = P_{\infty}(\theta) + \sum_n P_n(\theta)[1 - e^{-nh}]$$

$$= 1 - \sum_n P_n(\theta) e^{-nh}. \tag{55}$$

It is worth observing that the above expression is analytic in (e to the minus) h, at least when Re(h) is positive. Furthermore, $P_*(\theta,h) \downarrow P_\infty(\theta)$ as $h \downarrow 0$. Next, it is noticed that the derivitive, dP_*/dh, is given by

$$dP_*/dh = \sum_n nP_n(\theta) e^{-nh} , \tag{56}$$

which is none other than the expected size of the ghostless clusters. By analogy with magnetic systems, this is in fact the reason that the notation χ' ($\equiv \chi'(\theta,h)$) is used for the expected cluster size. Some immediate properties of the system, which can be gleaned from the above "generating functions," are

      i) monotonicity in h of $P_*(\theta,h)$

      ii) monotonicity (the other way) in h of $\chi'(\theta,h)$.

Indeed, $P_*(\theta,h)$ is completely monotone.

It is worth remarking, briefly, that the idea of "ghost sites" is due to Griffiths [21], in the context of Ising systems, and that the above formulation of a magnetic field percolation problem can be understood, to a reasonable degree of satisfaction, via the representation of Fortuin and Kasteleyn [22].

At the critical point, one expects (in most circumstances!) that $P_\infty(\theta_c) = 0$, and that, as a function of h, $P_*(\theta_c,h) \equiv P_*(h)$ tends to zero in a singular fashion. This "defines" the critical exponent δ:

$$P_*(h) \sim h^{1/\delta} . \tag{57}$$

In these last few paragraphs, we will introduce and obtain bounds on a new critical exponent which characterizes the magnetic behavior of the backbone clusters.

The backbone clusters also have a magnetic analogue. Indeed, here one simply defines $Q_*(\theta,h)$ to be the probability that there are two independent routes connecting the origin to the ghost site. It is not hard to show that $Q_*(\theta,h) \downarrow Q_\infty(\theta)$ as h tends to zero. Assuming, as is natural, that $Q_\infty(\theta_c) = 0$, we "define" a magnetic backbone exponent by

$$Q_*(\theta_c,h) \equiv Q_*(h) \sim h^{1/\delta_Q} . \tag{58}$$

As has been pointed out by Aizenman and Barsky [23], the remarkable aspect of these in-field problems is that, with suitable modifications, virtually all of the zero-field inequalities have magnetic analogues. As an example, they have shown that the magnetic version of (53) holds, which has the interpertation of a Burgers' inequality. According to the shock front analysis of Newman [24], this implies the exponent inequality

$$\beta(\delta-1) \geq 1 , \tag{59}$$

which, in combination with the results of Theorem 5, gives them the mean field bound $\delta \geq 2$ (although they reached the same conclusion earlier by different methods).

It turns out the above result result can be derived without any help from a Burgers' inequality. Here, we will derive an inequality which squeezes in between δ and 2, giving the mean field bound on the magnetic backbone exponent:

$$\delta \geq 2\delta_Q \geq 2 . \tag{60}$$

The upper bound is straightforward; it is an immediate consequence of the van den Berg - Kesten inequality applied to Q_* :

$$Q_*(h) \leq P_*^2(h) . \tag{61}$$

The lower bound follows from the inequality

$$P_*(h) \leq CX'Q_*(h) + (e^h - 1)X' , \tag{62}$$

which will be derived below. (As usual, C is nonsingular). Observe that (61) and (62) together imply the inequality $P_*(h) \leq CX'P_*^2(h) + (e^h - 1)X'$ -- the so-called master inequality of [23] -- from which the result $\delta \geq 2$ can be directly obtained.

Theorem 8. *For the percolation systems considered here, the inequalities (61) and (62) hold. In particular*

$$\limsup_{h \to 0} Q_*(h)/P_*^2(h) \leq 1$$

and

$$\limsup_{h \to 0} CQ_*(h)/h \geq 1 ,$$

so that if the critical exponents δ_Q and δ exist, then

$$\delta/2 \geq \delta_Q \geq 1.$$

Proof. The first "half" of this has already been described above. The inequality (62) is a generalization of Theorem 7 in an external field. Following Aizenman and Barsky [23], we first compute the contribution to $P_*(h)$ from configurations in which the origin is connected to a ghost via a single ghost bond. This is given by

$$\sum_n nP_n(\mathcal{O}) e^{-(n-1)h} (1 - e^{-h}) , \tag{63}$$

since, in a cluster of n sites, n-1 of them must be disconnected, and there are n ways of doing this. The above sum is simply

$$(1 - e^{-h})e^{+h}\sum_n nP_n(\vartheta) e^{-nh} = (e^h - 1)\chi' . \tag{64}$$

The rest of the rest of the derivation of (62) proceeds exactly along the lines of proof of Theorem 7, with the word "infinity" replaced by "the ghost."

To derive the mean field bound on δ_Q, we divide (61) by h and (62) by $\chi'h$, which gives

$$CP_*^2/h \geq CQ_*/h \geq P_*/\chi'h - 1 + O(h) . \tag{65}$$

Eliminating the middle term and solving (65) as though it were an equality would produce $P_*(h) \geq (1/\sqrt{C})h^{1/2}$, which up to o(h) provides a uniform lower bound on $P_*(h)$ [23]. Using this and l'Hôpital's rule, equation (65) implies

$$\limsup_{h \to 0} CQ_*/h \geq \limsup_{h \to 0} P_*/\chi'h - 1$$

$$\geq \limsup_{h \to 0} [\log P_*/\log h]^{-1} - 1$$

$$\geq 1 , \tag{66}$$

which is the desired result. ▨

Acknowledgments

We would like to thank H. Kesten for numerous discussions which were very useful in this work, and for having organized an enjoyble conference. We gratefully acknowledge the support of the IMA, which enabled us to participate in the conference. The research of JTC was supported by the NSF under grant no. DMR-83-14625, and that of LC by the DOE under grant no. DE-AC02-83-ER13044.

References

1. J. T. Chayes and L. Chayes, "Inequality for the infinite cluster density in Bernoulli percolation," *Phys. Rev. Lett.* **56**, 1619-1622 (1986).

2. G. R. Grimmett, M. Keane and J. M. Marstrand, "On the connectedness of a random graph," *Math. Proc. Camb. Phil. Soc.* **96**, 151-166 (1984).

3. L. Schulman, "Long-range percolation in one dimension," *J. Phys. A: Math. Gen.* **16**, L639-641 (1983).

4. C. M. Newman and L. Schulman, "One-dimensional 1/|j-i|^s percolation models: the existence of a transition for s ≤ 2," to appear in *Commun. Math. Phys.*

5. M. Aizenman and C. M. Newman, "Discontinuity of the percolation density in one-dimensional 1/|x-y|^2 percolation models," preprint.

6. M. Aizenman, J. T. Chayes, L. Chayes and C. M. Newman, in preparation.

7. M. Aizenman, J. T. Chayes, L. Chayes, J. Z. Imbrie and C. M. Newman, in preparation.

8. M. Aizenman, H. Kesten and C. M. Newman, in preparation.

9. L. Russo, "On the critical percolation probabilities," *Z. Wahrsh. verw Gebiete* **56**, 229-237 (1981).

10. J. van den Berg and H. Kesten, "Inequalities with applications to percolation and reliability theory," to appear in *Adv. Appl. Probab.*

11. J. van den Berg and U. Fiebig, "On a combinatorial conjecture concerning disjoint occurrence of events," to appear in *Ann. Probab.*

12. R. Peierls, "Ising's model of ferromagnetism," *Proc. Camb. Phil. Soc.* **32**, 477-481 (1936).

13. S. R. Broadbent and J. M. Hammersley, "Percolation processes I: crystals and mazes," *Proc. Camb. Phil. Soc.* **53**, 629-641 (1957).

14. L. Russo, "An approximate zero-one law," *Z. Wahrsh. verw Gebiete* **61**, 129-139 (1982).

15. R. Durrett, "Some general results concerning the critical exponents of percolation processes," *Z. Wahrsh. verw Gebiete* **69**, 421 (1984).

16. B. J. Last and D. J. Thouless, "Percolation theory and electrical conductivity," *Phys. Rev. Lett.* **27**, 1719-1721 (1971).

17. J. van den Berg and M. Keane, "On the continuity of the percolation probability function," *Contemp. Math.* **26**, 61-65 (1984).

18. J. T. Chayes and L. Chayes, "Bulk transport properties and exponent inequalities for random resistor and flow networks," to appear in *Commun. Math. Phys.*

19. H. Kesten, *Percolation Theory for Mathematicians*, (Birkhäuser, Boston 1982).

20. J. T. Chayes, L. Chayes and C. M. Newman, "Bernoulli percolation above threshold: an invasion percolation analysis," to appear in *Ann. Probab.*

21. R. B. Griffiths, "Correlations in Ising ferromagnets II: external magnetic fields," *J. Math. Phys.* **8**, 484-489 (1967).

22. C. M. Fortuin and P. W. Kasteleyn, "On the random cluster model I: introduction and relation to other models," *Physica* **57**, 536-564 (1971).

23. M. Aizenman and D. Barsky, in preparation.

24. C. M. Newman, lecture presented and announcement circulated at the 46th Statistical Mechanics Meeting, Rutgers University (December, 1981).

RECENT RESULTS FOR THE STEPPING STONE MODEL

by

J. Theodore Cox[1]
Syracuse University

and

David Griffeath[2]
University of Wisconsin

1. Ingredients

Let G be a graph, and $C = \{0,1,\ldots,\kappa-1\}$ a set of colors. We want to discuss a continuous time random process ζ_t with state space C^G = configurations of colors from C on the graph G. This system ζ_t, known as the <u>stepping stone model</u>, has very simple dynamics: in any state ζ and at any site x, the color at that site waits a mean $\frac{1}{2}$ exponential holding time and then paints a randomly chosen neighboring vertex with its color. We will also be treating a companion process ξ_t on {subsets of G} called <u>coalescing random walks</u>. As the name implies, ξ_t consists of rate $\frac{1}{2}$ continuous time random walks on G which coalesce when they collide. Let us write ξ_t^A to denote the evolution of those walks which start on a subset $A \subseteq G$. The graphs of principal interest to us are:

$$G_N = \text{the complete graph on } \{0,1,\ldots,N-1\},$$

$$Z_N^d = \text{the d-dimensional integers with period } N,$$

$$Z^d = \text{the d-dimensional integers.}$$

Tropical interpretations of the dimension d are good therapy for a Minnesota February: Baja California (d=1), Society Islands (d=2), Caribbean Marine World (d=3), and of course Mathematical Physics (d > 4). The cardinality κ of the color set C might be 2 or 32,768 or even ∞.

Our stepping stone process ζ_t has its source in mathematical genetics; the next section will review its origins. There sites of G represent individuals

---

[1]Partially supported by N.S.F. Grant DMS-841317.
[2]Partially supported by N.S.F. Grant DMS-830549.

(or more generally, colonies of individuals) and colors denote (allelic) types. ζ_t describes the evolution of types, while ξ_t gives the (reverse-time) evolution of geneologies. Of course other interpretations are possible. For $\kappa=2$, Williams and Bjerknes [37] considered biased versions of ζ as models for tumour growth (1=cancerous, 0=healthy), Clifford and Sudbury [8] named ζ the invasion process (two competing factions), and Holley and Liggett [20] called ζ the voter model (1="for", 0="against" on some issue).

2. Early History.

A discrete time, 2 color version of ζ on the complete graph G_N is one of the oldest models in mathematical population genetics. As W.J. Ewens writes in his comprehensive book [16],

"Naturally, biological reality should be the main criterion in our choice of model, but we shall also consider mathematical convenience in this choice. The model used implicitly by Fisher (1930) and explicitly by Wright (1931), and which we will call here the Wright-Fisher model, assumes that the genes in generation t+1 are derived by sampling with replacement from the genes of generation t."

Nontrivial spatial structure was introduced by Sewall Wright [39] about 10 years later when he described a discrete time, 2 color version of ζ on a finite lattice:

"At the opposite extreme from the [Fisher-Wright] model is that in which there is complete continuity of distribution, but interbreeding is restricted to small distances by the occurence of only short range means of dispersal. Remote populations may become differentiated merely from isolation by distance."

Wright's process was given a rigorous mathematical formulation in a beautiful little monograph by Malécot [28], who recognized that exact computations were possible using the theory of random walks. A few years later Kimura [23] dubbed ζ_t the "stepping stone model", and in 1964 Kimura and Weiss [24] published the first detailed analysis on the infinite lattice Z^d. They singled out $d = 2$ for biological applications:

"The two dimensional model can represent a population on a plane and cover the most important cases in nature."

Over the past twenty years the stepping stone model, in both its complete graph and integer lattice incarnations, has been exhaustively studied by a great many authors. Some of the more important contributions to stepping stone theory are listed at the end of this article. Probably the best general references are Ewens [16] (G_N) and Ligget [26] (Z^d), although Ewens deals with discrete time and Liggett only discusses the two color (voter model) case. The champion of many colors, in particular the case $\kappa = \infty$, is unquestionably Sawyer. We would like to single out his beautiful series of papers, especially [31], [32] and [33], which have been a major source of inspiration for out investigations.

The work which we will describe below extends the main theorems in [11] to models with more than two colors. Serious readers should probably look at [11] before proceeding, since our exposition there is more leisurely. Primary ojectives in this note are twofold: i) to document the early history of stepping stone type models, which was pointed out to us by a referee of [11], and ii) to announce our multi-color results. Proofs, which require some additional technical machinery, will appear elsewhere [12].

3. Duality.

The principal tool for analysis of stepping stone models is a <u>duality equation</u> connecting ζ_t with the coalescing walks ξ_t . If μ is an initial measure on C^G, then cylinder probabilities for the two models are related by the identity

(*) $$ P_\mu(\zeta_t(A_c) \equiv c; \; \epsilon \subset C) = E[\mu(\xi_t^{A_c}) \equiv c \; ; \; c \in C\}]. $$

(Both sides give the probability of the same event in a percolation scheme; stepping stone dynamics admit graphical representation (cf. [26]) for any κ.) Two general applications of (*) are as follows. First, let μ_θ denote Bernoulli product measure with $\mu_\theta(\zeta(x)=c) \equiv \theta_c$. Then

$$P_\theta(\zeta_t(x) = c) \equiv \theta_c \quad \text{(ζ is density preserving)}$$

and

$$P_\theta(\zeta_t(x) = c, \zeta_t(y) = c') = \theta_c \theta_{c'} P_x(\tau_y > t) \quad (x \neq y, c \neq c'),$$

where τ_y is the time a rate 1 random walk first hits y. So ζ_t exhibits absorption or clustering if $d = 1$ or 2, but is stable for $d \geqslant 3$. In any case, the two-point correlations can be analysed by means of random walk asymptotics. Higher order self correlations are expressed by

$$P_\theta(\zeta_t(A) \equiv c) = \sum_{k=1}^{\#A} \theta_c^k P(\#\xi_t^A = k),$$

and probabilities of more complex ξ events give the mixed correlations. As a second application, suppose each site of G starts with its own color. For the sake of notation say $C=G$ and start from $\mu = \delta_n$, where $n(x)=x$. In this case (*) yields

$$P_n(\zeta_t \text{ has } k \text{ colors on } A) = P(\#\xi_t^A = k) \quad (\#A < \infty).$$

4. Exact Results on G_N.

Exchangeability of ζ_t on the complete graph enables precise calculation of many probabilistic quantities. Clearly, if $\#A = n$, then $P(\#\xi_t^A = k) \equiv P_{n,k}(t)$. Moreover, $\#\xi^A$ is simply a pure death process with transitions

$$k \to k-1 \quad \text{at rate } \binom{k}{2}.$$

So $P_{n,k}(t)$ satisfies the backward equations

$$\begin{cases} \dfrac{d}{dt} P_{n,k}(t) = \binom{n}{2} [P_{n-1,k}(t) - P_{n,k}(t)], \\[2em] P_{n,k}(t)|_{t=0} = 1_{\{n=k\}}. \end{cases}$$

One can check that

$$P_{n,k}(t) = \sum_{j=k}^{n} c_n(j) \, a_k(j) \, \exp\{- \binom{j}{2} t\},$$

where

$$c_n(j) = \frac{\binom{n}{j}}{\binom{n+j-1}{j}} \quad \text{and} \quad a_k(j) = \frac{(-1)^{j+1}(2j-1)!\,(j+k-2!}{k!\,(k-1)!\,(j-k)!} .$$

Thus, two particularly simple explicit formulae are:

$$P_\eta(\zeta_t \text{ has } k \text{ colors}) = P_{N,k}(t) ,$$

$$P_\theta(\zeta_t \equiv c) = \sum_{k=1}^{N} \theta_c^k \, P_{N,k}(t).$$

$P_\theta(\zeta_t \text{ all one color})$ is obtained by summation, and $P_\theta(\zeta_t \text{ has } k \text{ colors})$ can be computed by inclusion exclusion. One can also compute the expected number of particles at time t, expected time until absorption (fixation), and other quantities of interest. Consult [16] or [35] for the details.

5. The Diffusion Limit.

Color proportions in the complete graph stepping stone model are asymptotically diffusive $(N\to\infty)$. Thus, let $X_t^{(N)}(c)$ be the proportion of sites on G_N with color c at time t. One easily verifies that

$$\lim_{h \to 0} \frac{1}{h} \, E\left[[X_{t+h}^{(N)}(c) - X_t^{(N)}(c)][X_{t+h}^{(N)}(c') - X_t^{(N)}(c')] \right] .$$

$$\frac{1}{2} X^{(N)}(c)]\delta_{cc'} - X_t^{(N)}(c')].$$

As $N \to \infty$, $(X_t^{(N)})$ converges to a diffusion (X_t) on the state space $\{X(c), c \in C: X(c) > 0, \sum x(c)=1\}$, with generator

$$A = \frac{1}{2} \sum_{c,c' \in C} x(c)[\delta_{cc'} - x(c')] \frac{\partial^2}{\partial x(c)\, \partial x(c')} .$$

Ethier and Kurtz [15] have shown that this generator (defined on a suitable domain) gives rise to a diffusion even in the case $\kappa = \infty$ (infinitely many alleles). If $\kappa = 2$ we get the (neutral) Fisher-Wright diffusion, which is absorbed at $(1,0)$ or $(0,1)$ after a finite time. For larger finite κ this process "falls" to lower order simplices until it is eventually trapped at one of the κ unit vectors. If $\kappa = \infty$ the diffusion instantly jumps to finite dimensions, and then proceeds as if κ were finite.

One can get duality equations for the moments of the limiting diffusion by passing to the limit in the duality between ζ and ξ. For simplicity, consider the case of two colors. Then

$$E_{(1-\theta,\theta)}[\{X_t(1)\}^n] = \lim_{N \to \infty} E_{(1-\theta,\theta)}[\{X_t^{(N)}(1)\}^n]$$

$$= P_{(1-\theta,\theta)}[\zeta_t \text{ has all 1's on } n \text{ sites}] = \sum_{k=1}^{n} \theta^k p_{n,k}(t) .$$

Note that by letting $n \to \infty$ we get the absorption probabilities:

$$P_{(1-\theta,\theta)}[X_t = (0,1)] = \sum_{k=1}^{\infty} \theta^k p_{\infty,k}(t) = \sum_{k=1}^{\infty} \theta^k \sum_{j=k}^{\infty} a_k(j) \exp\{-\left(\begin{smallmatrix}j\\2\end{smallmatrix}\right) t\} .$$

For more on genetics diffusion see, e.g., [15], [16], [34] and [36].

6. The Stepping Stone Model on Z^d.

The ergodic behavior of ζ_t is strongly dimension dependent. We are interested here in the case $d=2$, but before fixing on the planar model let us summarize the basic dynamics of other dimensions. Look at [26] and [11] for a more detailed survey. One should note in passing that the occupation times of ζ_t exhibit an even more elaborate dimension dependence (cf. [1], [4] and [9]).

d $\geq$ 3. The distribution of ζ_t approaches a steady state ν_θ as $t \to \infty$. If we let $S^c(x)$ denote the number of sites on a box of side n centered at x which have color c under ν_θ, then this field of block sums normalized by $n^{(d+2)/2}$ converges ([5], [27]) to a limiting Gaussian field with covariance

$$K_{d,\theta} \int \Lambda_1(x) \int \Lambda_1(y) \frac{1}{|u-v|^{d-2}} du \, dv,$$

where $\Lambda_n(x)$ is the box of side n centered at x. These are simple examples of strongly correlated equilibrium fields for which one can compute the renormalization limit. It would nice to know something about their geometry.

d=1. (from n) The clustering is easy to analyse due to the linear nature of the lattice. Boundaries between clusters execute coalescing random walks: colors disappear when they are swallowed by a surrounding pair of colors. Properly normalized, the edges converge to coalescing Brownian motions, one starting from every real position. Limit theorems ([6], [2]) can be read from this invariance principle. Clustering occurs at rate $\sqrt{t}$.

For the remainder of this paper the dimension is two.

d=2. The clustering is critical: there is no natural scale; rather the cluster "sizes" at time t are spread over all powers of t of order $< \frac{1}{2}$. We have made a micromovie on $Z^2_{512 \times 400}$ with $\kappa = 32{,}768$ which illustrates the effect nicely after 72 hours, say. [For postcards of this and other colorful particle system computer graphics, see the Note before the References.] Our principal objective in this paper is to announce rigorous results which quantify the critical clustering of ζ. Duality mandates study of coalescing random walks on Z^2. (With hindsight) the idea is to connect ξ on Z^2 with ξ on the complete graph $(G_\infty$???) and use the Fisher-Wright results. What follows is a very scant heuristic which may provide some insight; our actual proofs follow a rather different outline.

7. The Heuristic on Z^2.

Recall that the two point correlations of ζ_t are given by hitting probabilities for random walks. Asymptotics due to Erdös and Taylor [14] provide the key ingredient for our results. Using a "last time at 0" decomposition, for $0 < \alpha < \beta$ one gets

$$P_{(y-x)t^{\alpha/2}}(\tau_0 < t^\beta) = p_{t^\beta}(0,(y-x)t^{\alpha/2}) + \int_0^{t^\beta} p_u(0,(y-x)t^{\alpha/2})P_{(1,0)}(\tau_0 > t^\beta - u)du$$

$$\approx \int_1^{t^\beta} \frac{e^{-|y-x|^2 t^\alpha / u}}{\pi u} \; \frac{\pi}{\log(t^\beta - u)} \; du$$

$$\approx \int_0^\beta e^{-|y-x|^2 t^{\alpha - s}} \frac{\log t}{\log(t^\beta - t^s)} \; ds \; \overset{t \to \infty}{\to} \; 1 - \alpha/\beta \; .$$

Note that there is no dependence on $y-x$ in the limit! Now <u>fix</u> a large time t. For any α, the coalescing particles will be spread out on a $t^{\alpha/2}$ scale at time t^α. Considering this evolution as a process in a new variable u, given by $\alpha = e^{-u}$, the Erdös-Taylor asymptotics become $P(2 \text{ particles don't meet by } v \mid$ they don't meet by $u) \approx e^{-(v-u)}$, suggesting complete graph dynamics on this scale in the limit. So if $A_t = \{x_1 t^{\alpha/2}, \ldots, x_n t^{\alpha/2}\}$ (distinct x_i), we expect that

(#) $$\lim_{t \to \infty} P(\#\xi^{A_t}_{t^\beta} = k) = P_{n,k}(\log(\beta/\alpha)).$$

Equation (#) is part of our <u>Theorem 1</u>. To carry out the program for κ colors when $\kappa > 2$ it is necessary to compute probabilities concerning not just $\#\xi$, but also which particles collide with which. The asymptotics for such statistics are determined by a complete graph process known as <u>Kingman's coalescent</u> [25]. Theorem 1 states roughly that coalescing random walks ξ_t on Z^2 converge to the coalescent if particles are appropriately spread out and time is appropriately rescaled. The precise formulation of Theorem 1 is deferred to [12].

8. Exact Results on Z^2.

We are now prepared to state three theorems for the two dimensional stepping stone model. All follow from the convergence of ξ to Kingman's coalescent (Theorem 1) mentioned above. The first two are easy consequences of that result, whereas the third also relies on joint work with Maury Bramson [3].

<u>Theorem 2.</u> The $t^{\alpha/2}$ thinning of the two dimensional κ color stepping stone model converges as $t \to \infty$ to a limiting exchangeable random field with de Finetti mixture $F_{\theta,\alpha} = P_\theta(X_{\log(1/\alpha)} \in \cdot)$, where X is the κ color genetics diffusion. That is to say, as $t \to \infty$,

$$P_\theta((\zeta_t(xt^{\alpha/2}))_{x \in Z^2} \in \cdot) \rightarrow \int_\theta \mu_\theta \, dF_{\theta,\alpha} \, .$$

Theorem 3. As $t \rightarrow \infty$, the color densities in the two dimensional κ color stepping stone model on a box of side $t^{\alpha/2}$ converge as a process in $\alpha \in [0,1]$ to a time change of the κ color genetics diffusion. Specifically,

$$[B^c(\cdot)]_{c \in C} \overset{t \rightarrow \infty}{\Rightarrow} X_{\log(1/\cdot)} \, ,$$

with convergence in the sense of finite dimensional distributions.

Theorem 4. Let $N^t = \sup \{n: \zeta_t(\Lambda_n)) \equiv c$ for some $c\}$ be the width of the largest box centered at the origin which is all one color at time t. Then

$$\frac{\log N^t}{\log t} \Rightarrow L^\infty/2 \quad \text{as} \quad t \rightarrow \infty \, .$$

Starting from μ_θ ,

$$P_\theta(L^\infty < \alpha) = \sum_{k=1}^\infty \sum_{c \in C} \theta_c^k \, P_{\infty,k}(\log(1/\alpha)).$$

If each site of Z^2 starts with its own color (configuration η),

$$P_\eta(L^\infty < \alpha) = P_{\infty,1}(\log(1/\alpha)) = \{\prod_{k=1}^\infty (1 - \alpha^k)\}^3 \, .$$

(The last remarkable identity is due to Jacobi [22].)

NOTE

Computer graphics postcards in thousands of colors, which
illustrate the stepping stone model and other systems of
interacting particles, are available from David Griffeath,
Mathematics Dept., University of Wisconsin, Madison WI   53706

82

References

[1] Andjel, E. and Kipnis, C. (1985) Pointwise ergodic theorems for nonergodic interacting particle systems. Preprint.

[2] Arratia, R. (1982) Coalescing Brownian motions and the voter model on Z. Unpublished manuscript.

[3] Bramson, M., Cox, J.T. and Griffeath, D. (1986) Consolidation rates for two interacting systems in the plane. To appear.

[4] Bramson, M., Cox, J.T. and Griffeath, D. (1986) Occupation time large deviations of the voter model. To appear.

[5] Bramson, M. and Griffeath, D. (1979) Renormalizing the 3-dimensional voter model. Annals of Probability **7**, 418-432.

[6] Bramson, M. and Griffeath, D. (1980) Clustering and dispersion rates for some interacting particle systems on Z. Annals of Probability **8**, 183-213.

[7] Bramson, M. and Griffeath, D. (1980) Asymptotics for interacting particle systems on Z_d. Z. Wahrsch. verw. Gebiete **45**, 183-196.

[8] Clifford, P. and Sudbury, A. (1973) A model for spatial conflict. Biometrika **60**, 581-588.

[9] Cox, J.T. and Griffeath, D. (1983) Occupation time limit theorems for the voter model. Ann. Probability **11**, 876-893.

[10] Cox, J.T. and Griffeath, D. (1985) Large deviations for some infinite particle system occupation times. ContemporaryMath. **41**, 43-54.

[11] Cox, J.T. and Griffeath, D. (1986) Diffusive clustering in the two dimensional voter model. Ann. Probability **14**, 347-370.

[12] Cox, J.T. and Griffeath, D. (1987) Mean field asymptotics for the planar stepping stone model. In preparation.

[13] Donnelly, P. (1984) The transient behaviour of the Moran model in population genetics. Proc. Cambridge Phil. Soc. **95**, 349-358.

[14] Erdös, P. and Taylor, S.J. (1960) Some problems concerning the structure of random walk paths. Acta Math. Acad. Sci. Hungaricae **11**, 137-162.

[15] Ethier, S. and Kurtz, T. (1981) The infinitely-many-neutral alleles diffusion model. Adv. in Appl. Probab. **13**, 429-452.

[16] Ewens, W.J. (1979) Mathematical Population Genetics. Springer-Verlag, New York.

[17] Felsenstein, J. (1975) A pain in the torus: some difficulties with models of isolation by distance. American Naturalist **109**, 359-368.

[18] Fisher, R.A. (1930) The Genetical Theory of Natural Selection. Clarendon Press, Oxford.

[19] Griffiths, R.C. (1979) Exact sampling distributions from the infinite neutral alleles model. Adv. Appl. Prob. **11**, 326-354.

[20] Holley, R. and Liggett, T.M. (1975) Ergodic theorems for weakly interacting infinite systems and the voter model. Annals of Probability **3**, 643-663.

[21] Holley, R. and Stroock, D. (1979) Central limit phenomena of various interacting systems. Ann. Math. **110**, 333-393.

[22] Jacobi, C.G.J. (1829) Fundamenta Nova Theoriae Functionum Elipticarum, Regiomontis, fratum Borntraeger. Reprinted in Gesammelte Werke, vol. 1, Reimer, Berlin, 1881.

[23] Kimura, M. (1953) "Stepping stone" model of population. Ann. Rept. Nat. Inst. Gentics, Japan **3**, 62.

[24] Kimura, M. and Weiss, G. (1964) The stepping stone model of population structure and the decrease of genetic correlation with distance. Genetics **49**, 561-576.

[25] Kingman, J.F.C. (1982) The coalescent. Stoch. Proc. Applns. **13**, 235-248.

[26] Liggett, T.M. (1985) Interacting Particle Systems. Springer-Verlag, New York.

[27] Major, P. (1980) Renormalizing the voter model. Space and space-time renormalization. Studia Sci. Math. Hungar. **15**, 321-341.

[28] Malécot, G. (1948) The Mathematics of Heredity. English translation, W.H. Freeman, San Francisco, 1969.

[29] Moran, P. (1962) The Statistical Processes of Evolutionary Theory. Clarendon Press, Oxford.

[30] Presutti, E. and Spohn, H. (1983) Hydrodynamics of the voter model. Annals of Probability **11**, 867-875.

[31] Sawyer, S. (1976) Results for the stepping-stone model for migration in population genetics. Annals of Probability **4**, 699-728.

[32] Sawyer, S. (1977) Rates of consolidation in a selectively neutral migration model. Annals of Probability **5**, 486-493.

[33] Sawyer, S. (1979) A limit theorem for patch sizes in a selectively-neutral migration model. J. Appl. Probab. **16**, 482-495.

[34] Shiga, T. (1981) Diffusion processes in population genetics. J. Math. Kyoto Univ. **21**, 133-151.

[35] Tavaré, S. (1984) Line-of-descent and genealogical processes, and their applications in population genetics models. Theoretical Population Biology **26**, 119-164.

[36] Watterson, G.A. (1976) The stationary distribution of the infinitely-many neutral alleles diffusion model. J. Appl. Prob. **13**, 639-651.

[37] Williams, T. and Bjerknes, R. (1972) Stochastic model for abnormal clone spread through epithelial basal layer. Nature **236**, 19-21.

[38] Wright, S. (1931) Evolution in Mendelian populations. Genetics **16**, 97-159.

[39] Wright, S. (1943) Isolation by distance. Genetics **28**, 114-138.

DEPT. OF MATHEMATICS
SYRACUSE UNIVERSITY
SYRACUSE, NY 13210

DEPT. OF MATHEMATICS
UNIVERSITY OF WISCONSIN
MADISON, WI 53706

STOCHASTIC GROWTH MODELS

R. Durrett, Cornell[1]

R.H. Schonmann, Sao Paulo and Rutgers[2]

Introduction and Summary

1. Description of the models

2. Basic questions and set up

3. Edge speeds characterize p_c

4. A renormalized bond construction

5. The complete convergence theorem

6. Limit Laws for ξ_n^o

7. Results for $d > 1$

Introduction and Summary

This paper is based on a talk given by the first author at the I.M.A. in February, 1986 but incorporates improvements discovered during six later repititions. The second authour should not be held responsible for the style of presentation of the results but should be given credit for discovering the results independently in the Fall of 1985. The discussion below is equal to the talk with most of the details of the proofs filled in, but we have tried to preserve the informal style of the talk and concentrate on the "main ideas" rather than giving complete details of the proofs. If we forget about definitions then the results can be summed up in a few words "Everything Durrett and Griffeath (1983) proved for one-dimensional nearest neighbor additive groth models is true for the corresponding class of finite range models, i.e., those which can be constructed from a percolation structure."

[1] This author was partially supported by an NSF grant and an AMS "mid-career" fellowship.

[2] This author was partially supported by CNPq (Brazil) and NSF during the academic year 1985-86 which he spent at the Rutgers Math department. He will visit the Cornell Mathematical Sciences Institute for 1986-87.

We will describe the models we consider in a minute but even before we do this it is easy to see the main point of our generalization: the words nearest neighbor have been replaced by finite range. This generalization has two benefits. The first and most obvious it that it greatly increase the number of systems to which our results can be applied.

A second benefit is that we are able to improve what is known about the discrete time contact process (and other models) in Z^2. To be precise results which Durrett and Griffeath (1982) could only prove for $p > p_c(Z)$ the critical value for the process on the integers Z can now be shown for $p > p_c(Z \times \{-L,\ldots,L\})$ for any $L < \infty$. Presumably

$$\lim_{L \to \infty} p_c(Z \times \{-L,\ldots L\}) = p_c(Z^2).$$

(and then our results hold for all $p > p_c(Z^2)$) but we have no idea how to prove this, and in any case we are getting way ahead of ourselves. We will discuss the last topic in Section 6 but before this a number of other things must be done (e.g. defining $p_c(Z)$). In Section 1 we will describe the class of models for which we can prove our results. These "generalized percolation processes (g.p.p.s)" in Z^d are generalizations of oriented percolation in Z^{d+1} and have two special properties (additivity and duality) which make them easier to study than other discrete time growth models.

In Section 2 we will begin out study of g.p.p.'s by describing the questions we want to study, and the set up we will use to formulate our answers. The real work begins in Section 3 when we prove that "edge speeds characterize p_c". This is one of two keys to developments that follow, the other being the renormalized bond construction described in Section 4. That construction, in the words of Durrett and Griffeath (1983), "was inspired by work of Russo and Kesten and allows us to reduce results concerning supercritical contact processes to corresponding results about 1-dependent oriented percolation with p arbitrarily close to 1".

Once one has the two results in the last paragraph one can, following the pattern of Durrett (1984), obtain a large number of results. In Sections 5 and

6 we will prove two of the most important of these: the complete convergence theorem (which is called complete because it describes the limit in distribution starting from any initial configuration) and the strong law for $|\xi_n^0|$, the number of particles at time n starting from a single particle at 0. Exponential estimates and large deviations results like those in Sections 10-13 of Durrett (1984) could also be proved but no new ideas are needed so we will leave this as an exercise for the reader.

Given the dates of the papers with the two "keys" to the proof the reader may ask why he had to wait unitl 1986 for the results we have here. The answer is simple: the approach of Durrett (1980) relies on a "coupling" result which is a special feature of the nearest neighbor case (see Lemma 3.4 in Durrett (1980) or (6) in Section 3 below) and only recently did we have the idea to go around this step using the renormalized bond construction (Note: to close the circle, when we are done we can go back and prove that the coupling result is almost correct, see Section 6).

Having extolled the virtues of our results it is only fitting to close this introduction by listing their weaknesses. The first and most obvious is that we are able to prove our results only for generalized percolation processes and not for the more general class of monotone (or attractive) growth models. (If these terms are unfamiliar they will be defined in Section 1). Accomplishing that generalization will require a new idea and not just rearranging the old ones.

A second more technical defect is that we have only proved the result in discrete time. The reader will see the reason for this at the end of Section 4 when we use the green bonds to tie the blue paths together. This part of the argument can undoubtedly be done in continuous time hut would requires many more technical details, since continuous time paths can move arbitrarily fast while paths for a finite range discrete time system have a strict speed limit.

1. Description of the Models

In this section we will describe the various classes of models we will consider in this paper. In all cases the system will be a discrete time Markov chain whose state at time n is $\xi_n \subset Z^d$ and which evolves according to the following rules

(i) $\qquad\qquad P(x \in \xi_{n+1} | \xi_n) = g(\xi_n(x + y_1), \ldots \ \xi(x + y_k)).$

where $k < \infty$ $\{y_1, \ldots y_k\} \subset Z^d$ and we have used coordinate notation for the random set: $\xi_n(x) = 1$ if $x \in \xi_n$ and $\xi_n(x) = 0$ if $x \in \xi_n$.

(ii) given ξ_n, the state at time $n + 1$ is decided by flipping independent coins, i.e. for any j and $x_1, \ldots x_j \in Z^d$

$$p(x_i \in \xi_{n+1} \text{ for } 1 < i < j | \xi_n) = \prod_{i=1}^{j} p(x_i \in \xi_{n+1} | \xi_n).$$

Systems which satisfy (i) and (ii) are what we would call discrete time particle systems but are often referred to in the physics literature as stochastic cellular automata. (See Kinzel (1985)). If we impose the additional condition

(iii) g is monotone i.e. if $x < y$ coordinatewise then $g(x) < g(y)$
then we say the process is monotone or "attractive"

and if we insist in addition that

(iv) there is no creation from nothing, i.e. $g(0) = 0$

then we have the class of processes mentioned in the title of the paper: stochastic growth models.

If one thinks (as we do) of the points in ξ_n as being occupied by a particle (think of an animal or better yet a plant) then assumption (iv) is clearly natural. Assumption (iii) is also reasonable. The probability of a birth should be an increasing function of the occupancy of the neighbors (unless severe overcrowding cause higher death rates). In any case, assumption (iii) is very useful (see Liggett (1985), Chapter III, Section 2) and for most of our results we will have to restrict our attention to an even smaller class of processes which are the discrete time analogues of the additive process of Harris (1978) and Griffeath (1979).

These generalized percolation processes are constructed from a "graphical representation". Specifically, we make Z^2 into a random graph in which the oriented bond $(x - y,n) \rightarrow (x,n + 1)$ is open (resp. closed) with probability $f(y)$ (resp. $1 - f(y)$); bonds ending at different sites are independent; and the system is translation invariant (so the joint distribution of bonds ending at a given site is always the same).

To construct the process from this graphical representation we let

$$\xi_n^A = \{y: \text{ there is a path of open bonds from}$$
$$(x,0) \text{ to } (y,n) \text{ for some } x \in A\}.$$

The subscript and the superscript on ξ indicate that it is the state at time n when the initial state is A. To explain the name and the right hand side we observe that ξ_n^A is the set of wet sites at level n if we imagine there is a source of fluid at $(x,0)$ for each $x \in A$ and the fluid can travel only through open bonds.

A few examples should help clarify the definitions.

Example 1: <u>Oriented bond percolation.</u> In this model $f(y) = p$ if $y \in S$ where S is a finite set and all the bonds are independently open or closed. The name comes from the fact that in the special case $S = \{0,1\}$, ξ_n^0 is what results when we take the usual oriented bond percolation process in Z^2, map $(x,y) \rightarrow (x,x + y)$, and look at $\{z: (z,n) \text{ can be reached from } (0,0)\}$. For more on this see section 2 of Durrett (1984).

Example 2. <u>Oriented site percolation.</u> In this model $f(y) = p$ if $y \in S$ where S is a finite set (like the last model) but this time either all the bonds $(x - y,n) \rightarrow (x,n + 1)$ are open with probability p or all are closed with probability $1 - p$. Again the name comes from the fact that in the special case $S = \{0,1\}, \xi_n^0$ is what results if we consider the points in Z^2 (called sites) to be the objects which are open or closed, define a path to be open if it contains no closed sites, map $(x,y) \rightarrow (x,x + y)$, and look at $\{Z: (z,n) \text{ can be reached from } (0,0)\}$.

Examples 1 and 2 are extreme cases and a large number of examples can be constructed by combining these two. In the next two examples we will consider what happens when g depends on two or three values of $\xi_n(x + y)$ to try to convince the reader that "many interesting examples but by no means all growth models are g.p.p.".

Example 3: <u>Two-site g.p.p.</u> Consider systems in which

$$P(x \; \varepsilon \; \xi_{n+1} | \xi_n) = g(\xi_n(x + y_1), \; \xi_n(x + y_2))$$

where $y_1 \neq y_2$ are in Z. I claim that this model is a g.p.p. if and only if

$$g(0,0) = 0$$

$$g(0,1), g(1,0) < g(1,1) < g(0,1) + g(1,0).$$

To see this observe that if bonds $b_1 = (x - y_1, n) \to (x, n + 1)$ and $b_2 = (x - y_2, n) \to (x, n + 1)$ have

b_1	b_2	with probability
open	open	a
open	closed	b
closed	open	c
closed	closed	1 - a + b + c)

then

$$g(1,1) = a + b + c$$
$$g(1,0) = a + b$$
$$g(0,1) = b + c$$

so the conditions above are necessary and sufficient to have $a, b, c > 0$ $(a + b + c = g(1,1)$ so the sum is automatically $< 1)$.

Example 4. <u>A simple class of 3 site g.p.p.</u> If we look at the general model on three sites then we get a bewildering number of conditions so to simplify things we will only consider what we call sum rules

$$P(x \in \xi_{n+1} | \xi_n) = f_{|\xi_n \cap \{x - 1, x, x + 1\}|}$$

where $|A|$ = the number of points in A. (= the sum of the coordinates $\xi_n(x - 1) + \xi_n(x) + \xi_n(x + 1)$). Calculations similar to those in the last example show that these processes are g.p.p. if and only if

$$f_3 = a + 3b + 3c$$
$$f_2 = a + 3b + 2c$$
$$f_1 = a + 2b + c$$

for some $a, b, c > 0$ with $a + 3b + 3c < 1$. The last condition is automatic since $f_3 < 1$, and for the first three to hold we must have

$$c > 0: \quad f_3 > f_2$$
$$b > 0: \quad f_2 - f_1 > f_3 - f_2$$
$$a > 0: \quad (f_1 - 0) - 2(f_2 - f_1) + (f_3 - f_2) > 0.$$

The inequalities above imply

$$0 < f_1 < f_2 < f_3$$
$$f_1 - 0 > f_2 - f_1 > f_3 - f_2 > 0$$
$$(f_1 - 0) - 2(f_2 - f_1) + (f_3 - f_2) > 0.$$

in contrast to the conditions for two site sum rules:

$$0 < f_1 < f_2$$
$$f_1 - 0 > f_2 - f_1.$$

We leave it to the reader to find the general result (or see Harris (1978)).

In closing the discussion of the models we would like to note that although we have arrived at our conditions from a desire to use the graphical represen-tation, one can, after the fact, argue that the first two conditions, are not too bad biologically: increasing the number of occupied sites should increase the birth rate but each new individual should result in a smaller increase.

The third condition which says "the first difference is convex" is harder to defend but it is satisfied for three site bond percolation (where

$f_k = 1 - (1 - p)^k$. The number of unpleasant conditions we have to accept increases with the range but it is comforting to note that it is always an open set.

2. Basic Questions and Set Up

Having defined the models we want to study, the next thing to explain is what we want to prove about them. In the last section we mentioned the fact that we think of the points in ξ_n as occupied by plants or animals so it is natural to ask: Is $P(\xi_n^0 \neq \phi$ for all $n) > 0$? (i.e. does the species have positive probability of not dying out) and if the answer to the first question is yes, "what does ξ_n^0 look like on $\Omega_\infty = \{\xi_n^0 \neq \phi$ for all $n\}$?"

Most of the rest of the paper is devoted to answering the last two questions. We will not have much to say about the first but we will be able to give a fairly complete answer to the second question for all "supercritical" g.p.p. It will take a few minutes to explain what we mean by the word in quotation marks, so will postpone that for a moment and set the stage by describing what sorts of answers we have for the examples described in the last section. In oriented bond percolation (example 1) the fraction of open bonds increases to 1 as p does, so it is natural to let

$$p_c = \inf\{p: P(\xi_n^0 \neq \phi \text{ for all } n) > 0\}.$$

The first thing to be resolved is: "Is $p_c \in (0,1)$?" This question is answered by

Proposition 1. Let $|S|$ = the number of points in S.

If $|S| \geq 2$ then $\frac{1}{|S|} \leq p_c \leq \frac{8}{9}$.

Proof. For the left side compare with a branching process. For the right see e.g. Durrett (1984), Section 10.

Computing p_c has turned out to be a difficult problem (see Durrett (1984), Section 6) but somewhat surprisingly it is possible to prove results valid for all $p > p_c$ without knowing what p_c is. This was done in Durrett (1984) for the case $S = \{0,1\}$ (or $= \{-1,1\}$) and will be done for general finite S below.

Having heard us say "supercritical" above the reader has probably noticed that the results above are only stated for $p > p_c$ and no mention is made of the critical case $p = p_c$. Presumably $P(\Omega_\infty) = 0$ at p_c so the asymptotic behavior of ξ_n^0 is trivial there, but this has turned out to be highly nontrivial to prove (and is an important open problem).

The situation for site percolation is the same as for bond percolation so we turn out attention now to Example 3: two site models. Suppose for simplicity that $\{y_1, y_2\} = \{0,1\}$ and we have a sum rule, i.e. $g(1,0) = g(0,1) = p_1$, $g(1,1) = p_2$. By results in Section 1 this process is a g.p.p. if and only if $p_1 < p_2 < 2p_1$ or, geometrically, (p_1, p_2) lies in the triangle with vertices $(0,0)$, $(1,1)$ and $(\frac{1}{2},1)$ (see Figure 2.1).

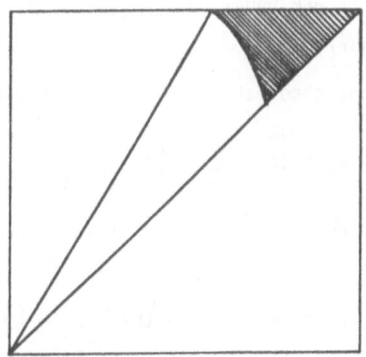

FIGURE 2.1.

The processes with $p_2 = 1$ and $p_1 = p$ are easy to understand. In this case if we draw a picture (see Figure 2.2) then it is easy to check that (for $0 < p < 1$) we have

$$0\ 0\ 0\ ?\ 1\ 1\ 1\ 1\ 1\ ?\ 0\ 0\ 0$$
$$0\ 0\ 0\ 1\ 1\ 1\ 1\ 1\ 1\ 0\ 0\ 0\ 0$$

$$? = 1 \text{ with prob } p$$
$$? = 0 \text{ with prob } 1-p.$$

Figure 2.2.

(1) ξ_n^o always equals $\{x: \ell_n^o < x < r_n^o\}$, where $\ell_n = \inf \xi_n^o$ and $r_n^o = \sup \xi_n^o$.

(2) r_n^o is a random walk which moves $x \to x + 1$ with probability p and $x \to x$ with probability $1 - p$.

(3) ℓ_n^o is a random walk which moves $x \to x$ with probability p and $x \to x + 1$ with probability $1 - p$

and

(4)   the increments $r_{n+1}^o - r_n^o$ and $\ell_{n+1}^o - \ell_n^o$ are independent on $\{\xi_n^o \neq \phi\}$.

For (4) observe that if $\xi_n^o = \{x\}$ then $\{x \in \xi_{n+1}^o\}$ and $\{x + 1 \in \xi_{n+1}^o\}$ are independent and these events are equal to $\{\ell_{n+1}^o - \ell_n^o = 0\}$ and $\{r_{n+1}^o - r_n^o = 1\}$. The case $|\xi_n^o| > 1$ is easier.

Combining the last four observation we see that the number of particles

$$Z_n = (1 + r_n^o - \ell_n^o) 1_{(\xi_n^o \neq \phi)}$$

is a random walk starting from 1 and run until it hits 0. The mean of r_1^o is p, the mean of ℓ_1^o is $1 - p$, and $p > 1 - p$ if and only if $p > 1/2$, so from the last three facts it is easy to see that $p_c = 1/2$, i.e., if $p < 1/2$ then the increments in Z_n have negative mean and $P(\xi_n^o \neq \phi$ for all $n) = 0$. On the other hand if $p > 1/2$ then $Er_1^o - E\ell_1^o = c > 0$ and we have $P(\Omega_\infty) > 0$.

For comparison with later results and an earlier conjecture, we would like the reader to observe that

$$P(\Omega_\infty) = 0 \quad \text{at} \quad p_c,$$

and if $p > p_c$ then on Ω_∞

$$\frac{r_n^o}{n} \to p \qquad \text{a.s.}$$

$$\frac{\ell_n^o}{n} \to 1 - p \qquad \text{a.s.}$$

$$\frac{|\xi_n|}{n} \to 2p - 1 \qquad \text{a.s.}$$

Having solved our problems when $p_2 = 1$ and $p_1 = p$ it is natural to make this a starting point for investigating the rest of the triangle $(0,0)$, $(1,1)$, $(\frac{1}{2},1)$. Let

$$P_{1,c}(\theta) = \inf \{p: P_{p,\theta}(\Omega_\infty) > 0\}$$

where $P_{p,\theta}$ is the probability measure for the system with $p_1 = p$ and $p_2 = \theta$. As θ decreases from 1, $P_{1,c}(\theta)$ increases (i.e. if $\theta_1 < \theta_2$ $P_{1,c}(\theta_1) > P_{1,c}(\theta_2)$) at least as long as $(P_{1,c}(\theta),\theta)$ stays in the set $\{(p_1,p_2): p_1 < p_2\}$ of attractive interactions. [Life below the diagonal will be the subject of a later paper].

The models with $p_2 = \theta$, $p_1 = P_{1,c}(\theta)$ are "critical" since they lie on the boundary of $\{(p_1,p_2): P(\Omega_\infty) > 0\}$ and we will have nothing to say about them (except that they presumably also have $P(\Omega_\infty) = 0$). We will however he able to prove fairly complete results about the models which live strictly above the critical curve (i.e. in the shaded region in Figure 2.1).

There is a similar but more complicated picture (which we leave for the reader to draw) for three site sum rules (example 4) or for the general case in example 3. In each situation the parameter space is three dimensional, there is a two dimensional critical surface, and we will be able to prove results about the "supercritical" models in the interior of $\{p: P(\Omega_\infty) > 0\}$.

To prove results for these supercritical models it is convenient to embed them in a one parameter family like site or bond percolation so we will assume that $(x - y,n) \to (x,n + 1)$ is open with probability $f(y,p)$. In order to prove our results we will, of course, head to make some assumptions about $f(y,p)$ and (even though the first author failed to mention this in his talk) also make assumptions about how the joint distributions change as p increases.

The first and most obvious of these is
(H1) the joint distribution of the bonds $(x - y_i,n) \to (x,n + 1)$ is stochastically increasing in p,
i.e., if $p' < p$ then the two systems can be constructed on the same space in such a way that if a bond is open in the p' system it is also open in the p system.

We will need a little more than this at two points below: we will need to know that the systems are strictly increasing in p. The technical assumptions · required will be obvious when we get there so we will introduce them as they are needed. The reader can be assured that site and bond percoaltion and the models above with $p_2 = \theta$ and $p_1 = p$ will always be included.

3. Edge Speeds Characterize p_c.

The monotonicity assumption we just made allows us to define a critical value

$$p_c = \inf\{p: P\{\xi_n^0 \neq \phi \text{ for all } n\} > 0\}$$

which has the property that

$$\text{if } p < p_c \quad \text{then} \quad P(\Omega_\infty) = 0$$
$$\text{if } P > p_c \quad \text{then} \quad P(\Omega_\infty) > 0.$$

The key to being able to prove results for all $p > p_c$ without knowing what p_c is, is finding a way to characterize p_c. The answer is hinted at in the title of the section and described below. It, like everything else in this section, is from Durrett (1980).

The first step in our analysis is to define the "right edge" by

$$r_n = \sup(\xi_n^{(-\infty,0]})$$

and embed r_n into a two parameter process by setting

$$r_{m,n} = \sup\{y: \text{ there is an open path from}$$
$$(x,m) \text{ to } (y + r_{m,n}) \text{ for some } x < r_m\}.$$

In words, $r_m + r_{m,n}$ is the rightmost site we can reach at time n if we pretend all the sites to the left of r_m are occupied at time m, so it is clear that

$$r_m + r_{m,n} > r_n$$

$$r_{m,n} \overset{d}{=} r_{n-m} \quad \text{and is independent of} \quad r_m.$$

Combining the last two observations with some ideas from the proof of Kingman's subadditive ergodic theorem one can show

(1)   As $n \to \infty$ $\quad r_n/n \to \alpha$ almost surely where

$$\alpha = \inf_{m \geq 1} E r_m/m.$$

For a proof see Durrett (1980), 893-896 or for a better proof see Liggett (1985), Chapter VI, Section 2.

Looking at the last argument in the mirror gives

(1')   Let $\ell_n = \inf\{\xi_n^{[0,\infty)}\}$. As $n \to \infty$ $\quad \ell_n/n \to \beta$ almost surely

where $\beta = \sup_{m \geq 1} E \ell_m/m.$

From (1) and (1') it follows immediately that we have

(2)   if $\alpha < \beta$ then $P(\xi_n^0 \neq \phi$ for all $n) = 0.$

Proof: Let $r_n^0 = \sup \xi_n^0$. Since $\xi_n^0 \subset \xi_n^{(-\infty,0]}$ (recall that the graphical representation defines the process simultaneously for all initial states) we have $r_n^0 \leq r_n$ and since $r_n/n \to \alpha$ it follows that

$$\limsup_{n \to \infty} r_n^0/n \leq \alpha.$$

and looking in the mirror again we see

$$\liminf_{n \to \infty} \ell_n^0/n \geq \beta.$$

Combining the last two observatons it follows that if n is large then

$$\sup \xi_n^0 = r_n^0 < \ell_n^0 = \inf \xi_n^0$$

with high probability, but the only set A with $\sup A < \inf A$ is the empty set ($\sup \phi = -\infty < \inf \phi = +\infty$) so the proof is complete.

Remark. By observing that

$$r_{kN} < r_N + r_{N,2n} + \cdots + r_{(k-1)N,N}$$

and that if $\alpha > a$ and N is large the right hand side is a random walk with drift $< a$. (and $E \exp(\theta r_N) < \infty$ for all $\theta > 0$) it is easy to see that if $a > \alpha$ then there are constants $C, \gamma \in (0,\infty)$ so that

$$P(r_n > an) < Ce^{-\gamma n}$$

(see Durrett (1984), Section 7 for details. We will need this fact in the next section).

Having seen that $\alpha < \beta$ implies that the process dies out it is natural to ask if there is a converse. This is true but much harder to prove.

Theorem 1. If (H1)-(H3) are satisfied then

$$p_c = \inf\{p: \alpha(p) > \beta(p)\}$$
$$= \sup\{p: \alpha(p) < \beta(p)\}.$$

(Note: as we mentioned above we will have to make two technical assumptions to make this a correct statement but we will introduce them as they are needed in the proof. (H2) is given in the proof of (4) below, (H3) at the very end of Section 4.)

The first step in the proof of Theorem 1 is to prove the following fact which is a generalization of an observation due to Tom Liggett. (see Durrett (1980), Lemma 4.1).

(3)   If $B \subset \{-1,-2,\ldots\}$ is an infinite set and we let $r_n^B = \sup \xi_n^B$ then

$$E(r_n^{B \cup \{0\}} - r_n^B) > 1.$$

The proof is the same since all that it uses is the <u>additivity property</u>

$$\xi_n^{A \cup B} = \xi_n^A \cup \xi_n^B$$

of processes defined on graphical representations.

With this result in hand we can repeat the proof of Lemma 4.2 of Durrett (1980) to conclude that $p \to \alpha(p)$ is strictly increasing. If we let $R = \sup\{y: f(y,p) > 0\}$ and suppose

(H2) If $f(y,p) > 0$ for some $p > 0$ then

$$\frac{\partial f(y,p)}{\partial p} > C > 0 \ ,$$

then what we obtain is

(4) $$E(r_n^{p+\delta} - r_n^p) > C\delta n.$$

Proof: Since this result is different from the one given in the talk we will supply a few details. As in Durrett (1980) if we let $\tau = \inf\{s > 0: r_s^{p+\delta} > r_s^p\}$ then using the Markov property and (3) we conclude

$$E(r_t^{p+\delta} - r_t^p) > P(\tau < t)$$

but this time

$$P(\tau > n) < (1 - [f(R,p+\delta) - f(R,p)])^n$$

since if the $p + \delta$ process jumps by R and the other one doesn't the $p + \delta$ process gets ahead by 1. Dividing the interval $[p,p+\delta]$ into M pieces, using the inequality above, and letting $M \to \infty$ gives (4). For more details see Durrett (1980), p. 901.

Looking in the mirror again we have

(4') $$E(\ell_n^{p+\delta} - \ell_n^p) < -C\delta n$$

for the same constant. If we combine the last result with (4), let $p_\alpha = \sup\{p: \alpha(p) < \beta(p)\}$, and let $n \to \infty$ we get

(5) $$\text{if } p > p_\alpha \text{ then } \alpha(p) - \beta(p) > C(p - p_\alpha) > 0 \ .$$

The last result implies

$$\sup\{p: \alpha(p) < 0\} = \inf\{p: \alpha(p) > 0\}$$

so the "only" thing that remains is to show that if $\alpha(p) > \beta(p)$ then
$P(\xi_n^0 \neq \phi$ for all $n) > 0$.

In Durrett (1980) this was done for Example 3 by using a coupling result which is a special feature of that case:

(6)   if $\ell_n = \inf \xi_n^{[0,\infty)}$ and $r_n = \sup \xi_n^{(-\infty,0]}$ then

     on $\{\ell_m < r_m$ for all $m < n\}$

$$\xi_n^0 = \xi_n^Z \cap [\ell_n, r_n]$$

and consequently $\ell_n^0 = \ell_n$, $r_n^0 = r_n$, $\xi_n^0 \neq \phi$.

NOTE: this coupling property is NOT true in the three site case. See Durrett (1980), Section 6 for a discussion or draw a random picture.

With (6) it was easy to prove what we wanted to (see Durrett (1980) for details) but until recently it was not clear how to do without (6). A large part of the solution it turns out was in Durrett and Griffeath (1983) so we turn to describing those results now.

4. A Renormalized Bond Construction

To almost quote Durrett (1984), p. 1023. "In this section we will introduce a construction which will allow us to reduce questions about supercritical finite range g.p.p. to corresponding questions about a k-dependent nearest neighbor site percolation process with p arbitrarily close to 1." The argument given here like the last quote is a simple modification of the corresponding thing in Durrett (1984) so we will start by describing the argument in that special case: oriented bond percolation $S = \{-1,1\}$; and then describe the changes which are necessary for finite range g.p.p.

The first thing to do is to define the site percolation process and its relationship to the original process. Let $\mathcal{L}$ be the graph with vertices $V = \{(m,n) \in Z^2 \quad m + n$ is even, $n \geqslant 0\}$ and oriented bonds connecting each $(m,n) \in V$ to $(m + 1, n + 1)$ and to $(m - 1, n + 1)$. Stealing a term from the physics literature we call $\mathcal{L}$ the renormalized lattice. To explain the name (and the idea behind the construction) the reader should imagine $\mathcal{L}$ mapped into the upper half plane $R \times [0,\infty)$ by $\phi(x,y) = (aLx,Ly)$ where a is a special constant and L is a large number to be chosen below.

We will define the site (m,n) in V to be open if a "good event" happens in the graphical representation near $z_{m,n} \in R \times [0,\infty)$ and we will do this in such a way that

(i)   the random variables $\eta(v)$ $v \in V$ which indicate whether the sites are open or not are k-dependent (i.e. if the distance from x to y on the graph $>k$ they are independent).

(ii)   if L is large the probability $\eta(v) = 1$ is close to 1.

(iii)   if percolation occurs starting from 0 on the renormalized lattice then it does starting from some point near $z_{0,0}$ in the original percolation process.

It is by now well known that (i) and (ii) imply that if L is large then the probability of percolation is positive (for more on this see Durrett (1984), Section 10) so once (i)-(iii) are demonstrated we can conclude that if $\alpha(p) > \beta(p)$ then $P(\xi_n^0 \neq \phi$ for all $n) > 0$ completing the proof of Theorem 1 in Section 2. We will see below that the construction can be used to prove a number of other things about percolation processes so if the reader gets bored or confused by the details of the construction, he/she/it should skip ahead to the next two sections to see part of what it is good for: the complete convergence theorem and the strong law for ξ_n^0.

Details of the construction (oriented percolation $S = \{-1,1\}$).

We begin by describing the fundamental building blocks: the renormalized bonds which appear in the title of the section. Let A be the parallelogram with vertices $(-2\epsilon L,0)$, $(2\epsilon L,0)$, $((\alpha-2\epsilon)L,L)$, and $((\alpha + 2\epsilon)L,L)$. From (1) in Section 3 we know that $r_L/L \to \alpha$ as $L \to \infty$, so if $\delta > 0$ and $L > L_0(\delta)$ then

$$P(r_L \in ((\alpha - \epsilon)L,(\alpha + \epsilon)L)) > 1 - \delta$$

When $r_L \epsilon ((\alpha - \epsilon)L, (\alpha + \epsilon)L))$, we know that there is a path from $(-\infty, 0] \times \{0\}$ to $((\alpha - \epsilon)L, (\alpha + \epsilon)L) \times \{L\}$ but it might look like the dotted line in Figure 4.1. Our next task is to show that it looks like the solid line. i.e. it stays

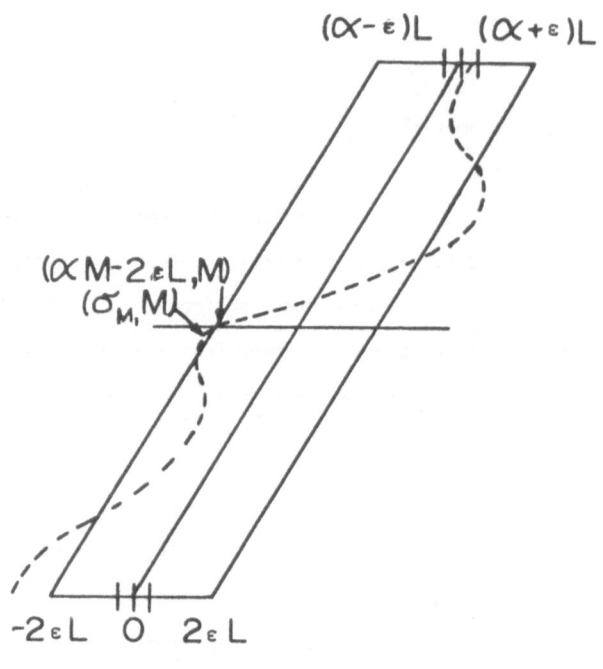

$(\alpha - \epsilon)L \quad (\alpha + \epsilon)L$

$(\alpha M - 2\epsilon L, M)$
(σ_M, M)

$-2\epsilon L \quad 0 \quad 2\epsilon L$

FIGURE 4.1.

in the parallelogram. To prove that it doesn't hit the right side we observe that if N is large $Er_N < (\alpha + \epsilon)N$ and

$$r_{kN} \leqslant r_N + r_{N,2N} + \cdots + r_{(k-1)N,kN}$$

with the right hand side being a random walk, so the simple argument we used in the last section shows that there are constants $C, \gamma \epsilon (0, \infty)$ such that

(1) $$P(r_m > (\alpha + \epsilon)m) \leqslant Ce^{-\gamma m}$$

Now if $R = \sup \{y: f(y,p) > 0\}$ (which is independent of $p > 0$ by (H2)) the right edge can increase by at most R per jump so

$$P(r_m \text{ exits right side of } A) \; < \; \sum_{m=2\,\varepsilon L/R}^{\infty} P(r_m > (\alpha + \varepsilon)m)$$

$$< \; Ce^{-\gamma L}$$

(where $C, \gamma \in (0, \infty)$ are new constants and will continue to change as we go along.)

To estimate the probability that the path escapes from the left side of the box we use an observation due to Larry Gray which allows us to turn our upper bound into a lower bound. Let σ_t $0 < t < L$ be a path from $(-\infty, 0] \times \{0\}$ to $((\alpha - \varepsilon)L, (\alpha + \varepsilon)L) \times \{L\}$ in the graphical representation used to construct the g.p.p. Let $M = \sup\{m : (\sigma_m, m) \notin A\}$ Larry's simple but useful insight is (see Figure 4.1 again) that the line from (σ_M, M) to (σ_L, L) has slope $> (\alpha + \varepsilon)$ so if $M = m$ then the right edge of the process starting from $(-\infty, -2\varepsilon L + \alpha m]$ at time m must be $> (\alpha + \varepsilon)(L - m)$ at time L and summing the estimate in (1) we conclude that

$$P(\sigma_m \text{ exits left side of } A) \; < \; Ce^{-\gamma L}$$

Combining the last three estimates shows that if $\varepsilon > 0$ and $L > L_1(\delta)$ then with probability $> 1 - 3\delta$ there is a path lying in A. These events are the raw material for the construction that follows. The next step in carrying it out is to associate the sites in the renormalized lattice with translates of A in the percolation structure.

Drawing a picture (see Figure 4.2) motivates letting

$$z_{m,n} = ((\alpha - 4\varepsilon)m, n) \quad (m,n) \; \varepsilon \; V$$

be the points of the renormalized lattice, defining translates of A by

$$A_{m,n} = (z_{m,n} + (-4\varepsilon L, 0)) + A$$
$$B_{m,n} = (z_{m,n} + (4\varepsilon L, 0)) - A$$

(where $x - A = \{x - y : y \varepsilon A\}$ etc.) and declaring that the site (m,n) is open if there are paths in $A_{m,n}$ and in $B_{m,n}$.

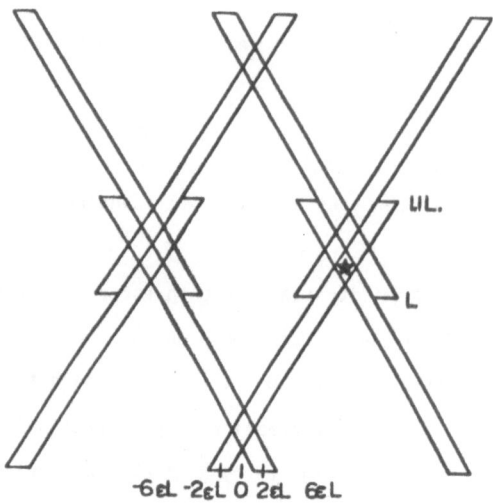

Figure 4.2.

From the definition it is clear that we have property (i) and the arguments above show that (ii) holds. To check (iii) we observe that in the case under consideration paths cannot "jump over each other" so the arrangement of the $A_{m,n}$ and $B_{m,n}$ guarantees that if say $(m-1, n-1)$ and (m,n) are open then there is a path from $z_{m-1,n-1} + (-6\varepsilon L, -2\varepsilon L)$ to $z_{m+1,n+1} + (2\varepsilon L, 6\varepsilon L)$ and to $z_{m-1,n+1} + (-6\varepsilon L, -2\varepsilon L)$ (see Figure 4.3). From this it follows easily that (iii) holds. (for more details consult Durrett (1984), p. 1025).

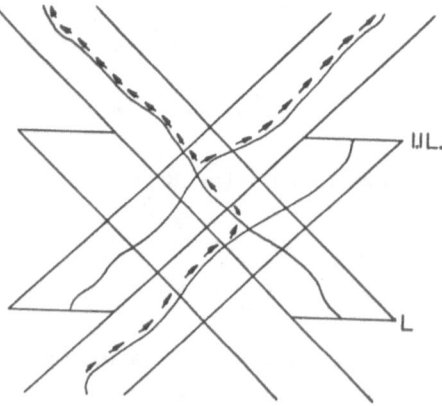

Fig. 4.3

Details of the construction (finite range g.p.p)

The last argument does not work outside the case $S = \{-1,1\}$ (or $S = \{0,1\}$) because paths which cross do not need to intersect but (here finally is our new idea) paths which cross will intersect with probability $> \eta > 0$ so if we use a zillion little paths to try to connect two of the (long) paths used in the construction then there will be a success with high probability. As the reader can probably guess carrying out this idea requires a little ingenuity and a large number of unpleasant details. To keep things as simple as possible we will first give the details for oriented bond percolation with $S = \{x: |x| < R\}$ and then treat the general case.

The first step in the argument is to make the tubes smaller. Let A' be the left half of A, i.e. the parallelogram with vertices $(-2\epsilon L,0)$, $(0,0)$, $((\alpha - 2\epsilon)L,L)$ and $(\alpha L,L)$. We keep the renormalized lattice the same

$$z_{m,n} = ((\alpha - 4\epsilon)Lm, Ln) \quad (m,n) \ \epsilon \ V$$

define translates of A' as before by

$$A'_{m,n} = (z_{m,n} + (-4\epsilon L,0)) + A'$$

$$B'_{m,n} = (z_{m,n} + (4\epsilon L,0)) - A'.$$

Thinning the tubes creates space near each point of the renormalized lattice (see Figure 4.4) and into this space which has width $4\epsilon L$ we put $[4/\epsilon]$ tubes of width $\epsilon^2 L$ and length $2L$ in the manner indicated in the picture and then (for reasons that will become clear in a minute) we remove every other one. Paths in the new smaller tubes will be used to connect the paths in the four large tubes.

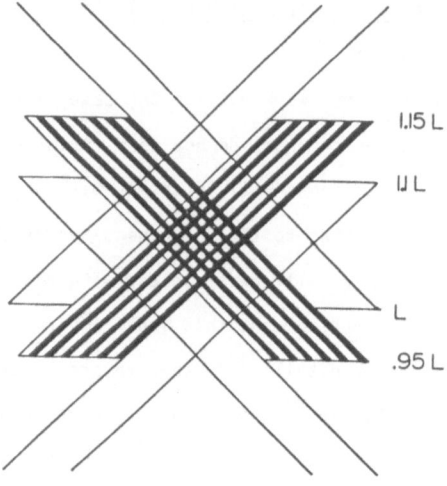

1.15 L

1.1 L

L

.95 L

Figure 4.4

Tying paths together is delicate because of "conditioning problems" - i.e. picking a path by some algorithm makes the conditional distribution of bonds near the chosen path different from the original one. To avoid difficulties of this type we "save a little randomness to make connections at the end." To be precise:

(a) We pick $p' < p$ with $\alpha(p') > 0$ (and observe that in the proof of $p_c < p_\alpha$ - which is what we're doing now! - this can be done without loss of generality.)

(b) Construct the processes with parameters p and p' on the same space by assigning independent uniformly distributed random variables $U(b)$ to each bond $b = (x-y, n) \rightarrow (x, n+1)$ where $|y| \leq R$ and declaring b to be open for the p' (resp. p) system if $U(b) < p'$ (resp. $U(b) < p$).

(c) Do the renormalized bond construction for the p' system (with the corresponding $\alpha(p') > 0$) and call one of the large or small tubes in the construction good if it has a path in the p' percolation structure from one end to the other which stays in the tube.

Since all the paths in the little tubes must pass within R of a path in the large tube and the number of little tubes is large then it is clear that if ε is small then the situation drawn in Figure 4.5 will occur with high probability i.e. there are paths in the small tubes which intersect and intersect the paths in the four large tubes.

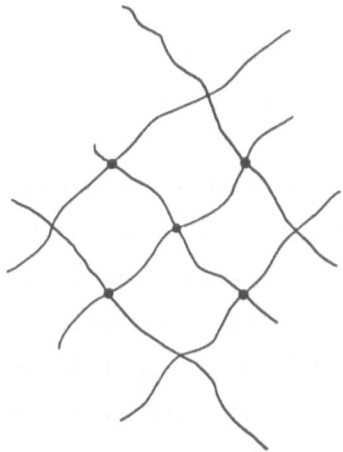

FIGURE 4.5.

To prove this we pick for each tube which was called good in (c), a path in the p' percolation structure with the desired properties and for ease of reference later, we will say that these paths are drawn in blue. Now each pair of blue paths σ, τ that we want to connect must come within a distance R of each other at some point, i.e. there are integers x,y and n with $\sigma_n = x$, $\tau_{n+1} = y$ and $|x - y| < R$.

Now if we condition on the value of $U(b) \wedge p'$ for all the bonds b then there is still probabilty $\geqslant (p - p')/(1 - p') > 0$ that $U((x,n) \to (y,n + 1)) < p$ and hence open in the p percolation structure. When present the "green bond" $(x,n) \to (y,n+1)$ (its color intended to signify its unconditioned state) allows us to connect the blue paths. Since we have arranged for there to be lots of little tubes and we have separated them by removing every other one to make the connection events independent, it follows that if ε is small and L is large then all the desired connections happen with high probability.

Given the argument for oriented percolation on S = {-1,1} the denouement
should be clear at this point. We declare a site in the renormalized lattice to
be open if there are paths in the four large tubes near it and the green bond
construction above succeeds in connecting them as indicated in Figure 4.5. From
the definition it is clear that we have property (i) listed in the first version
of the proof and the arguments above show that (ii) holds. To check (iii)
observe that above we have been careful to choose one path in each large tube
and then connect these paths so, having worked harder to get here, the last step
is now trivial.

With (i)-(iii) verified the rest follows as before and we have completed
the proof for oriented bond percolation with S = {x: |x| < R}. In tackling the
general g.p.p the first (trivial) extension to be considered is what happens for
other oriented percolation processes, e.g. bond percolation process with
S = {1,2,3,4}, S = {-2,2}, S = {-51,50},... In the first case mentioned we just
need to slant the construction: if (m,n) ε V and ℓ = n - m then

$$z_{m,n} = (n - \ell)(\alpha - 4\varepsilon) + \ell(\beta - 4\varepsilon)$$

(for more details in the nearest neighbor case see Schonmann (1986)). In the
second case (like S= {-1,1}) restricting to a sublattice gives a problem to
which the results for solid intervals can be applied. Last but not least when S
is not a solid interval but the group it generates is all of Z, a finite number
of iterates allow us to reach all points in an interval and blah, blah, blah.

The generalization mentioned in the last paragraph are, like the extension
of Markov Chain results from the case of a positive matrix to that of an irreducible,
one, routine although somewhat tedious and hence are left as an exercise for
an energetic reader. We turn now to the last important item of business:
proving the result for a general g.p.p. Having discussed the asymmetric and
non-interval cases above we will assume that model is symmetric and f(y,p) > 0
if and only if |y| < R.

Looking back at the proof now it is clear that special properties of
oriented bond percolation were only used in the (p',p) property of the

construction above and for this the important point was

(H3). If we condition on the state of all bonds in the p'-system then for any x,n

and $|y| < R$ there is always conditional probability $> \delta(p,p') > 0$ that

$(x - y,n) \rightarrow (x,n + 1)$ is open.

This is our last hypothesis that "the models increase strictly with p and with

it made it is trivial to complete the proof.

5. **The Complete Convergence Theorem**

   In this section we will prove a result which allows us to determine the

limiting distribution of ξ_n^A for any A when $p > p_c$. The first step in doing

this is to describe the process which appears in the limit theorem.

   Let $\tilde{\xi}_n$ be the process generated by the graphical representation in which

the oriented bond $(x,n) \rightarrow (x - y,n + 1)$ is open (closed) with probability

$f(y)$ (resp. $1 - f(y)$); bonds beginning at different sites are independent; and

the joint distribution of the bonds $(x,n) \rightarrow (x - y,n + 1)$ $y \in Z$ is the same

as that of $(x - y,n - 1) \rightarrow (x,n)$ in ξ_n. Comparing the last paragraph with

the definition in Section 1 it should be clear that the new graphical represen-

tation can be obtained by reversing time (and the direction of the arrows) in

the old one and a little more thought leads us to the following important

conclusion

(1) $$P(\xi_n^A \cap B \neq \phi) = P(A \cap \tilde{\xi}_n^B \neq \phi).$$

Proof: From the definition of ξ_n^A we see that $\{\xi_n^A \cap B \neq \phi\}$ = {there is an

open path from $(x,0)$ to (y,n) for some $x \in A, y \in B$ and from the discussion

above we see that the right hand side is equal to the probability of a path down

from (y,n) to $(x,0)$ in the same percolation structure.

   Taking $A = Z$ in (1) we see that $P(\xi_n^Z \cap B \neq \phi) = P(\tilde{\xi}_n^B \neq \phi)$ which

decreases to a limit as $n \rightarrow \infty$ (since ϕ is an absorbing set for $\tilde{\xi}_n$). The

inclusion-exclusion formula allows us to write all probabilities of the form

$$P(\xi_n^Z(x_1) = i_1, \ldots, \xi_n^Z(x_k) = i_k)$$

where $\{x_1, \ldots, x_k\} \subset Z$ and $i_1, \ldots, i_k \in \{0,1\}$ in terms of $P(\xi_n^Z \cap R = \phi)$ so it follows that we have

(2) As $n \to \infty$ $\xi_n^Z \Rightarrow$ to a limit ξ_∞^Z,

where $\Rightarrow$ denotes weak convergence of probability measures on $\{0,1\}^Z$ (which in this setting is = convergence of finite dimensional distributions).

Having defined the limit we can now state our convergence result

Theorem 2. If $\alpha(p) > 0 > \beta(p)$ then as $n \to \infty$

$$\xi_n^A \Rightarrow \delta_\phi P(\tau^A < \infty) + \xi_\infty^Z P(\tau^A = \infty).$$

where $\tau^A = \inf\{m > 0: \xi_m^A = \phi\}$,

$$\delta_\phi = \text{the point mass on the empty set } \phi ,$$

and we use ξ_∞^Z to denote the limit distribution starting from $\xi_0^Z = Z$.

The first part of the right hand side is easy to see: on $\{\tau^A < \infty\}$ we have $\xi_n^A = \phi$ for $n > \tau^A$. The second part is much harder to prove: it says that if ξ_n^A does not die out (i.e. $\tau^A = \infty$) and n is large then ξ_n^A looks like ξ_n^Z with high probability on any (fixed) finite set. (We will prove a sharper version of this in the next section).

An immediate consequence of Theorem 2 is that all stationary distributions have the form $\theta\delta_\phi + (1 - \theta)\xi_\infty^Z$ for some $\theta \in [0,1]$. When confronted with the last observation the reader should ask: is $\xi_\infty^Z \neq \delta_\phi$ for $p > p_c$? The density of particles in ξ_∞^Z can be read off from the duality equation (1):

$$P(x \in \xi_\infty^Z) = P(\tilde{\xi}_n^0 \neq \phi \text{ for all } n).$$

so if we let

$$p_e = \inf\{p: \xi_\infty^Z \neq \delta_\phi\}$$

(where e is equilibrium) then it is clear that we have

$$p_e = \tilde{p}_c = \inf\{p: P(\tilde{\xi}_n^0 \neq \phi \text{ for all } n) > 0\},$$

and the question becomes "$p_c = \tilde{p}_c$" ? The answer as we will see in the proof is: Yes.

The first step in proving Theorem 2 is to observe that if ξ_t^A and $\tilde{\xi}_t^B$ are independent

$$(3) \qquad P(\xi_{2t}^A \cap B \neq \phi) = P(\xi_t^A \cap \tilde{\xi}_t^B \neq \phi).$$

(the first event being the probability of a path from $(x,0)$ to $(y,2t)$ for some $x \in A$, $y \in B$ while the second = {there are $x \in A$, $y \in B$, and $z \in Z$ so that $(x,0) \to (z,t) \to (y,2t)$}.) Now

$$P(\xi_t^A \cap \tilde{\xi}_t^B \neq \phi) = P(\xi_t^A \neq \phi, \tilde{\xi}_t^B \neq \phi) - P(\xi_t^A \neq \phi, \tilde{\xi}_t^B \neq \phi, \xi_t^A \cap \tilde{\xi}_t^B = \phi)$$

and the first term $= P(\xi_t^A \neq \phi)P(\tilde{\xi}_t^B \neq \phi)$ which converges to $P(\tau^A = \infty)P(\xi_\infty^Z \cap B \neq \phi)$ as $t \to \infty$ so to prove the theorem it suffices to show

$$(4) \qquad P(\xi_t^A \neq \phi, \tilde{\xi}_t^B \neq \phi, \xi_t^A \cap \tilde{\xi}_t^B = \phi) \to 0.$$

The proof of (4) requires one new bit of inspiration (followed by quite a bit of perspiration) so we will start by stating the new idea: "when the renormalized bond construction works it produces points at a positive density of sites between the left and right edges so if $\beta(p) < 0 < \alpha(p)$ and the time is large t then ξ_t^A and $\tilde{\xi}_t^B$ will intersect with high probability (if both are nonempty)."

With this idea in mind the rest is routine following the proof of similar results in Durrett (1984) so we will just give an outline.

The sentences in quotation marks below were the ones I said during my talk. In between them I have tried to supply enough details so that the reader (with the help of the paper cited above) can fill in the rest. As in the last section the summary becomes tedious or confusing the reader can safely skip to the beginning of the next section where we will start to consider $|\xi_n^0| = $ the number of occupied sites at time n.

1. "After a geometric number of trials either (a) $\xi_n^A = \phi$ or (b) the renormalized bond construction works and on the renormalized lattice ξ_n^A dominates oriented percolation with p close to 1."

The proof is a "restart argument" following Durrett (1980), 903-904 and/or Durrett (1984), 1031-1032. The proof is based on a simple idea: "if at first you don't succeed try, try again" but requires a depressing number of definitions to carry out.

Let $x_0 = \sup A$. If $M > 6\epsilon L$ and we are very lucky then (i) $[-6\epsilon L + x_0, 6\epsilon L + x_0] \subset \xi_M^{x_0}$ and (ii) we get a path to ∞ on the renormalized lattice when we try the renormalized bond construction translated by x_0. These are the two things we dream about and they have positive probability of happening on the first try.

When they don't then we have to go to work: if (i) does not occur and $\xi_M^A = \phi$ then we are happy since all we have to do is show that things are OK when $\xi_n^A \neq \phi$. If (i) does not occur and $\xi_M^A \neq \phi$ we let $x_1 = \sup \xi_M^A$ and look M units of time later to see if $[-6\epsilon L + x_1, 6\epsilon L + x_1] \subset \xi^{(x_1, M)}(2M)$ where the superscript indicates we are looking at the process starting from $\{x_1\}$ at time M. Each time we repeat the last step we have a positive probability of success so after a geometric number of failures we get to try the renormalized bond construction.

As the reader has probably already anticipated the renormalized bond construction may fail but if it does we try again: we wait until the process on the renormalized lattice dies out (and then wait 1.1L units of time more for "good luck", i.e. so that the death of the construction does not adversely effect the future development of the graphical representation) and then start again to look for an interval of length $12\epsilon L$ to try the construction again.

2. "$\tilde{\alpha}(p) = \beta(p)$ and $\tilde{\beta}(p) = \alpha(p)$ so the same construction can be used on the dual." As observed on p. 9 of Durrett and Griffeath (1983)

$$P(r_m > k) = P(\text{there is a path from } (-\infty,0] \times \{0\} \text{ to } [k,\infty) \times \{m\})$$

$$= P(\text{there is a dual path from } [k,\infty) \times \{m\} \text{ to } (-\infty,0] \times \{0\})$$

$$= P(\text{there is a dual path from } [0,\infty) \times \{m\} \text{ to } (-\infty,-k] \times \{0\})$$

$$= P(\tilde{\ell}_m < -k).$$

The last identity shows $r_m \overset{d}{=} -\tilde{\ell}_m$ from which it follows immediately that $\tilde{\beta}(p) = \alpha(p)$. From this the rest of the statement in quotes follows immediately and using results from Section 4 shows $p_c = \tilde{p}_c$

3. "When the renormalized bond construction works for the process and its dual then ξ_n^A and $\tilde{\xi}_n^B$ intersect with high probability."

In this case a picture is worth (and probably replaces) a thousand words. (see Figure 5.1). The squiggly line above A and below B indicates that with high probability we have to wait at most 100 years (i.e. a time independent of t) before the renormalized bond construction works and the rest of the picture is meant to suggest that when it does then the process on the renormalized lattice dominates (k-dependent) oriented site percolation with p close to 1. Results of Section 10-11 of Durrett (1984) show that if the probabilty of a site being open is close to 1 then the set of occupied sites for oriented percolation on the renormalized lattice

$$\liminf_{k \to \infty} |\stackrel{0}{\ell_k}|/k > 1 - \delta \quad \text{on} \quad \{\stackrel{0}{\ell_k} \neq \phi \text{ for all } k\}$$

almost surely. Since $\stackrel{0}{\ell_k} \subset \{-k,\ldots,k\}$ fills up a fraction $(1-\delta)$ of the available space. The last observation implies that if we pick δ much smaller than the minimum of α and β, and ξ_n^A and ξ_n^B are $\neq \phi$ then there will be a large number of pairs (x_i,y_i) with $x_i \in \xi_\ell^A$, $y_i \in \tilde{\xi}_\ell^B$ and $|x_i - y_i| < L$ and running things for $2L$ more units of time we conclude

$$P(\xi_n^A \neq \phi, \tilde{\xi}_n^B \neq \phi, \xi_n^A \cap \tilde{\xi}_n^B = \phi) \to 0$$

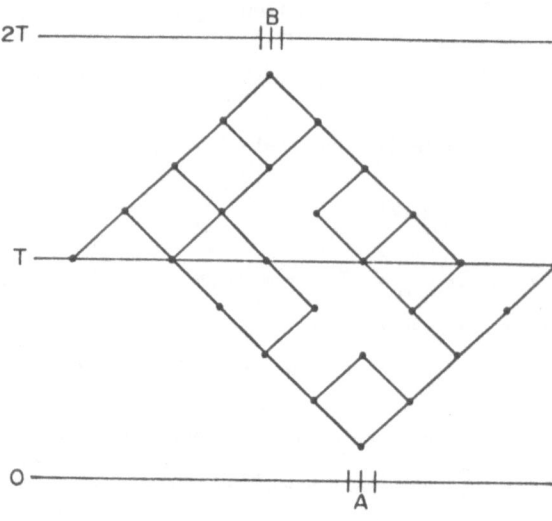

FIGURE 5.I.

6. Limit Laws for ξ_n^0

In this section we will take a closer look at the behavior of the system starting from a single particle at 0. The first step is to state a result which follows from the construction in the last section.

(1)   If $\epsilon > 0$ then there are constants $C, \gamma \in (0, \infty)$ so that if $(\beta + \epsilon)n < x < (\alpha - \epsilon)n$ then

$$P(\xi_n^0 \neq \phi, \ \xi_n^0(x) \neq \xi_n^Z(x)) < Ce^{-\gamma n}.$$

Proof: As enunciated on p. 1031 of Durrett (1984) the proof is based on two simple ideas.

(i) If you have a sequence of independent events with probability p then K, then the number of failures before the first success has $P(K = n) = p(1 - p)^n$ $n = 0,1,2,\ldots$ and

(ii) If X_i is a sequence of independent random variables with $P(X_i > m) < c \exp(-\gamma_m)$ so that $(X_1, \ldots, X_k)$ is independent of $\{K = k\}$ then

$$P(X_1 + \ldots + X_k > m) < C' \exp(-\gamma'm).$$

where C', γ' are new constants ϵ $(0,\infty)$. To use these ideas to prove (1) you have to check that when the renormalized bond construction fails it only lasts an amount of time T with $P(T > t) < C \exp(-\gamma t)$ but this is true, see Durrett (1984), p. 1031-1032.

From (1) it follows immediately that we have

(2) For any $\epsilon > 0$, on $\Omega_\infty = \{\xi_n^0 \neq \phi$ for all $n\}$ we have

$$\{x: \xi_n^0(x) = \xi_n^Z(x)\} \supset [(\beta + \epsilon)n, (\alpha - \epsilon)n] \cap Z$$

for all n sufficiently large. With the coupling result ((6) in section 3) recaptured one can repeat arguments from Section 13 of Durrett (1984) now to show

Theorem 3. On Ω_∞ we have

$$|\xi_n^0|/n \rightarrow (\alpha - \beta) \text{ almost surely as } n \rightarrow \infty.$$

where $\rho = P(z \in \xi_\infty^Z)$ and α, β are the by now familiar limits of $r_n/n, \ell_n/n$.

The last result has a simple explanation: the distance between the left and right particles is $\sim(\alpha - \beta)n$ and this interval is filled with particles at density ρ.

7. Results for $d > 1$

Last but not least we come to the original motivation for doing this paper: to improve what is known in higher dimensions. We begin by "recalling" the results proved by Durrett and Griffeath (1982). We have put the word recalling in quotation marks because those results were proved for a class of models in continuous time (called permanent one-sided growth processes there) and we will have to ask the reader to believe that the analogous results are true in discrete time. First some notation:

$$H_n = \bigcup_{m \leq n} \xi_m^0 = \text{sites \underline{hit} by time } n$$

$$K_n = \{x: \xi_n^0(x) = \xi_n^Z(x)\} = \text{sites \underline{coupled} at time } n.$$

For a variety of reasons it is convenient to enlarge the last two sets by replacing each point x by a cube of side 1 centered at that point:

$$\overline{H}_n = \bigcup_{x \in H_n} x + [-\tfrac{1}{2}, \tfrac{1}{2}]^d$$

$$\overline{K}_n = \bigcup_{x \in K_n} x + [-\tfrac{1}{2}, \tfrac{1}{2}]^d$$

(the notation is meant to suggest closure).

With this notation introduced we are now (almost) ready to state the result of Durrett and Griffeath (1982). For simplicity and concreteness we will restrict our attention to oriented bond percolation in Z^3: i.e. the process defined from the percolation structure in which all bonds are independent and $(x,n) \to (x + y, n + 1)$ is open with probabilty p if (and only if) $|y| = 1$.

(1) Suppose p is large enough so that the process restricted to $Z \times \{0\}$ has positive probability of survival for all time. There is a (non random) convex set U so that on $\Omega_\infty = \{\xi_n^0 \neq \phi \text{ for all } n\}$ we have

$$\tfrac{1}{n}(\overline{H}_n \cap \overline{K}_n) \to U \quad \text{a.s.} \quad \text{on } \Omega_\infty$$

as $n \to \infty$ i.e. for any $\varepsilon > 0$ and $\omega \in \Omega_\infty$

$$(1 - \varepsilon) n U \subset (\overline{H}_n \cap \overline{K}_n) \subset (1 + \varepsilon) nU$$

for all n sufficiently large.

Roughly speaking (1) says that ξ_n^0 looks like $\xi_\infty^Z \cap nU$ on Ω_∞ or even rougher it is a "blob in equilibrium" in the terminology of Durrett and Griffeath (1982). The statement of (1) is made contorted by the fact that $K_n \supset \{x: \xi_n(x) = 0\}$ for trivial reasons so we have to intersect with H_n to get the interesting part. The strength of this is the fact that the theorem says "almost everywhere we have hit we are in equilibrium" and has as a consequence the complete convergence theorem

(2) For any A, as $n \to \infty$

$$\xi_n^A \implies P(\tau^A < \infty)\delta_\phi + P(\tau^A = \infty)\xi_\infty^Z.$$

So much for the virtues of (1). Its shortcoming is obvious: the result is only for $p > p_c(Z)$ the critical value for the process on Z and not for $p > p_c(Z^2)$. The next result, our last theorem improves this but does not yet complete the story. Let p_c^L be the critical value for oriented percolation in $Z^2 \times \{-L, \ldots, L\}$. It is easy to see that as $L \to \infty$ p_c^L decreases to a limit we call p_c^∞ and it is natural (if somewhat optimistic) to conjecture that $p_c^\infty = p_c(Z^3)$. In any case the next result improves on (1) but is not the last word. The reader should note that the complete convergence theorem is again a consequence.

Theorem 4. If $p > p_c^\infty$ then there is a nonrandom convex set U so that on $\Omega_\infty = \{\xi_n^0 \ne \phi$ for all $n\}$ we have

$$\frac{1}{n}(\overline{H_n} \cap \overline{K_n}) \to U \quad \text{a.s.}$$

as $n \to \infty$.

This result can be proved by using an abstract theorem (see Durrett and Griffeath (1982), p. 529) which was designed five years ago for the application we are making today: all we have to do is check that the three conditions of the theorem hold and then Theorem 3 follows. If we let $\tau = \inf\{n: \xi_n^0 = \phi\}$ then what we need to show is that there are constants $\delta, C, \gamma \in (0, \infty)$ so that

(a) $$P(n < \tau < \infty) \le Ce^{-\gamma n}$$

(b) $$P(x \in H_n, \tau = \infty) \le Ce^{-\gamma n} \quad \text{if} \quad |x| < \delta n$$

(c) $$P(x \in K_n, \tau = \infty) \le Ce^{-\gamma n} \quad \text{if} \quad |x| < \delta n.$$

Checking (a), (b), and (c) is neither trivial nor pleasant but following the argument on p. 545-550 in Durrett and Griffeath (1982) and using the renormalized bond construction one can do this. Details of the proof of this will be the subject of a future publication.

Acknowledgements

RHS would like to thank the Rutgers mathematics department for the warm hospitality it showed him during his stay there.

References

Amati, D., et al (1976) Expanding disc as dynamic vacuum instability in Reggeon field theory. Nucl. Phys. B 114, 483-504.

Brower, R.C., Furman, M.A., and Moshe, M. (1978) Critical exponents for the Reggeon quantum spin model. Phys. Lett B. 76, 213-219.

Cardy, J.L. and Sugar, R.L. (1980) Directed percolation and Reggeon field theory. J. Phys. A 13, L423-L427.

Dhar, D. and Barma (1981) Monte Carlo simulation of directed percolation on the square lattice. J. Phys. C., L1-6.

Domany, E. and Kinzel, W. (1981) Directed percolation in two dimensions: numerical analysis and an exact solution Phys. Rev. lett. 47, 1238-1241.

Durrett, R. (1980) On the growth of one dimensional contact processes. Ann. Prob. 8, 890-907.

Durrett, R. (1984) Oriented percolation in two dimensions. Ann. Prob. 12, 999-1040.

Durrett, R.,and Griffeath, D. (1982) Contact processes in several dimensions, Z. fur Wahr. 59, 535-552.

Durrett, R. and Griffeath, D. (1983) Supercritical contact processes on Z. Ann. Prob. 11, 1-15.

Grassberger, P. and de la Torre, A. (1979) Reggeon field theory (Shlogl's first model) on a lattice: Monte Carlo calculations of critical behavior. Ann. Phys. 122, 373-396.

Griffeath, D. (1979) Additive and Cancellative Interacting Particle Systems Springer Lecture Notes in Math, Vol. 724.

Griffeath, D. (1981) The basic contact process. Stoch. Pro. Appl. 11, 151-168.

Harris, T.E. (1974) Contact interactions on a lattice. Ann. Prob. 2, 969-988.

Harris, T.E. (1978) Additive set valued processes and graphical methods. Ann. Prob. 6, 355-378.

Kertesz, J. and Vicsek, T. (1980) Oriented bond percolation. J. Phys. C. 13, L343-348.

Kinzel, W. and Yeomans, J. (1981) Directed percolation: a finite size renormalization approach. J. Phys. A14, L163-168.

Kinzel, W. (1985) Phase transitions of cellular automata. Z. Phys. B. 58, 229-244.

Liggett, T. Interacting Particle Systems. Springer-Verlag, New York.

Schonmann, R.H. (1985) Metastability for the contact process. J. Stat. Phys. 41, 445-464.

Schonmann, R.H. (1986a) A new proof of the complete convergence theorem for contact processes in several dimensions with large infection parameter. Ann. Prob., to appear.

Schonmann, R.H. (1986b) The asymmetric contact process. J. Stat. Phys., to appear.

Wolfram, S. (1983) Statistical mechanics of cellular automata. Rev. Mod. Phys. 55, 601-650.

Wolfram, S. (1984) Universality and complexity in cellular automata. Phsica 10 D, 1-35.

RANDOM WALKS AND DIFFUSIONS ON FRACTALS

Sheldon Goldstein
Department of Mathematics
Rutgers University
New Brunswick, N.J. 08903

Abstract

We investigate the asymptotic motion of a random walker, which at time n is at $X(n)$, on certain "fractal lattices". For the "Sierpinski lattice" in dimension d we show that as $\ell \to \infty$, the process $Y_\ell(t) \equiv X([(d+3)^\ell t])/2^\ell$ converges in distribution (so that, in particular, $|X(n)| \sim n^\gamma$, where $\gamma = (\ln 2)/\ln(d+3)$) to a diffusion on the Sierpinski gasket, a Cantor set of Lebesgue measure zero. The analysis is based on a simple "renormalization group" type argument, involving self-similarity and "decimation invariance".

1. Introduction

Some of the results described here have been independently obtained by Kusuoka [1] and, more recently, by Barlow and Perkins.

I wish to discuss two related questions involving two related fractal structures: what I will call the Sierpinski lattice Γ_0 and the Sierpinski gasket Γ.

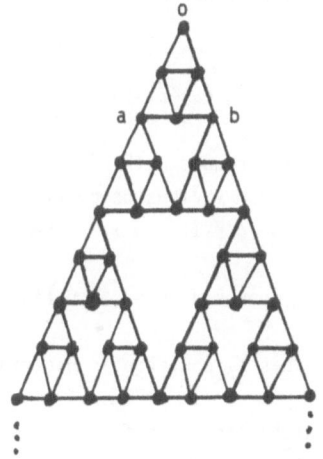

Figure 1. The Sierpinski lattice Γ_0.

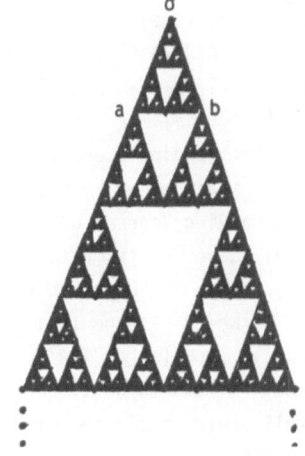

Figure 2. The Sierpinski gasket Γ.

The Sierpinski lattice Γ_0, a fragment of which is shown above for dimension $d = 2$, can be recursively constructed by adjoining to the triangular (tetrahedral) structure arising at any generation, d copies of that structure, attaching to each bottom vertex of the original the upper vertex of one of the copies, to obtain the triangular structure of the next generation. The Sierpinski gasket Γ can be recursively constructed by removing from each elementary triangle (tetrahedron) present at any stage its open middle triangle ("middle polyhedron"). Γ_0 and Γ are defined for any $d = 1, 2, 3, \ldots$. (For $d = 1$, Γ_0 may be identified with $\mathbb{Z}_+$ and Γ with R_+.) Γ is a connected continuum of Hausdorff dimension $(\ln(d+1))/\ln 2$. For $d > 1$, it is a "Cantor set" of d-dimensional Lebesgue measure zero.

I will discuss the following two questions:

(1) Consider the symmetric nearest neighbor random walk $X(n)$ on Γ_0, in which nearest neighbor transitions occur with equal probability. What is the asymptotic behavior of $X(n)$, $n \to \infty$? (A more precise description of the problem will soon be given.)

(2) How can Brownian motion on Γ be defined? (Note that this question is not natural for the usual Cantor set since it is totally disconnected.)

As one should expect, (2) will be (partially) answered when (1) is, so I will concentrate on (1).

Before giving a more precise formulation of (1), I wish to review the situation for the symmetric n.n. random walk $X(n)$ on $\mathbb{Z}^d$ [2]:

(i) $|X(n)| \sim n^{1/2}$ (e.g., $<X(n)^2> \sim n$).

(ii) limit law for random variable $X(n)$, suitably rescaled:
$$X(n)/n^{1/2} \implies \text{Normal}.$$
(Here "$\implies$" means convergence in distribution.)

(iii) limit law for process $(X(n))_{n \geq 0}$ - asymptotic behavior of process on macroscipic length and time scales:
$$(X([At])/A^{1/2})_{t \geq 0} \implies (W(t))_{t \geq 0}$$
where $W(t)$ is standard Brownian motion and $[\cdot]$ is the greatest integer function. Note that (iii) more or less implies (ii) and (i).

The corresponding properties and or questions for the random walk $X(n)$ on Γ_0 are:

(i) $|X(n)| \sim n^\gamma$ where $\gamma = \dfrac{\ln 2}{\ln(d+3)}$. This is known from various computations with varying degrees of rigor [3, 4]. Note that for $d = 1$, $\gamma = 1/2$, the classical value, as it must.

(ii) $X(n)/n^\gamma \implies$? Does the rescaled random variable converge in distribution? If so, to what? Is the limit Gaussian?

(iii) $(X_A(t))_{t \geq 0} \equiv (X([At])/A^\gamma)_{t \geq 0} \implies$? Does the process X_A converge in distribution? If so, to what?

I will sketch a simple, direct derivation of γ which also gives the existence of the asymptotic process. It turns out, in fact, that (ii) and (iii) are false for the random walk on Γ_0, but for trivial reasons which I will come to later. We have, however, the following

Proposition: Let $X(n)$ be the symmetric nearest neighbor random walk on Γ_0, and let

$$Y_\ell(t) = X([(d+3)^\ell \, t])/2^\ell$$

Then $(Y_\ell(t))_{t>0} \xrightarrow[\ell \to \infty]{} (Y(t))_{t>0}$, where $Y(t)$ is a Markov process on Γ with continuous sample paths, in fact, a Feller process.

$Y(t)$ should be regarded as the (standard) Brownian motion on Γ. "$\Longrightarrow$" in the
proposition refers to weak convergence for the Skorohod topology on path space [2].
Note that the formula for $\gamma = \frac{\ln 2}{\ln(d+3)}$ follows immediately, as does the fact
that $Y(t)$ is itself invariant under the rescaling $x \to 2x$, $t \to (d+3)t$. Note also
that the proposition asserts that a limit law exists only for the discrete set of
rescalings $A = (d+3)^\ell$, $\ell = 1, 2, 3, \ldots$.

I conclude this section with some general remarks comparing the proofs for $\mathbb{Z}^d$
and Γ_0: On $\mathbb{Z}^d$, $X(n)$ is the sum of independent, identically distributed random variables. Thus the central limit theorem applies, and asymptotically $X(n)$ has independent Gaussian increments, in fact, becomes a Brownian motion. In proving
convergence, the limit is explicitly exhibited. Now Γ_0 is not homogeneous - not
all points are equivalent - so that for the random walk on Γ_0, $X(n)$ cannot be decomposed into a sum of independent random variables in any obvious useful way. But
for $X(n)$ on Γ_0 it is easy to establish "self-similarity" - the decimation invariance
described in the next section - and this leads to a simple argument for the existence of the asymptotic process; however, the argument doesn't provide a great deal
of information about the limit.

2. Decimation Invariance

Focus on the sublattice of even sites $2\Gamma_0 \subset \Gamma_0$, the even sublattice, indicated
below by "●"'s:

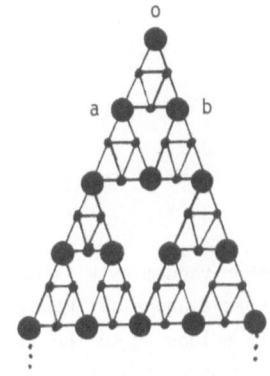

Figure 3. The even sublattice.

Let X(n) be any nearest neighbor random walk on Γ_0. From each even site let the
random walk X(n) run until a <u>new</u> even site is reached. The successive even sites
reached in this way define a random walk on $2\Gamma_0$ and hence, since $2\Gamma_0$ may be identified with Γ_0 in the obvious way, a random walk on Γ_0 which we call the <u>decimation</u>
of X(n). If X(n) is described by the transition matrix P, then the decimation of
X(n) is described by RP. If RP = P, then the random walk X(n), or P, is <u>decimation
invariant</u>. Similarly, decimation invariance can be defined for random walks on any
lattice Γ_0, not necessarily the Sierpinski lattice, satisfying $k\Gamma_0 \subset \Gamma_0$ for some
integer $k \geq 2$.

By symmetry, it is clear that the symmetric nearest neighbor random walk on
the Sierpinski lattice Γ_0 (or on $\mathbb{Z}$) is decimation invariant. On $\mathbb{Z}^d$ for $d \geq 2$, however, it is not decimation invariant, since in this case RP has "infinite range".
A more interesting example is provided by the lattice (k = 3) of Fig. 4.

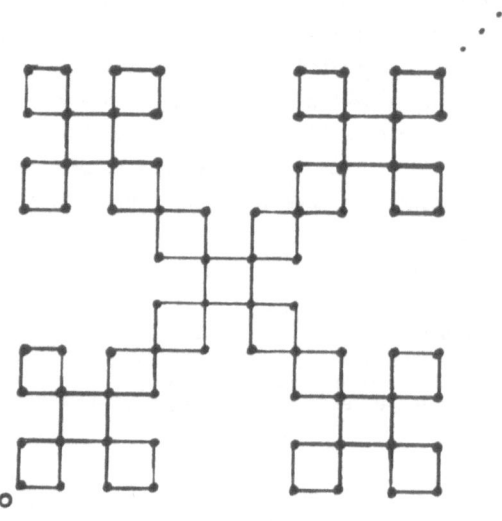

Figure 4.

Here the symmetric nearest neighbor random walk is not decimation invariant, since
RP allows direct transitions across diagonals. However, if all direct transitions
from a vertex to any other vertex belonging to the same elementary square occur with
equal probability (so that, for example, from a vertex belonging to two elementary
squares six direct transitions, each with probability 1/6, are allowed) i.e., if we

consider the symmetric nearest neighbor random walk on the graph obtained by replacing each elementary square by ⊠ , then, by symmetry, we have decimation invariance.

Another key ingredient in our analysis, which goes hand in hand with decimation invariance, is the average decimation time, which determines the appropriate relationship between the macroscopic length and time scales, for which the random walk has nontrivial asymptotics. For the Sierpinski lattice the _decimation time_ τ is the time required to reach a new even site from a point p on the even sublattice. For the symmetric random walk the distribution of τ clearly does not depend upon p, and for d = 2 the average decimation time $\langle\tau\rangle$ is just the expected hitting time, starting from o, of the set {a, b} in Figure 1. An easy computation shows that $\langle\tau\rangle = d + 3$.

Thus, for the symmetric n.n. random walk on the Sierpinski lattice, spatial rescaling by 2 is associated with a time rescaling by d + 3, which suggests asymptotic symmetry under $x \to 2x$, $t \to (d+3)t$ and that $\gamma = (\ln 2)/\ln(d+3)$. Similarly, for a general decimation invariant random walk, we should have $\gamma = (\ln k)/\ln\langle\tau\rangle$. In particular, for the decimation invariant random walk, on the lattice of Figure 4, based on ⊠ , $\langle\tau\rangle = 15$ so that we should have $\gamma = (\ln 3)/\ln 15$.

3. Asymptotics

The preceding section motivates our definition

$$Y_\ell(t) = X([(d+3)^\ell t])/2^\ell$$

of the rescaled process for the symmetric n.n. random walk on the Sierpinski lattice. Note that Y_ℓ lives on $\Gamma_\ell \equiv 2^{-\ell}\Gamma_0$, which may be regarded as the lattice of vertices arising from Γ_0 by constructing ℓ generations of interior middle triangles (polyhedra). Note also that

$$\Gamma_\ell \subset \overline{\underset{j}{\cup}\Gamma_j} = \Gamma , \tag{1}$$

the Sierpinski gasket.

Let $X_\ell(n)$ be the symmetric n.n. random walk on Γ_ℓ. Suppose, for simplicity, that $X(0) = X_\ell(0) = o$. Then the process $Y_\ell(t)$ may be identified with $X_\ell([(d+3)^\ell t])$, i.e., instead of spatially rescaling the random walk on Γ_0, we may equivalently consider the random walk on Γ_ℓ. Since Γ_ℓ is the even sublattice of $\Gamma_{\ell+1}$, decimation invariance now immediately gives

$$(X_{\ell+1}(T_n))_{n\geq 0} \overset{\text{law}}{=\!=} (X_\ell(n))_{n\geq 0} \tag{2}$$

where $T_n = \tau_1 + \tau_2 + \ldots + \tau_n$ with τ_i the time required for the ith "jump" in the decimation of $X_{\ell+1}$. Moreover, the τ_i's are independent (of everything) with distribution that of τ. Thus, for large n, $T_n \cong n\langle\tau\rangle = (d+3)n$ with error $\sim \sqrt{n}$.

If T_n were exactly equal to $(d+3)n$, we would have from (2), because $Y_{\ell+1}$

involves one more factor of $d + 3$ than Y_ℓ, that

$$(Y_{\ell+1}(t))_{t\geq 0} \overset{\text{law}}{=} (Y_\ell(t))_{t\geq 0}$$

on the discrete time skeleton. In fact, we have that

$$Y_{\ell+1} \overset{\text{law}}{\cong} Y_\ell$$

with exponentially small error (since $(d + 3)^\ell$ grows exponentially and the "time error" / total time $\sim$ (total time)$^{-1/2}$). Thus $(Y_\ell(t))_{t\geq 0}$ is Cauchy in distribution (in, say, the Vasserstein norm based on the Skorohod metric on path space) and hence converges in distribution to a process which we denote by $Y(t)$. (In fact, it is not difficult to simultaneously realize all of the processes Y_ℓ on the same probability space in such a way that the paths of Y_ℓ, $\ell \to \infty$, converge uniformly on bounded time intervals.)

So far we have considered only starting at o. The analysis applies, however, almost verbatim to random walks starting at any vertex of $\bigcup_\ell \Gamma_\ell$, a dense subset of the Sierpinski gasket Γ. This suggests that the asymptotic process $Y(t) = Y^x(t)$ with $Y(0) = x$ can be naturally defined for any $x \in \Gamma$. This, in fact, can be done, in such a way that we obtain a Markov process with continuous sample paths, in fact, a Feller process. Moreover, since Y_ℓ lives on $\Gamma_\ell \subset \Gamma$, a closed set, the limit process Y lives on Γ.

Note that the random variable $Y(t)$, $t > 0$, cannot be Gaussian, since its distribution is supported by Γ which has Lebesgue measure 0.

4. Brownian Motion on the Sierpinski Gasket

The process $Y(t)$, the proof of whose existence has just been sketched, should naturally be regarded as the generalization to the Sierpinski gasket Γ of Brownian motion on $\mathbb{R}^d$. In this short section we mention some of its basic properties, which support this generalization.

$Y(t)$ has a stationary measure, the "Cantor measure" μ on Γ, which distributes mass "uniformly" over Γ. More precisely, we identify a "unit triangle" of Γ with the set of sequences $\xi_1, \xi_2, \ldots$ with $\xi_i = 1, 2,$ or 3 according to which of the three (surviving) i-th generation triangles contain the corresponding point x. (This correspondence is one-to-one unless x is a vertex.) Then μ (restricted to the unit triangle) is the image of the $(\frac{1}{3}, \frac{1}{3}, \frac{1}{3})$-product measure on sequence space under this correspondence. μ is the (scaled) limit of the (modified) counting measures μ_ℓ on Γ_ℓ (with $\mu_\ell(\{o\}) = 1/2$). The Cantor measure μ is, in fact, reversible for Y, since μ_ℓ is reversible for X_ℓ.

Brownian motion on $\mathbb{R}^d$ is uniquely determined, among time-homogeneous Markov processes on $\mathbb{R}^d$ with continuous sample paths (the "diffusions" on $\mathbb{R}^d$), by Euclidean invariance, i.e., by (spatial) homogeneity and isotropy. Similarly, up to a uniform speed change, $Y(t)$ is uniquely determined, among time-homogeneous Markov processes on Γ with continuous sample paths (the "diffusions" on Γ), by "local homogeneity and isotropy", the natural symmetries of Γ, which will not be defined here more precisely.

Modulo a linear transformation, there are only two scale invariant homogeneous diffusions on $\mathbb{R}^d$ - Brownian motion and uniform motion (pure, constant drift). It follows from its construction as a limit that the Brownian motion $Y^o(t)$ on Γ is scale invariant, under $x \to 2x$, $t \to (d+3)t$. Are there other scale invariant, locally homogeneous diffusions on Γ? Yes, in fact, up to a uniform speed change the scale invariant locally homogeneous diffusions on Γ are in one-to-one correspondence with the "homogeneous" decimation invariant nearest neighbor random walks on Γ_0. (The homogeneous n.n. random walks on Γ_0 are determined, for $d = 2$, by the four probability vectors describing transitions from the four types of vertices, namely o, $\angle$, $\swarrow$, and $\rightarrow$.)

5. Other Fixed Points of R

Any upper triangle (of side 2^n) in Γ_0 has three axes of symmetry, L_0, L_1, and L_2, the perpendicular bisectors through the vertices o, 1, and 2, respectively, where 1 is the bottom left vertex of the triangle and 2 the bottom right. The symmetric n.n. random walk on Γ_0, which we will now denote by P_{symm}, is invariant under the group D_3 generated by reflections in these three axes (ignoring the transitions from the special vertex o). Another decimation invariant homogeneous n.n. random walk on Γ_0, also invariant under D_3, is P_{out}, whose transition probabilities are $\angle$, $\swarrow$, and $\rightarrow$. (Again, we ignore the transitions from o; note, however, that any transition probability from o is consistent with decimation invariance for P_{out}.) What about P_{in}, which is also invariant under D_3? (P_{in} is obtained by interchanging 0's and 1/2's in the definition of P_{out}.)

There are at least six other decimation invariant homogeneous n.n. random walks on Γ_0, two for each of the axes of symmetry: For each $i = o, 1, 2$ we have $P_{\to i}$ and $P_{\leftarrow i}$ ("towards vertex i" and "away from vertex i") which are invariant under reflection in L_i. Whenever possible these random walks do what their description suggests. When a choice must be made, its statistics are determined by symmetry. (If the origin o presents any problem, make it an absorbing point.) We will describe the asymptotics for one of these, $P_{\leftarrow o}$, which we will also call P_{down}, in some detail.

The fixed point P_{down}, $RP_{down} = P_{down}$, has transitions

(a) $\angle$, (b) $\swarrow$, (c) $\rightarrow$,

(and $\wedge$). From a vertex of type (b) the expected decimation time is 4, while from a vertex of type (a) or (c) the expected decimation time is 5/2. Therefore the random walk P_{down} spends most, i.e., a fraction approaching unity, of its time in vertices of type (b). Hence, the average decimation time $\langle\tau\rangle$ = 4. The appropriate scale is thus the Brownian, i.e., we consider

$$Y(t) \quad = \quad \lim_{\ell\to\infty} X([4^{\ell}\,t])/2^{\ell} \ .$$

For $Y(t)$, which lives on Γ, we presumably have the following behavior. On horizontals $Y(t)$ is ordinary Brownian motion, until an endpoint of the horizontal is reached, at which time $Y(t)$ goes down. But $Y(t)$ spends no time going down, and it can't jump - this would not be compatible with scale invariance. $Y(t)$, in fact, spends almost all of its time undergoing Brownian motion on horizontals. On a set D of times of Lebesgue measure 0, the complement of the union of time intervals in which horizontal motion is occurring, $Y(t)$ makes its descent. Let $Z(t)$ be the "distance down". Then $Z(t)$ is increasing and continuous, with $Z'(t) = 0$ for $t \notin D$, i.e., $Z(t)$ is a Cantor function or devil's staircase.

6. General Fractal Lattice

The basic ingredient of our analysis for a random walk P on a lattice Γ_0 is the existence of an integer $k \geq 2$ such that

1) $k\Gamma_0 \subset \Gamma_0$, so that decimation R can be defined, and
2) P is decimation invariant: RP = P.

If 2) is not satisfied, the asymptotic behavior of the random walk should be governed by the fixed point $\hat{P}$ of R in whose domain of attraction P is situated:

$$\hat{P} \quad = \quad \lim_{\ell\to\infty} R^{\ell}\,P \ .$$

For example, for the symmetric n.n. random walk P on the lattice in Figure 4, k = 3 and we presumably have $\hat{P}$ = "$(\frac{1}{3},\frac{1}{3},\frac{1}{3})$", the symmetric n.n. random walk when ☐ is replaced by ☒ . Thus γ = ln 3 / ln 15 (see section 2), for P as well as for $\hat{P}$.

Acknowledgements

I would like to thank Itzhak Webman, Joel Lebowitz, David Wick, and Yves Elskins for valuable discussions. I also wish to gratefully acknowledge the warm hospitality of the IMA, University of Minnesota. This work was supported in part by NSF Grants No. DMR 81-14726 and DMS 85-12505.

References

[1] S. Kusuoka, A diffusion process on a fractal, preprint.
[2] P. Billingsley, <u>Convergence of Probability Measures</u>, Wiley, New York (1968).
[3] R.A. Guyer, Phys. Rev. A <u>29</u>, 2751 (1984).
[4] R. Rammal and G. Toulouse, J. Physique-Lettres <u>44</u>, L-13 (1983).

THE BEHAVIOR OF PROCESSES WITH STATISTICAL MECHANICAL PROPERTIES

Lawrence Gray
School of Mathematics
University of Minnesota
Minneapolis, MN 55455

1. Introduction.

Ever since Spitzer's famous paper in 1970, there has been interest in a class of Markov processes which have as time-reversible stationary measures certain special distributions from the theory of statistical mechanics. The state space for these processes is $\Xi = \{-1, +1\}^{\mathbf{Z}^d}$, which is the space of configurations of $+$ and $-$ spins on the sites of the lattice $\mathbf{Z}^d$. Transitions occur when there is a "flip" at a site $x \in \mathbf{Z}^d$, or in other words, a change of sign in the spin at x. The probability that a flip occurs at x in a short time interval $(t, t + h]$, given the history of the process up to time t, is $c_x(\xi_t)h + o(h)$, where ξ_t is the state of the process at time t, and c_x is a non-negative function defined on Ξ, called the <u>flip rate at x</u>. Simultaneous flips at two different sites do not occur. A system of Markov processes with this description, one process for each possible initial state, is often called a "spin-flip system" with rates $\{ c_x \}$. Spitzer pointed out that for certain kinds of interaction potentials commonly used in statistical mechanics, one can always find a set of rates $\{ c_x \}$ such that the corresponding spin-flip system has as time-reversible equilibria the Gibbs states that correspond to the interaction potential. (Spitzer's results required a certain uniqueness hypothesis that was later verified for a large class of systems by Liggett (1972).) This is illustrated in the following example:

Example 1. (The stochastic Ising model) We start with a <u>pair potential</u> with <u>range R</u>: let $J: (0,\infty) \to [0,\infty)$ be a non-increasing function such that $J(r) = 0$ for $r > R$, and then we define the <u>energy at x in state ξ</u> by

$$E_x(\xi) = -\sum_{y:\, y \neq x} \xi(x)\xi(y) J(\|x - y\|) \ .$$

We have used $\|\cdot\|$ for the usual Euclidean distance in $\mathbb{Z}^d$, and $\xi(x)$ stands for the value of the spin (+1 or −1) at x in the configuration ξ. We have restricted our attention to pair potentials whose strength decreases with distance merely for convenience. The assumption that J is non–negative is more significant. It ensures certain monotonicity properties that seem to be almost indispensable in the study of spin–flip systems.

Next we define a <u>Gibbs state with potential J</u> to be any probability measure μ on the Borel sets of Ξ such that for all $\eta \in \Xi$ and all $x \in \mathbb{Z}^d$,

$$\mu\left(\xi\colon \xi(x) = \eta(x) \mid \xi(y) = \eta(y) \text{ for } y \neq x\right) = Z^{-1}(\eta)\exp(-E_x(\eta)), \tag{1}$$

where $Z^{-1}(\cdot)$ is the normalizing constant that makes the total conditional probability equal to one,

$$Z^{-1}(\eta) = \exp(-E_x(\eta)) + \exp(E_x(\eta)) \ .$$

To get a spin system that has all the Gibbs states with potential J as its time–reversible equilibria, we have many choices for the rates, but we choose $c_x(\xi) = \exp(E_x(\xi))$, which is one of the choices that makes the system symmetric with respect to interchange of + and − spins. For a detailed explanation as to why a system with these rates has time–reversible equilibria as claimed, see Spitzer's original paper, or more recently, Ligget's book on interacting particle systems, both listed in the references.

Next we introduce a parameter $\varepsilon > 0$ (corresponding to a constant multiple of the temperature in statistical mechanics) which we call the <u>noise</u> parameter. Let

$$c_x^\varepsilon(\xi) = \exp(E_x(\xi)/\varepsilon),$$

which defines rates for a system corresponding to the interaction potential $J^\varepsilon = J/\varepsilon$. The idea is that as the noise decreases, the interaction becomes stronger.

One final parameter: let H be a real constant, which we call the <u>bias parameter</u> (the external field strength in statistical mechanics), and define flip rates for a biased system by

$$c_x^{\varepsilon,H}(\xi) = \exp(E_x(\xi)/\varepsilon)\exp(-H\xi(x)). \qquad (2)$$

These biased systems also have Gibbs states as their time-reversible equilibria. They are defined by replacing $E_x(\xi)$ in (1) by $(E_x(\xi)/\varepsilon) - H\xi(x)$.

The expression in (2) ensures that the flip rate at x is high if E_x would be lowered by a flip, and the flip rate is low if the energy would be raised. A quick glance at the expression for the energy shows that there is a positive contribution to it for each site y within range R of x such that the spins at x and y disagree. In biased systems, there is also a positive contribution to the energy if the spin at x disagrees with the sign of H. Conversely, sign agreement between such x-y spin pairs or between H and the spin at x contribute negatively to the energy. Thus the rates are set up so that the system tends to spend more time in configurations of low overall energy. In the unbiased case, there are two configurations of minimum energy, which we call ξ^+ and ξ^-, the two configurations in which all the spins have the same sign. These are called

<u>ground states.</u> In the biased case, only one of these states has minimum energy, namely the one that agrees in sign with H. Under certain circumstances which will be discussed in the next section, the extreme equilibria of the system become more and more concentrated near the ground states as the noise parameter decreases to 0. This might be expected from the form of the flip rates.□

One of the purposes in defining a Markovian system with Gibbs states as equilibria is that it makes available the tools and techniques of Markov processes to the study of statistical mechanics. Semigroups, generators, martingales, coupling methods, etc., can now be brought to bear on problems that originally had no time dependence. Furthermore, information about rates of convergence to equilibrium, dependence of the process on the initial state and other details about the time evolution can be translated into information about uniqueness and mixing properties of the Gibbs states. As a simple example of how this can work, we take one of the most basic questions in the study of Gibbs states, namely the question of whether there exists more than one Gibbs state corresponding to a given potential J. If the corresponding spin system is <u>ergodic</u> (or in other words, has a unique equilibrium), then one can conclude that there is only one Gibbs state for J, and if the system converges exponentially fast to this equilibrium, then the Gibbs state must have exponential mixing properties (see Liggett's book for these and other results).

To a certain limited extent, this carryover from the study of spin-flip to statistical mechanics has occurred. Some nice contributions to various questions are to be found in the work of Holley and Stroock (1976a and 1976b).   There has also been some interesting work about the increase of free energy in the time evolution which adds to one's understanding of the concept of free energy in statistical mechanics —

see Holley and Stroock (1977). But these instances are the exceptions, rather than the rule. Most of what is known about the equilibrium behavior of the stochastic Ising model is derived from already known facts about Gibbs states. This is putting the cart before the horse if one is trying to find applications for spin-flip systems. (Of course, a great deal of interesting and important work has been done on other kinds of spin-flip systems which have nothing directly to do with statistical mechanics, such as those which are related to population or genetics models like the contact process or the voter model.)

Based on what has been said so far, the problem seems to be that most techniques for studying the time evolution of the stochastic Ising model depend on *a priori* information about the equilibria. It is proposed here that methods should be developed which deal directly with the time evolution. These methods should be robust, in the sense that they should apply to systems whose rates are qualitatively like the rates defined in Example 1, but which do not necessarily have the same precise form, systems which do not necessarily have Gibbs states as equilibria. Once such methods are developed, then it should be possible to isolate those essential properties of a time evolution that lead to "statistical mechanical-like" behavior (we will call this SML behavior for short), thus leading to a better understanding of time evolutions of actual statistical mechanical models.

In the next section, we will propose a program of research that has these goals in mind. We will do this by first defining a class of flip rates that we conjecture have the features necessary for SML behavior, then by giving some examples of such rates, and finally by listing some of the properties that we feel are a part of typical SML behavior. In the third section of the paper, we will "set a good example" by outlining a proof that a certain well-known system which is not a stochastic Ising model (the majority vote model) has SML behavior in one dimension for all sufficiently small values of the noise parameter ε. (The SML behavior I have in mind here is ergodicity, or as the people in statistical mechanics would say, "no phase transitions in one-dimensional, finite range systems".) This is a partial solution to an open problem of several years (one would like to remove the restriction that ε be small). Since this is only a research report, the proof will only be indicated for the nearest neighbor case. An extension to the arbitrary finite range case is forthcoming in a paper which is in preparation (Gray (1986)). It should even be possible to extend the proof to cover the wider class of systems defined in Section 2; the proof can certainly be adapted to cover one-dimensional stochastic Ising models. In Section 3 we will also discuss briefly an unexpected benefit of the proof, namely a sort of invariance principle for one-dimensional systems as ε → 0. At the time of the writing of this report, I don't quite know what to make of this last material, but it is hoped that it has some consequences for the theory of statistical mechanics.

2. SML flip rates and SML behavior.

It seems (to the author at least) that there is nothing special about the exponential function in (1) and (2). Its presence is dictated by certain considerations from statistical mechanics, but from a purely dynamical viewpoint, it is hard to see why its use in the definition of the flip rates should be crucial. To best see what are the essential properties of this function, we consider a state in which there is a flat inter-face between the two types of spins. To simplify things, let $H = 0$. Let π be any $(d-1)$-dimensional hyperplane that does not contain any of the lattice points in $\mathbf{Z}^d$. Assume that a normal vector is attached to π to give it some orientation, and let ξ^π be the state in which the sites on the positive side of π have spin +1 and the sites on the negative side have spin −1. In this configuration, the energy $E_x(\xi)$ at any x is less than or equal to 0. If we change the sign of the spin at some site x, then the energy at x becomes strictly positive. Thus, if for any configuration ξ, we define the configuration $^x\xi$ by

$$^x\xi(y) = \xi(y) \quad \text{for } y \neq x,$$
$$-\xi(y) \quad \text{for } y = x,$$

then the ratio $c_x(\xi^\pi)/c_x(^x\xi^\pi)$ is strictly less than 1. When we introduce the noise parameter, we find that

$$c_x^\varepsilon(\xi^\pi)/c_x^\varepsilon(^x\xi^\pi) \to 0 \quad \text{as } \varepsilon \to 0. \tag{3}$$

(Note that it is the ratio of the rates that is important, since we can always rescale time without changing the equilibrium behavior. Also note that (3) still holds for the biased rates $c_x^{\varepsilon,H}$.) Thus, the smaller the noise, the more the system attempts to maintain a flat interface where one exists. If we define $\varphi^\varepsilon(E) = \exp(E/\varepsilon)$, then for all $E > 0$, $\varphi^\varepsilon(E)/\varphi^\varepsilon(0)$ converges to ∞ as ε goes to 0, and it is this property of the exponential that implies (3). We feel that this property is also the reason that stochastic Ising models behave the way that they do, and that other non-decreasing positive functions φ^ε with this property would do just as well. This leads us to the following:

Definition of SML flip rates. Let $E_x(\xi)$ and $J(\cdot)$ be as in Example 1. We will say that a parameterized family $\{c_x^{\varepsilon,H}, \varepsilon > 0\}$ is a family of SML flip rates with potential J if there are positive, non-decreasing functions $\{\varphi^\varepsilon, \varepsilon > 0\}$ defined on $\mathbb{R}$ such that

$$c_x^{\varepsilon,H}(\xi) = \varphi^\varepsilon(E_x(\xi)) \exp(-H\xi(x)), \tag{4}$$

and such that for all E > 0,

$$\varphi^\varepsilon(E)/\varphi^\varepsilon(0) \rightarrow \infty \quad \text{as} \quad \varepsilon \rightarrow 0. \tag{5}$$

We further impose the regularity condition that the ratios $\varphi^\varepsilon(E)/\varphi^\varepsilon(0)$ be non-decreasing for E > 0 and non-increasing for E < 0 as ε decreases to 0, so that a decrease in the noise parameter corresponds to an increase in the strength of the potential. □

In our definition, we have retained the way in which the rates depend on H and we have not done anything to generalize the energy function. Presumably more generality is possible, but we feel that we have done enough violence to the original model already. Actually, the class of SML rates is even larger than it first appears, due to the flexibility with which φ_ε and J may be chosen. This point will be partially illustrated by the following examples.

Example 2. Perturbations of the stochastic Ising model. We simply let $\varphi(E)$ be some increasing positive function that is uniformly close to exp(x) and define $\varphi^\varepsilon(E) = \varphi(E/\varepsilon)$. It seems incredible, but it is true, that as soon as a perturbation like this is made, virtually all the facts known about the stochastic Ising model become open questions.

Example 3. Sums of Ising model rates. Let $\{a_x^{\varepsilon,H}\}$ and $\{b_x^{\varepsilon,H}\}$ be two sets of rates for stochastic Ising models, corresponding to two different potential energies, and define $c_x^{\varepsilon,H} = a_x^{\varepsilon,H} + b_x^{\varepsilon,H}$. As in Example 2, most of the usual results are no longer known to apply. For example, suppose both of the original systems have phase transitions (non-ergodicity for small ε). It is not known whether the hybrid system with rates $\{c_x^{\varepsilon,H}\}$ has a phase transition. (It is not immediately clear that this example fits into our class of systems with SML flip rates. It turns out, however, that by choosing J and φ^ε properly, the rates in this example can always be realized as SML flip rates. I am indebted to R. Schonmann and J. Lebowitz for pointing out this example.)

Example 4. The majority vote model (continuous time). Define $J \equiv 1$ and

$$\varphi^\varepsilon(E) = \varepsilon \qquad \text{if } E \leq 0$$

$$= 1 \qquad \text{otherwise.}$$

With H = 0, the flip rate at a site x is 1 if $\xi(x) \neq \xi(y)$ for more than half the sites y within range R of x (these sites are called the neighbors of x). When the spin at x agrees with at least half the spins at neighbors of x, the rate is ε. This is the simplest continuous time example of SML rates, yet its behavior is only known in the one-dimensional case with R = 1, in which case it just happens to be a stochastic Ising model! □

We are also interested in models in discrete time. Such models have often been avoided by people interested in statistical mechanics, because there is no simple way to define such a model so that it has Gibbs states as equilibria, as one can in continuous time. From our point of view, however, they are just as worthy of study as their continuous time relatives.

In a discrete time model, there are no rates. Instead, if the systems is in state ξ_t at time $t \in \{0, 1, 2, \ldots\}$, there is a flip probability $c_x(\xi_t)$ that the spin at x changes sign during the next time unit, in which case it takes on the new value at time $t + 1$. During any single step of the time evolution, the flips are made independently at the different sites. We will use the same notation for flip probabilities as we use for flip rates (of course, flip probabilities must be between 0 and 1). An analogue to the notion of SML flip rates is:

Definition of SML flip probabilities. Let $E_x(\xi)$ and $J(\cdot)$ be as in Example 1. We will say that a parameterized family $\{c_x^{\varepsilon,H}, \varepsilon > 0\}$ is a family of SML flip probabilities with potential J if there are non-decreasing, positive functions $\{\varphi^\varepsilon, \varepsilon > 0\}$ defined on $(-\infty, 0]$ such that

$$
\begin{aligned}
c_x^{\varepsilon,H}(\xi) &= \varphi^\varepsilon(E_x(\xi))\exp(-H\xi(x)) &&\text{if } E_x(\xi) \leq 0 &&(4') \\
&= 1 - \left(\varphi^\varepsilon(-E_x(\xi))\exp(H\xi(x))\right) &&\text{otherwise,}
\end{aligned}
$$

and such that for all $E < 0$,

$$
\varphi^\varepsilon(E) \to 0 \quad \text{as} \quad \varepsilon \to 0. \tag{5'}
$$

As before, we make a regularity assumption, namely that $\varphi^\varepsilon(E)$ be non-increasing for all $E < 0$ as ε decreases. Note that we have formulated (4') in such a way that for all states ξ, $c_x^{\varepsilon,H}(\xi) + c_x^{\varepsilon,H}(^x\xi) = 1$. This is not a necessary assumption, but it does save us from being forced to pay attention to several annoying details that come up in discrete time models. We are now ready to define what we consider to be the simplest SML model of all:

Example 5. The majority vote model (discrete time). Define $J \equiv 1$ and
$$
\varphi^\varepsilon(E) = \varepsilon \quad \text{for all } E \leq 0.
$$
Thus the spin at x changes with probability ε if it is in agreement with at least half the spins at neighbors of x, otherwise it changes with probability $1-\varepsilon$. Until now, nothing was known about this model for small $\varepsilon > 0$, even in the one-dimensional, nearest neighbor case (compare with the continuous time model above). It seems that the simpler the model gets, the less one knows about its SML behavior. In the next section,

we will alleviate this situation a little by discussing the very simplest of all models, namely the one-dimensional nearest neighbor discrete time majority vote model. For this model, we will sketch the proof that the equilibrium is always unique for sufficiently small $\varepsilon > 0$. Our methods can be adapted to the arbitrary finite range case in both discrete and continuous time (see Gray (1986)) , and probably even to the general one-dimensional SML model. $\square$

We conclude this section with a discussion of what, in our view, constitutes basic SML behavior. We have picked out a few properties that we feel characterize the behavior of a fairly general class of statistical mechanical systems. We are aware that there are many interesting examples of systems which do not conform to this picture (which is one reason that they are so interesting). For example, we are ignoring the variety of behavior that occurs in systems with infinite range interactions, or in systems with infinitely many "spin" values possible at each site.

We start by defining the __magnetization at x__, which is merely the expected value of $\xi_t(x)$. For a fixed family of SML flip rates, this quantity depends on the initial state ξ_0, the site x, the time t, and on the parameters ε and H. We will only be interested in the initial states ξ^+, ξ^-, and ξ^π defined earlier. Let $M_t^+(\varepsilon, H)$, $M_t^-(\varepsilon, H)$, and $M_t^\pi(x, \varepsilon, H)$ be the corresponding magnetizations at x.

Note that by translation invariance, $M_t^+(\varepsilon, H)$ and $M_t^-(\varepsilon, H)$ do not depend on x, and that by symmetry, $M_t^+(\varepsilon, H) = -M_t^-(\varepsilon, H)$. It follows from our assumption that J be a non-negative function that the flip rates are "attractive" (see Liggett's book in the references), which in turn implies that $M_t^+(\varepsilon, H)$ decreases as $t \to \infty$. We will write $M_\infty^+(\varepsilon, H)$ and $M_\infty^-(\varepsilon, H)$ for the limits of $M_t^+(\varepsilon, H)$ and $M_t^-(\varepsilon, H)$ as $t \to \infty$. It is not clear that $M_t^\pi(x, \varepsilon, H)$ has a limit as $t \to \infty$, so we will use the notation $M_\infty^\pi(x, \varepsilon, H)$ for the lim sup if x is on the positive side of π, otherwise it will stand for the lim inf.

Using the notion of magnetization, we can now list four kinds of behavior that we conjecture are exhibited by all systems with SML flip rates in which the range R is finite:

SML Behavior.

(i) Monotonicity in the parameters: for any fixed potential J and for any time $0 \leq t \leq \infty$, the magnetizations $M_t^+(\varepsilon, H)$, $M_t^-(\varepsilon, H)$, and $M_t^\pi(x, \varepsilon, H)$ are monotone functions of the range R, the dimension d, the noise ε and the bias H. Note: the monotonicity in H — like the monotonicity in t —is a consequence of the fact that the kinds of flip rates that we are concerned with are attractive. Monotonicity in the remaining parameters does not come so easily.

(ii) No spontaneous magnetization in biased systems: If $H \neq 0$, then the limits $M_\infty^+(\varepsilon, H)$, $M_\infty^-(\varepsilon, H)$, and $M_\infty^\pi(x, \varepsilon, H)$ are all equal for all $\varepsilon > 0$.

(iii) No spontaneous magnetization in one dimension: If $d = 1$, then the limits $M_\infty^+(\varepsilon, 0)$, $M_\infty^-(\varepsilon, 0)$, and $M_\infty^\pi(x, \varepsilon, 0)$ are all equal to 0 for all $\varepsilon > 0$. Furthermore, the convergence is exponentially fast in t.

(iv) Spontaneous magnetization in dimensions greater than 1: If $d > 1$, then $M_\infty^+(\varepsilon, 0) > M_\infty^-(\varepsilon, 0)$ for all sufficiently small $\varepsilon > 0$.

(v) No tight interface in one or two dimensions: If $d = 1$ or 2, then $M_\infty^\pi(x, \varepsilon, 0) = 0$ for all $\varepsilon > 0$ and all sites x.

(vi) Existence of tight interfaces in dimensions greater than 2: If $d > 2$, then for all sufficiently small $\varepsilon > 0$, $M_\infty^\pi(x, \varepsilon, 0)$ is dependent on x. □

This list is based on known behavior of Gibbs states. There is a great deal that can be said about these properties — a proper discussion of them is beyond the scope of this report. The interested reader should consult Liggett's book, and also the very readable monograph of Kindermann and Snell (1980). We will merely note here that we feel that property (i) is of the most immediate importance. It corresponds to various correlation inequalities in statistical mechanics which are used over and over again (for example, the monotonicity in ε corresponds to the "Griffith's inequalities"). In particular, it allows one to talk about "critical" values of the various parameters.

3. SML behavior in the one—dimensional majority vote model.

In this section, we will outline the proof that property (iii) of SML behavior holds for the one—dimensional nearest neighbor discrete time majority vote model, at least for all sufficiently small ε > 0. If one could prove that property (i) also holds for this model, then we could remove the restriction that ε be sufficiently small.

Theorem. Let (ξ_t), t = 0, 1, 2, . . . , be the discrete time system in Example 5, with H = 0 and R = 1. There exists $\varepsilon_0 > 0$ such that for all ε ∈ $(0, \varepsilon_0]$, $M_t^+(\varepsilon, 0)$, $M_t^-(\varepsilon, 0)$ and $M_t^\pi(x, \varepsilon, 0)$ converge exponentially fast to 0 as t → ∞. (As noted earlier, this result will be extended, using essentially the same proof, to all finite range majority vote models in discrete and continuous time in Gray (1986).)

Outline of Proof. This is not a complete proof. We will be very detailed for certain parts of the argument and very sketchy in others. Our choice will be based on whether we consider the part of the argument to contain new ideas or not. See Gray (1986) for more details.

We start by giving an explicit construction of the process. For integer times t ≥ 0 and sites x ∈ **Z**, let e(x, t) be i.i.d. random variables, with distribution determined by

$$P(e(x, t) = +1) = P(e(x, t) = -1) = \varepsilon \quad \text{and} \quad P(e(x, t) = 0) = 1 - 2\varepsilon.$$

We will construct the entire system of processes (one process for each initial state) on the probability space associated with the random variables e(x, t). Assuming that an initial state ξ_0 has been chosen, then we inductively define ξ_t for times t > 0 by

$$\xi_t(x) = e(x, t) \qquad\qquad \text{if } e(x, t) \neq 0$$

$$= \text{the majority spin value in}$$

$$\text{the set } \{\xi_{t-1}(y), |y - x| \leq 1\} \qquad \text{if } e(x, t) = 0.$$

Thus, to determine the value at x at time t, a vote is taken at time t − 1 at the sites within distance R = 1 of x (including x itself). The outcome of this vote determines $\xi_t(x)$, provided the "error variable" $e(x, t)$ is 0. If this error variable is not 0, then the vote is ignored, and $\xi_t(x)$ is determined by $e(t, x)$. This clearly agrees with the description of the system in Example 5. Note that if $e(x, t)$ happens to agree with the outcome of the majority vote, then the value of $\xi_t(x)$ is the same as if $e(x, t)$ had been 0. We will say that <u>an error occurs at x at time t</u> if $e(x, t)$ is not 0 and if $e(x, t)$ does not agree with the outcome of the majority vote taken at time t − 1 at the sites within distance 1 of x. Thus, an error occurs independently with probability ε at any given point (x, t) in space–time.

We can now describe the general strategy of the proof. Let ξ_t^+ and ξ_t^- be the processes with initial states ξ^+ and ξ^- respectively (these are the "all +1's" and "all −1's" initial states defined earlier). As noted in the previous paragraph, these two processes are jointly defined on the same probability space, so we can define the joint process $y_t = (\xi_t^+, \xi_t^-)$. It is easy to check that $\xi_t^+(x) \geq \xi_t^-(x)$ for all t and x, so that $y_t(x)$ can only take three values: (+1, +1), (+1, −1), and (−1, −1). We will call these values +, 0, and − respectively. Note that

$$\tfrac{1}{2} P(y_t(0) = 0) = M_t^+(\varepsilon, 0) - M_t^-(\varepsilon, 0).$$

Since M_∞^π is sandwiched between M_∞^+ and M_∞^-, it is enough to prove that

For all sufficiently small $\varepsilon > 0$, $P(y_t(0) = 0) \to 0$ exponentially fast as $t \to \infty$. (6)

(Of course, the exponential rate is allowed to depend on ε.)

The basic idea behind the proof of (6) is simple. Since there is a positive probability ϵ of an error occurring at any given site, and since these errors occur independently, during each time step there will be infinitely many places where at least 2
consecutive sites have simultaneous errors of the same sign. Such a block will show
up as a string of consecutive +'s or consecutive −'s in the process y_t. As long as no
errors occur near the endpoints of such a string, no 0's can appear within the block, so
we have a somewhat stable string of non-zeroes. This string of non-zeroes will change
in size as errors occur near its endpoints. If ϵ is small, the most likely event is that a
single error will eventually occur near an endpoint (rather than simultaneous multiple
errors), and the symmetry of the model ensures that, for strings of length 4 or more,
such an error is equally likely to result in an increase of the size of the string by 1
unit as it is to result in a decrease by 1 unit. One should try to envision infinitely
many such strings of non-zeroes whose endpoints are essentially doing independent
simple random walks. Of course, this picture is only approximate, for several reasons.
First, simultaneous multiple errors do occur. Secondly, a single error can simultaneously affect both endpoints of a short string (of length 3 or less). Finally, when the
endpoints of different strings get close to each other, various movements can occur
which are not due to errors. If we could ignore these difficulties, the rest of the
proof would be easy. Simple computations would show that with infinitely many
strings of non-zeroes appearing in each time step, all of which change size like random walks, the probability would go to 1 exponentially fast as $t \rightarrow \infty$ that a given site
be contained in such a string, implying (6).

The crux of our proof is to show that for small enough ϵ, the behavior of strings of
non-zeroes in the process y_t is sufficiently like the naive description given in the preceding paragraph to obtain (6). We begin by investigating the behavior of a single

string of +'s. Let $\zeta_t^n = (\xi_t^n, \xi_t^+)$, where ξ_t^n is the process with initial state defined by

$$\xi_0^n(x) = +1 \quad \text{if} \quad |x| \le n$$

$$= -1 \quad \text{otherwise.}$$

As with the process y_t, $\zeta_t^n(x)$ can take three values, +, 0, and −. It starts with the value + at x inside the interval [−n, n] and 0 at all other x. We wish to analyze the movement of the endpoints of this interval of non−zero values due to two different causes, namely due to errors and due to collisions with other intervals of non−zero values.

There is one more point that we wish to make before we start our analysis. It will be seen when we make our estimates that events that have probability $O(\varepsilon^3)$ can be completely ignored. Thus we do not need to worry about the occurrence of clusters of more than two errors in space−time. More precisely, we will eventually be working on some block of space−time [−N, N] × [0, T], where N is $O(1/\sqrt{\varepsilon})$ and T is $O(\varepsilon^{-2})$. It follows that the probability that somewhere in the [−N, N] × [0, T] rectangle, three or more errors occur in any space−time block of size 10 × 10, say, will be small as $\varepsilon \to$ 0. One should keep this in mind when checking the various claims that will be made — in particular, we will never concern ourselves with checking for the effects of clusters of three or more errors.

Movement of the endpoints due to errors. The most important effects arise from a single error. Let us assume that at time 1 a single error occurs at some site x within the interval [n −1 , n + 2], and that no other errors occur anywhere near the site n for several time units. Also, for simplicity, assume that n is at least 2. If $x \le n$ (the error will produce a − at x in this case) then it is easy to check that at time 2, there will be a 0 at the site n in the process ζ_t, while all other sites near n will retain their original value. Thus, the right endpoint of the interval of +'s has moved one unit to the left. Similarly, if $x > n$, and if the error at x produces a $-$, then at time 2 there will be a $+1$ at the site $n + 1$ and the endpoint will have moved 1 unit to the right. Examples of this type of movement at the right endpoint are illustrated below in "Movie #1". An analogous description applies to the left endpoint of the string, with all the directions reversed.

It is possible to give a similar but more involved analysis of the possible effects of two simultaneous errors. The result is that there is again a balance between movement to the right and movement to the left. However, in the case with $R = 1$, there is no need to make the effort. We will find that the direct effects of two simultaneous errors are negligible, as far as the movement of endpoints is concerned. Of course, the occurrence of two simultaneous errors at some distance from the endpoint can indirectly affect the movement of the endpoints through eventual collisions, as will be seen below. Also, remember that we are ignoring clusters of three or more errors, as explained earlier.

Thus we see that, at least to the extent that we can ignore interactions with other strings and the occurrence of two simultaneous errors near the endpoints of the string of $+$'s, they are seen to behave like independent symmetric random walks. Each endpoint moves 1 unit in either direction with probability ε during each unit of time. Of course, the same is true of a string of $-$'s. (The situation in the general finite range case is very similar. The proper statements are somewhat more difficult to formulate and require more machinery to justify, but behind all the technicalities lie just two factors: symmetry and the fact that we can ignore clusters of more than R errors.)

Collisions. Let us suppose that we have in some way defined the positions of the endpoints of an interval of $+$'s or $-$'s up to some time t. Let E_t^l and E_t^r stand for the positions of the left and right endpoints at time t. Initially, each endpoint behaves like a random walk as described above. Now we consider the problem of what happens when one endpoint comes close to the endpoint of another interval. For example, suppose that we are dealing with a string of +'s, and that somewhere between E_t^l and E_t^r, two errors that produce -1's occur simultaneously at neighboring sites. This produces an "inner" string of $-$'s which will move about with the same kind of random walk behavior exhibited by the "outer" string. After some time, its right endpoint may come within three units of E_t^r. Let the three sites that separate E_t^r from the right endpoint of the interval of $-$'s be called x, y, and z. The process ζ_t^n has +'s at these sites. If an error now occurs at y that produces a $-$, then after the dust settles (in two time units), the process ζ_t^n will have values $-$, 0, and 0 at the sites x, y, and z respectively (see Movie #2; note in the movie that a single inner string can be the cause of two collisions — once when the string first gets close to the endpoint, and then later when the inner string becomes too short). Such collisions with inner strings cause us a few problems. One is that the value + at the end of the outer string seems to have changed to $-$. Up until now, E_t^r has always marked a dividing line between +'s and 0's in the process ζ_t^n . We will be forced to drop this interpretation and will have to move E_t^r two sites to the left when such a collision occurs (for a more precise definition, see below). Our interpretation is now that the endpoints mark positions between which there are essentially no zeroes in the process ζ_t^n . We will be more explicit later.

There is still another more serious difficulty that the collision causes for us. The endpoint has moved 2 units to the left, but only a single error has occurred. One may say that the error caused a movement of 1 unit and that the collision caused a further movement of 1 unit. This introduces a drift into the random walk behavior of the endpoints. Of course, one might hope that this drift is balanced out by the possibility that a symmetrically opposite collision occurs on the right side of E_t^r. That is, two consecutive errors that produce +'s could occur to the right of E_t^r, then this string of +'s could eventually come within 3 units of the endpoint, then an error could occur at the middle site separating the string of +'s from E_t^r, and then E_t^r would move at least 2 units to the right. This is probably correct, but rigorous proof seems very difficult. For example, it is hard to say anything about the independence of increments due to collisions in the movement of E_t^r. These collisions are events that involve large chunks of space–time and are highly dependent. They constitute the main problem that has prevented any progress on this model for so many years.

The way out of the difficulty is to prove that the effects of collisions can be ignored when ε is sufficiently small. We will show that when ε is sufficiently small, E_t^l and E_t^r behave enough like a random walks so that the heuristics given near the beginning of the proof can be carried out. Before we can do this, we must get the right definition of E_t^l and E_t^r. But first, here are the movies promised earlier.

Note: In all movies, time progresses down the page.

Movie #1

$\ldots$ + + + + + + + + 0 0 0 0 0 $\ldots$	These are the sites near the right endpoint.
$\ldots$ + + + + + + + + 0 $\oplus$ 0 0 0 $\ldots$	An error (circled) occurs $\ldots$
$\ldots$ + + + + + + + + + 0 0 0 0 $\ldots$	moving the endpoint to the right.
$\ldots$ + + + + + + + + $\ominus$ 0 0 0 0 $\ldots$	Another error occurs $\ldots$
$\ldots$ + + + + + + + + 0 0 0 0 0 $\ldots$	and the endpoint moves left.

Movie #2

$\ldots$ + + + + + + + + + 0 0 0 0 $\ldots$	The sites near the endpoint.
$\ldots$ + + + − − + + + + 0 0 0 0 $\ldots$	An inner string of −'s appears.
$\ldots$ + + + − − − + + + 0 0 0 0 $\ldots$	Later, the inner string has moved.

```
...+++---+⊖+0000....          An error occurs and . . .

...+++----+00000....          in two time steps . . .

...+++----000000....          the 0's have gained two sites.

...++++---000000....          Later, the string of -'s shortens.

...++++-⊕-000000....          An error occurs and . . .

...+++++00000000....          two time units later, the 0's gain
```
$$\text{two more sites}$$

Definition of the endpoints. We will explain how to construct the right end-point process E_t^r. The left endpoint is defined analogously. We start by defining $E_0^r = n + 1/2$. This position marks the midpoint between the rightmost $+$ and its neighboring 0 in the initial state of the process ζ_t^n. Next let us assume inductively that E_t^r has been defined to be some half integer value for some $t \geq 0$. We will include in our inductive assumption the requirement that the value assigned by ζ_t^n to the site $E_t^r - 3/2$ is non-zero and agrees either with the value assigned to the site $E_t^r - 1/2$ or with the value assigned to $E_t^r - 5/2$. Thus we assume that there is an interval of at least two consecutive $+$'s or two consecutive $-$'s which includes the site $E_t^r - 3/2$. Note that this assumption is satisfied at $t = 0$ as long as $n > 0$. We call the common value at the two consecutive sites the (right) endpoint sign. In the case that the endpoint sign does not agree with the value assigned by ζ_t^n to the site $E_t^r - 1/2$, we make the further assumption that value assigned to the site $E_t^r + 1/2$ does agree with the endpoint sign. To make it easier to visualize all this, we note that if the endpoint sign is $+$, then according to the assumptions made, the values at the four sites in $[E_t^r - 3, E_t^r + 1]$ are either $? + + ?$ or $+ + ? +$, where the symbol "?" stands for any of the three possible values $+$, $-$, or 0.

We will now define the increment $E_{t+1}^r - E_t^r$. Following our previous discussion, this increment is built from two pieces, the first of which may be considered to be the result of errors, and the second the result of collisions. The part due to errors will be called X_{t+1}^r and the part due to collisions will be called Y_{t+1}^r, with the increment being the sum of the two.

We start by defining X_{t+1}^r. Consider the values of the error variables $e(x, t+1)$ for sites x in the interval $[E_t^r - 2, E_t^r + 2]$. If all of these have the value 0, then set X_{t+1}^r equal to 0. Next we define X_{t+1}^r in the case that exactly one of these ⌐ ⌐ variables is not 0. Let x' be the site where this occurs. If $x' <$ ⌐⌐ ... if $e(x', t+1)$ does not agree with the endpoint sign, then let x^r . -1. If $x' > E_t^r$ and if $e(x', t+1)$ does agree with the endpoint ˙ .. let X_{t+1}^r equal $+1$. In all other cases, let X_{t+1}^r equal 0. It ˙ ˙ ... that this definition of X_{t+1}^r exactly corresponds to the jumps described in the discussion of movement of the endpoint due to the occurrence of a single error.

Now that X_{t+1}^r has been defined, we let Y_{t+1}^r be the largest integer ≤ 0 such that the choice

$$E_{t+1}^r = E_t^r + X_{t+1}^r + Y_{t+1}^r$$

satisfies the inductive assumption at time $t+1$. (If no such integer Y_{t+1}^r exists, we let $E_u^r = -\infty$ for all $u \geq t+1$. It is easy to see that if $\varepsilon > 0$, such an integer will exist with probability one at all times.) We may describe our definition of the movement of the right endpoint in this way: first move the endpoint in accordance with the description given in the paragraphs which discuss movement due to a single error; then, if neces-sary, move it to the left until the values assigned by ζ_{t+1}^n to the four sites in $[E_{t+1}^r - 3, E_{t+1}^r + 1]$ are $? + + ?$ or $? - - ?$ or $+ + ? +$ or $- - ? -$, with "?" representing any of the three possible values 0, + or − as before. A little investigation of the possibili-ties will reveal that the extra movement to the left represented by Y_{t+1}^r can only occur when there is a string of sites close to the endpoint which have the opposite value from the endpoint sign (a collision occurs), or when multiple simultaneous errors occur near the endpoint.

Let E_t^s, X_t^s and Y_t^s be defined in the obvious way. Then it is easy to check inductively that

All sites in the interval $[E_t^s + 1, E_t^r - 1]$ are assigned non-zero values by ζ_t^n. (7)

In fact, the statement in (7) is essentially true for the larger interval $[E_t^s, E_t^r]$: any zeroes that occur at the end sites of this interval are temporary (they last only one time unit) because of the assumption made about the signs at sites near the endpoints. (Such zeroes can arise at, say, the right endpoint, when the right endpoint moves one unit to the right due to an error that occurs at the site $E_t^r + \frac{3}{2}$ at time t+1.) Thus for all practical purposes, we may consider the interval $[E_t^s, E_t^r]$ to be free from 0's in the process ζ_t^n.

Inner strings. One of the goals in this proof is to obtain good upper bounds on the amount that the endpoint moves due to collisions with inner strings. Thus we need to analyze the way in which these strings move about and interact with one another. We will separate the inner strings into two types. The first type, called a <u>simple string</u>, is an inner string which has never joined together with another inner string which has the same sign and which was originally disjoint from it. When two simple strings with the same sign join together, the larger string formed is called a <u>complex string</u>. We will also insist that the occurrence of multiple simultaneous errors not figure into the movements of the endpoints of a simple string. If more than one error occurs simultaneously within four sites of either side of the endpoint of a simple string, it becomes a complex string.

Let us describe the typical life of a simple string. It is born when two simultaneous errors with the same sign occur at adjacent sites at some time t in the interval
$[E_t^l, E_t^r]$, forming a block of two neighboring sites which are assigned the same non-zero
value by ζ_t^n. As single errors occur near the endpoints of this block, it grows or
shrinks as described in the paragraphs on the movement of endpoints due to errors. If
at any time one of these endpoints comes too close to another block of sites containing the same sign as the endpoint of the simple string, or if multiple simultaneous errors occur near one of these endpoints, then its life as a simple string ends and it becomes a complex string (see Movie # 3 below). It can also happen that an "inner,
inner" string is formed within the first simple string, with the opposite sign. If, say,
the right endpoint of this inner, inner string eventually joins up with the right endpoint of the simple string, then we will consider the simple string to have shrunk by
an amount equal to the width of the inner, inner string (move the right endpoint of the
simple string to coincide with the left endpoint of the inner, inner string). The inner,
inner string then "breaks out" of the simple string and is then absorbed, so to speak,
by whatever is outside of the original simple string (see Movie #4). Note that we are
treating simple inner strings differently here than we treated the large interval $[E_t^l,$
$E_t^r]$. A simple inner string always retains its integrity as a string of +'s or a string of
−'s. It does not change its endpoint sign like the interval $[E_t^l, E_t^r]$ does. Incidentally, an
inner, inner string is considered to be a separate inner string itself, which may be either simple or complex, and which may cause its own collisions with the endpoints E_t^l
and E_t^r. Once it breaks out, it either disappears as a string, or joins together with
another string with the same sign, so in any case, it will at that time cease its life as
a separate string. To summarize, the endpoints of simple strings move in two ways:
the usual random walk kind of movement due to single errors, and larger jumps that

shrink the size of the simple string and allow inner, inner strings to break out.

Movie #3

...+++++++++++++...	Some sites in the interval $[E_t^l, E_t^r]$.
...+++++--++++++...	An inner string is born.
...+++---+++++++...	The inner string has grown and moved.
...+++---+++--++...	A second inner string appears.
...+++---+⊖+--++...	An error (circled) occurs between the strings . . .
...+++----------++...	and a complex string is formed in 2 time units.

Movie #4

...+++++++++++++...	Some sites in the interval $[E_t^l, E_t^r]$.
...+++++--++++++...	An inner string appears and . . .
...++--------+++...	grows after some time has passed.
...++---++---+++...	An inner, inner string appears.
...++-⊕-++---+++...	An error (circled) occurs, allowing. . .
...++++++++---+++...	the inner, inner string to break out, causing the string of −'s to be diminished after 2 time units.

We are interested in upper bounds on the number of times that a collision occurs between one of the endpoints E_t^l or E_t^r and one of the inner strings. We will be able to estimate the probability that such a collision occurs when the inner string remains simple, but estimates for complex strings are troublesome (they _can_ be done, but we prefer not). We find that it is sufficient to be quite crude with complex strings: it is enough to note that the number of collisions between complex strings and E_t^l or E_t^r is bounded by the twice number of times that a new complex string is formed (recall from Movie #2 that an inner string can cause two collisions). In other words, once a complex string is first formed, we will count it as if it has already collided twice with E_t^ℓ or E_t^r, whether it actually eventually does so or not. The formation of complex strings in the interval $[E_t^\ell, E_t^r]$ is rare enough that this overestimate does not get us into trouble. Note that we do not need to keep track of instances where simple strings join up with complex strings, because such events do not increase the number of complex strings. We simply count complex strings once they are formed, and then we ignore them. Note also that we do not exclude the possibility that the same set of simultaneous errors that destroyed one simple string could also start an new simple string. Our estimates will allow for this possibility. To summarize, we have the following:

Assume that $E_t^r > E_t^\ell + 2$ for all $t \in [0, T]$. Then the number of collisions between inner strings and E_t^ℓ or E_t^r during the time interval $[0, T]$ is bounded above by twice the number of simple inner strings that collide with E_t^ℓ or E_t^r during $[0, T]$ plus twice the number of (8) collisions between two disjoint simple inner strings during $[0, T]$ plus twice the number of simple strings that become complex through the occurrence simultaneous errors.

Thus, according to (8), we only need to work with simple strings in our estimates. This fact is quite useful, because for all practical purposes, we can treat simple strings as if their endpoints moved like random walks. The deviation from random walk behavior occurs when an inner, inner string breaks out. This causes a decrease in the size of the simple string, making it harder for it to collide with another disjoint simple string or with the endpoints E_t^ℓ or E_t^r. If we ignore such decreases, we are only making it easier for complex strings to be formed and for collisions to occur between simple strings and E_t^l or E_t^r. Thus we are justified in our estimates in treating simple strings as if the movements of their endpoints were only of the random walk type that results from single errors.

Estimates. Our goal is to show that the endpoint processes E_t^l and E_t^r behave like random walks for small $\varepsilon > 0$. Define

$$\sum_{s \leq t} Y_s^r = C_t^r \qquad \text{and} \qquad \sum_{s \leq t} X_s^r = B_t^r$$

so that $E_t^r - E_0^r = B_t^r + C_t^r$. The increments X_t^r were constructed to be independent identically distributed symmetric random variables, so the B_t^r process is just a symmetric random walk. Thus we would like to show that the decreasing process C_t^r is negligible in comparison to B_t^r as $\varepsilon \to 0$ (and similarly for the left endpoint process). Recall that Y_t^r is 0 unless there is some movement of the right endpoint E_t^r which is not due to the occurrence of a single error. Multiple simultaneous errors near E_t^r and collisions between an inner string and the right endpoint can cause Y_t^r to be negative. When such events occur, $-Y_t^r$ measures the distance that the right endpoint must move toward the left in order that the assumption about the values assigned by the process ζ_t^n to sites near the right endpoint remain valid. This distance is roughly the distance to the nearest pair of neighboring sites that are both assigned the same non-zero value. Recall that none of the sites in the interval $[E_t^l + 1, E_t^r - 1]$ can be assigned the value 0, so as long as E_t^r remains larger than $E_t^l + 2$, the sites in the interval $[E_t^r + Y_t^r$, $E_t^r - 1]$ must contain alternating +'s and −'s. It is not hard to check that intervals of alternating +'s and −'s of length N in the process ζ_t^n can only be produced by clusters of at least $N/2 - 1$ errors occurring near each other in space-time. As mentioned earlier, we can assume that clusters of three or more errors do not occur (they can be ignored if ε is small), so it is safe to assume that $-Y_t^r$ is always less than or equal to 6. If we combine this bound with (8), we obtain the following, which is valid for small $\varepsilon > 0$:

Assume that $E_s^r > E_s^\ell + 2$ for all $s \in [0, t]$. Then

$$C_t^r \leq 6 \, [(\text{the number of collisions between the right endpoint } E_s^r \text{ and}$$

inner strings during $[0, t])$ +

(the number of times $s \in [0, t]$ that two simultaneous   (9)

errors occur near E_s^r)]

$$\leq 12 \, [(\text{the number of simple inner strings that collide}$$

with the right endpoint E_s^r during $[0, t])$ +

(the number of complex strings formed between

E_s^ℓ and E_s^r at some time $s \in [0, t])$ +

(the number of times $s \in [0, t]$ that two simultaneous

errors occur near E_s^r)].

We will estimate the expected values of each of the three terms on the right of (9). It is easy to see that

The expected number of times $s \in [0, t]$ that two simultaneous   (10)

errors occur near E_s^r is $O(t\varepsilon^2)$,

provided we understand "near" to mean "within some fixed number of sites". Estimates for the other two terms will take more effort. Let

$F(x, s)$ = the event that $x \in [E_s^\ell, E_s^r]$ and $e(x, s) = e(x + 1, s) \neq 0$.

In other words, $F(x, s)$ is the event that an inner string of two +'s or two −'s starts at time s at the sites x and $x + 1$. We are ignoring strings that start in other ways, since all other ways require clusters of three or more errors, which we have claimed all along are negligible. We are also including some situations in which a true string is not formed, namely those cases where the signs of the error random variables $e(x, s)$ and $e(x + 1, s)$ agree with the values of the process ζ^n_{s-1} at the sites x and $x + 1$. This overcounting of strings will only increase our estimates, so it is justified. Next let

$G(x, r, s)$ = the event that $F(x, r)$ occurs , and the resulting inner string
eventually collides <u>as a simple string</u> at time s with the right endpoint E^r_s

$H(x, r, s)$ = the event that $F(x, r)$ occurs , and the resulting inner string
becomes a complex string at time s, either by colliding with
another simple string or through the occurrence of multiple
simultaneous errors near one of its ends.

$I(N, t)$ = the event that $[-N, N] \supset [E^s_{s}, E^r_{s}]$ and $E^r_s \geq E^s_s + 2$ for all $s \in [0, t]$.

According to (9) and (10),

$$E(C^r_1; I(N, t)) \leq 12 \sum_{r < s < t} \sum_{x \in [-N, N]} (P(G(x, r, s)) + P(H(x, r, s))) + O(t\varepsilon^2) \qquad (11)$$

We will first obtain an estimate for $P(H(x, r, s))$. There are two ways in which $H(x, r, s)$ can occur. One is that an inner string appears at the sites x and $x + 1$ at time r, this inner string survives as a simple string until time s, and then at time s, two simultaneous errors occur near one of its ends. The probability that the inner string appears is ε^2. The length of a simple inner string behaves essentially like a symmetric random walk which starts at 2 and has an absorbing barrier at 1, and which has jumps at rate 2ε. Thus the probability that it survives for $s - r$ time units is bounded above by a constant times $1/\sqrt{(s - r)\varepsilon}$. The probability that two errors occur simultaneous near one of its ends at time s is a constant times ε^2. There is enough independence around so that we can multiply these probabilities, yielding a bound of $C(\sqrt{(s-r)\varepsilon})^{-1}\varepsilon^4$ for the probability that H(x, r, s) occurs in this way, where C is some constant independent of ε. The second way that H(x, r, s) can occur is that the inner string appears at time r and survives for $s - r$ time units, as before, and then a second inner string appears at time s at a distance d from the first inner string, and the two inner strings collide __as simple strings__ some time later. We claim that the probability that this later collision occurs is bounded by a constant times d^{-1}. Again there is enough independence so that we are justified in multiplying probabilities, so we obtain a bound of $C\varepsilon^4(d\sqrt{(s-r)\varepsilon})^{-1}$, where the factor of ε^4 comes from the probabilities of the appearances of the two inner strings, and the factor of $(\sqrt{(s-r)\varepsilon})^{-1}$ arises as before. Of course, we have ignored the possibility that the collision occurred with a second inner string which appeared __before__ time s. However, this only introduces a factor of 2 in our estimates (the collision will be counted in H(x', r', s) for some other site x' and some time r' < r), so we will not worry about it. Since all inner strings must appear somewhere in the interval [-N, N], d must be less than 2N. Summing over all such d we find that P(H(x, r, s)) is O($(s-r)^{-1/2}\varepsilon^{7/2}\ln N$), so that

$$\sum_{r \leq s \leq t} \sum_{x \in [-N, N]} P(H(x, r, s)) \quad \text{is} \quad O(t^{3/2}\varepsilon^{7/2}N\ln N). \qquad (12)$$

We will now justify our claim about the probability of collision being a constant times d^{-1}. In order to picture the situation, let us assume that the second inner string appears to the right of the first. We will concentrate on the movements of the right endpoint of the first inner string and both endpoints of the second inner string. They essentially all move like independent random walks until two of them come close enough together to collide. If the two endpoints of the second inner string come too close to one another, the second inner string will disappear and no collision between the two inner strings will occur. Thus the two inner strings can only collide if the right endpoint of the first inner string and the left endpoint of the second inner string come close to each other before the two endpoints of the second inner string get too close together. This is a variation on the classical gambler's ruin problem: the right endpoints of the two strings play the roles of the absorbing barriers, and the left endpoint of the second inner string plays the role of the gambler's fortune. This comparison makes the claim quite plausible. One can easily rigorously justify the claim by using the same kind of martingale arguments that one uses in the gambler's ruin problem. In fact, with a little work, one can prove the following: let $x_1(t)$, $x_2(t)$ and $x_3(t)$ be martingales with respect to the same sequence of σ-algebras (the martingales are not assumed to be independent). Assume that $x_1(0) = x_2(0) = x_3(0) = 0$, and for simplicity also assume that the increments of each of the three martingales are bounded in size by some fixed constant. Finally assume that the expected value of either $x_1(t)^2$ or $x_2(t)^2$ grows at a non-zero rate. Let τ_1 be the first time t that $x_1(t) > x_2(t) + 1$ and let τ_2 be the first time t that $x_2(t) > x_3(t) + d$. Then $P(\tau_1 > \tau_2) < Cd^{-1}$ for some constant C independent of d.

We now turn to the estimate of $P(G(x, r, s))$. The event $G(x, r, s)$ occurs as follows. First, the inner string appears at x and x + 1 at time r. This happens with probability ε^2. Then the two ends of this inner string move about like symmetric random walks, jumping at rate ε. At time s, the right one of these two random walks collides with E_s^r. The two endpoints of the inner string must not collide during [r, s], since the inner string would disappear if they did. Thus we can consider them to be moving like independent random walks during that time. These two random walks are also moving independently of the process B_u^r, which is itself moving like a random walk. If we could identify E_u^r with B_u^r, then we would again have a gambler's ruin type problem involving the three random walks. There would be a probability of the form $p(r, s)$ that the collision at time s would occur between E_s^r and the right end of the inner string before the inner string disappeared. The sum over $s > r$ of these probabilities would be the probability that the inner string would eventually collide with E_u^r before disappearing. If we let d be the distance between x and the right endpoint E_r^r, then as before, we would obtain an upper bound for this sum of the form Cd^{-1}. It is easy to see that the probabilities $p(r, s)$ depend only on $s - r$, so we would also have the same bound for the sum over all values of r less than s. Summing over $d < 2N$, we would then have

$$\sum_{r: r < s} \sum_{x \in [-N, N]} P(G(x, r, s)) \quad \text{is} \quad O(\varepsilon^2 \ln N) .$$

Unfortunately, we cannot identify E_u^r with B_u^r. We must take into account the effect of the process C_u^r. The simplest way to do this is to replace the event $G(x, r, s)$ by a different event. Let $x(t)$ be the position of the right endpoint of the inner string formed at x and $x + 1$ at time r and define

$\tilde{G}(x, r, s) =$ the event that $F(x, r)$ occurs and the resulting inner string is

still alive and simple at time s, and s is the smallest time such that

either $B_s^r - B_r^r + E_r^r - x(s) < (E_r^r - x - 2)/2$ or $-(C_s^r - C_r^r) > (E_r^r - x - 2)/2$.

The event $\tilde{G}(x, r, s)$ does not contain the event $G(x, r, s)$, but the union over x, r, s of the $\tilde{G}$ events does contain the union of the G events, and more importantly, the number of $\tilde{G}$ events that occur is greater than or equal to the number of G events that occur. The reason for this is that in order for a collision between an inner string and E_u^r to occur, the (distance $- 2$) between the two must be reduced at least half way either through the drift caused by the process C_u^r or by the wandering about of the random walk components of the various endpoints, and all this must happen before the inner string disappears. If we let d equal this distance (i.e., $d = (E_r^r - x - 2)$), then the probability that half this distance is covered by the wandering of the random walks is bounded above by a quantity $p(r, s)$ which sums to Cd^{-1} as before. The probability that it is covered by the process C_u^r is equal to $P(C_r^r - C_u^r \leq d/2$ for all $u \in [r, s-1]$; $C_r^r - C_s^r > d/2)$. Since the process C_u^r is decreasing, we have a bound of $P(-C_s^r > d/2)$ for the sum of these terms over $r < s$. Summing over $d < 2N$, we obtain

$$\sum_{r: r < s} \sum_{x \in [-N, N]} P(\tilde{G}(x, r, s)) < 2\varepsilon^2 E(-C_s^r) + O(\varepsilon^2 \ln N).$$

Thus we have the following inequality:

$$E(-C_t^r; I(t, N)) \leq C\varepsilon^2 (t\ln N + \varepsilon^{3/2} t^{3/2} N \ln N + \sum_{s \leq t} E(-C_s^r; I(s, N))) \qquad (13)$$

(We have glossed over a technical point here, namely the appearance of the event $I(s, N)$ on the right side of (13). This can be justified by using a stopping time that stops the process E_u^r at the first time that the inequalities in the definition of $I(s, N)$ become violated. We will spare the reader the details.) The constant C in (13) is independent of ε, t and N. Let us assume that $t < c/\varepsilon^2$ and $N < c/\sqrt{\varepsilon}$ for some constant c independent of ε, and define $f^\varepsilon(s) = \sqrt{\varepsilon} E(-C_s^r; I(s, N))$. Then (13) becomes

$$f^\varepsilon(t) < C(\sqrt{\varepsilon} |\ln \varepsilon| \varepsilon^2 t + \varepsilon^2 \sum_{s \leq t} f^\varepsilon(s)). \qquad (14)$$

Again, C is a constant which is independent of ε, N and t. It follows from Gronwall's inequality (see Coddington and Levinson, Chapter 1, Exercise 1) that

$$f^\varepsilon(t) < C\sqrt{\varepsilon} |\ln \varepsilon| \varepsilon^2 t \qquad \text{for all } \varepsilon > 0, t < c/\varepsilon^2 \text{ and } N < c/\sqrt{\varepsilon}. \qquad (15)$$

This last inequality has been the main goal of all our estimates, so let us take a moment here to understand what it gives us. By the Central Limit Theorem , $\sqrt{\varepsilon}\, B_t^r$ is approximately normally distributed, with mean 0 and variance equal to a constant times $\varepsilon^2 t$. On the other hand, we have just shown that if $I(t, N)$ occurs, $-\sqrt{\varepsilon}\, C_t^r$ has an expected value which is small in comparison with $\varepsilon^2 t$, at least when ε is small. Typically, the size of $\sqrt{\varepsilon}\, B_t^r$ will be the same order of magnitude as the square root of its variance, while $-\sqrt{\varepsilon}\, C_t^r$ will be comparable to its expected value, so for $t < c/\varepsilon^2$ and $\varepsilon > 0$ sufficiently small, the drift part of E_t^r, namely C_t^r, will be small in comparison to the random walk part, which is B_t^r . Of course, there are still some technicalities to overcome, but the worst is over. We will not give these details here, but the remaining work on this part of the proof involves using the fact that there is sufficient independence built into the $\tilde{G}$ and H events to make the phrases "typically" , "comparable to" and "small in comparison to" sufficiently precise to prove the following statement:

Let $E_t = E_t^r - E_t^\ell$ and define $\tau = \inf\{t > 0 : E_t < 2$ or $E_t \geq d/\sqrt{\varepsilon}\}$,

for some $d > 0$. Then there is a constant C independent of d such that (16)

for all sufficiently small $\varepsilon > 0$ (depending on d)

$$P(E_\tau \geq d/\sqrt{\varepsilon}\,) > C\sqrt{\varepsilon}\, n/d.$$

(Recall that n is the number that determines the length of the original interval of +'s in the process ζ_t^n , so that $E_0 = 2n + 1$.) The lower bound on $P(E_\tau \geq d)$ is just what we would expect if there were no movement of the endpoints E_t^r and E_t^ℓ due to collisions. In other words, if C_t^r and C_t^ℓ were both 0, then E_t would equal $B_t^r - B_t^\ell$, which is just a random walk. If such a random walk starts at $2n + 1$, then the probability is $\sim n/d$ that it will hit d before hitting 1. Thus (16) is a way of expressing the statement that the movement of endpoints due to collisions is negligible in comparison with the movement of endpoints due to single errors. We will use (16) to complete the proof in the next subsection.

Conclusion of proof. After our long excursion concerning the movement of end-points, we are finally ready to return to an analysis of the process y_t and the proof of (6). Recall that the initial state of y_t has 0's at all sites. There is a probability of ε^2 that simultaneous errors with the same sign will occur at any fixed pair of neigh-boring sites at any given time. One can imagine that an interval of non-zeroes has ap-peared, with endpoints 2 units apart. One can define the subsequent positions of these endpoints in the same manner that E_t^r and E_t^l were defined (the values at sites outside this interval are not relevant — we never assumed anything about these values in our actual definitions of the endpoints). Estimate (16) will apply. In particular, the probability that the interval will achieve a length of at least $d/\sqrt{\varepsilon}$ before disappear-ing is greater than $C\sqrt{\varepsilon}/d$, as long as $\varepsilon > 0$ is sufficiently small (depending on d).

Now consider a $c/\sqrt{\varepsilon} \times c^2/\varepsilon^2$ rectangle of space-time. Within such a rectangle, we would expect about $c^3/\sqrt{\varepsilon}$ intervals to appear as described in the last paragraph. Thus the expected number of these intervals to achieve a length of $d/\sqrt{\varepsilon}$ before disap-pearing is greater than Cc^3/d, at least for sufficiently small $\varepsilon > 0$. Choose c large enough so that $Cc^2 > 1$ and set d = c. Then we have the following:

The expected number of intervals originating in a $c/\sqrt{\varepsilon} \times c^2/\varepsilon^2$
space-time rectangle which achieve length $c/\sqrt{\varepsilon}$ before disappearing   (17)
is at least 1 for sufficiently small $\varepsilon > 0$, where c is independent of ε.

There are now two remaining steps. These are both non-trivial, but they contain no

new ideas, so we will only briefly describe them and refer the reader to the paper referred to at the beginning of the proof for details. The first step is to convert (17)
into the following:

Let c be as above. Then there exists $\delta > 0$ such that for sufficiently

small $\varepsilon > 0$ and for any $c/\sqrt{\varepsilon} \times c^2/\varepsilon^2$ space–time rectangle R,   (18)

P(an interval originates in R which achieves length $c/\sqrt{\varepsilon}$ in time c^2/ε^2)

$$> 1 - \delta.$$

As in other parts of the proof, we claim that there is sufficient independence around
to get from (17) to (18). The dimension of the rectangles in the time direction has
been chosen so that any interval that achieves length $c/\sqrt{\varepsilon}$ has a minimum positive
probability of doing it within a time interval equal to the length of the time side of
the rectangle. This is important in order to maintain a certain amount of independence
for the rest of the argument.

In the final step of the proof, we finally use a fact that we have carefully avoided
until now, namely the fact that two intervals of non–zeroes can join to form a larger
interval. The statement in (18) guarantees that there is a minimum positive density
of intervals of non–zeroes with a certain minimum length in rescaled space–time. If
we focus on one of these intervals, we find that its endpoints will continue to move
about like random walks (or Brownian motion as $\varepsilon \to 0$). There will be a certain rate
at which the interval will join with another interval. The endpoints of this new longer
interval will also move about like a random walk until a further coalescence with yet
another interval. It is not too hard to show that we therefore will get an interval
which grows linearly in length. By choosing ε sufficiently small, this description can

be made accurate for arbitrarily long periods of <u>rescaled</u> time. There are certain percolation techniques (fast becoming standard in this area) which can now be used to
show that for all sufficiently small $\varepsilon > 0$, with probability 1 these growing intervals
will link together to form an infinite region in space-time, and every site will eventually become a part of this region forever. The statement in (6) follows easily from
this. We do not have space here to give the details of the percolation argument. For
an example, see Durrett and Griffeath (1983). This and other missing parts of our
proof will be found in Gray (1986). □

A concluding note. In the proof we rescaled space and time in such a way that
the endpoint processes behaved like Brownian motions as $\varepsilon \to 0$. The same thing can
be done in other models with SML flip probabilities or flip rates. It is natural to ask
whether there is some limiting distribution in space-time which is independent of the
particular model. I believe that there is. In other words, I conjecture that there is an
invariance principle in operation which applies to all finite range SML models as $\varepsilon \to 0$.
The chief difficulty with this conjecture is not to prove that it applies to a large
class of models. Instead, I have difficulty trying to make sense out of it even in the
special case of the nearest neighbor majority vote model. The problem lies in trying
to identify the limiting object. One must try to imagine a process which lives on **R** in
which swarms of infinitesimally small intervals of two different colors are appearing.
The endpoints of these intervals move like Brownian motions, so most of them disappear immediately. However, so many of them are produced that some of them become long enough to be "visible". Collisions produce no effects except to join two intervals into one. One is vaguely reminded here of the production of so-called virtual
particles in physics, most of which immediately disappear, but a few of which have a
relatively long life. It is hoped that if it is really possible to make sense out of such

a process, some insights into the nature of Gibbs states at low temperatures will be attained. Then our goal of applying the stochastic theory to equilibrium theory would be realized.

References

1. Coddington, E. and Levinson, N. (1955). Theory of Ordinary Differential Equations. McGraw–Hill. New York.

2. Durrett, R. and Griffeath, D. (1983). Supercritical contact processes on **Z**. Z. Wahrsch. verw. Gebiete **61** 389–404.

3. Gray, L. (1986). Finite range majority vote models in one dimension. In preparation.

4. Holley, R. and Stroock, D. (1976). Applications of the stochastic Ising model to the Gibbs states. Communications in Math. Physics **48** 249–266.

5. Holley, R. and Stroock, D. (1976). L_2 Theory for the stochastic Ising model. Z. Wahrsch. verw. Gebiete **35** 87–101.

6. Holley, R. and Stroock, D. (1977). In one and two dimensions, every stationary measure for a stochastic Ising model is a Gibbs state. Communications in Math. Physics **55** 37–45.

7. Kindermann, R. and Snell, J. L. (1980). Markov Random Fields and their Applications. Contemporary Mathematics Volume 1. Amer. Math. Soc.

8. Liggett, T. (1972). Existence theorems for infinite particle systems. T.A.M.S. **165** 471–481.

9. Liggett, T. (1985). Interacting Particle Systems. Springer–Verlag. New York.

10. Spitzer, F. (1970). Interaction of Markov processes. Adv. in Math. **5** 246–290.

STIFF CHAINS AND LEVY FLIGHT: TWO SELF AVOIDING WALK MODELS AND THE USES OF THEIR STATISTICAL MECHANICAL REPRESENTATIONS

J.W.Halley

School of Physics and Astronomy
University of Minnesota
Minneapolis, Minnesota 55455

Two self-avoiding walk models are described: In the stiff chain model, useful in the description of some polymers, the walk "prefers" to continue in the direction of the last step with probability 1-p. The limit of small p and a large number of steps N is of particular interest. We present numerical results indicating the nature of the crossover from "ballistic" to random walk behavior. In two dimensions, the walk is asymptotically behaving like a self avoiding walk but in three dimensions, the numerical results suggest a random (not self-avoiding) behavior for a wide range of N. We describe this walk in terms of a vector spin model in the limit that the number of components n -> 0 and use this formulation to account for the difference between two and three dimensions observed numerically. Secondly, we consider self-avoiding Levy flight: the step length distribution is of the form $P(x) = C/x^{\mu+1}$. We study a type of Levy flight with a self-avoiding constraint called node-avoiding Levy flight here. This node-avoiding case is shown to be obtained from the n -> 0 limit of the statistical mechanics of another kind of vector spin model. The critical properties of this model were previously studied by Fisher, Nickel and Saks using the ϵ expansion. By comparing numerical simulations of node-avoiding Levy flight with their results, we can obtain information about the ϵ expansion in statistical mechanics at very small values of ϵ.

1. Introduction

In this paper we illustrate two themes with two examples: The first theme is that by simple modifications of the well known self-avoiding random walk model we obtain numerical evidence of qualitatively new properties. The second theme is that the use of associations of these models with lattice n-vector spin models in the $n \rightarrow 0$ limit permits some understanding of these properties.

We [1,2] call the first model a "stiff" self-avoiding walk. By this we mean an equilibrium (*not* "true"[3]) self-avoiding walk in which, at each step, if there were no self-avoiding constraint, the probability of stepping in the same direction as the previous step would be $1-p$ while the probabilities of stepping in any other direction would all be equal and sum to p. To avoid ambiguity we state the algorithm for generating the ensemble of walks of interest:

At each step (after the first step, which is in an arbitrary direction) in a nearest neighbor walk on the sites of a hypercubic lattice of dimension d, choose a random number r evenly distributed between 0 and 1. If r lies between 0 and $1-p$, then attempt to step in the same direction as the direction of the previous step of the walk. If r lies between $(1-p)$ and $(1-p)+p/2(d-1)$ then attempt to step in the first of the remaining $2(d-1)$ directions on the lattice which are neither forward nor back along the direction of the previous step. If r lies between $(1-p)+p/2(d-1)$ and $(1-p)+p/(d-1)$ then attempt to step in the second of these directions and so forth. When an attempted direction has been selected, check to see if stepping to the nearest neighbor in that direction results in self intersection with the path of the walk. If it does, then do not include this realization in the ensemble of stiff walks and begin a new walk from the origin. If no self-intersection results, then take the step and continue. (In our earlier papers[1,2], this kind of walk is called a "biased" self-avoiding walk, but this is an unfortunate name because other authors [4] use the term "biased" walk to describe the completely different model in which a particular direction *fixed in the lattice* is preferred at each step.)

The physical motivation for studying the stiff self avoiding walk just defined is that it provides an approximate description of the fact that, in polymer chains, bonding an additional atom to a hydrocarbon chain costs less energy if the resulting bond causes the chain to grow in the same direction in which it grew when the last atom was added [5]. From the point of view of the theoretical physics community interested in phase transitions, the stiff walk has often been assumed [6] to be uninteresting because it was believed that, with an appropriate rescaling of lengths, its asymptotic properties at large numbers of steps N would turn out to be equivalent to those of the ordinary self avoiding walk While this is almost certainly true, in the limit $N \to \infty$, p finite, we have found strong numerical evidence of nonanalyticity at the point $p=0$, $1/N=0$ at least in three dimensions.

The second model to be discussed here is called self-avoiding Levy flight. The concept of Levy flight was first introduced by Mandelbrot [7] without any self-avoiding constraint. That model described a random walk with variable step length l in which the probability distribution for l was algebraically decreasing with increasing l. By adding a self avoiding constraint to be described in section 3. below we [8] find a model that is closely related to magnetic models of phase transitions which had previously [9] been studied by use of the ϵ expansion. Thus it turns out that by study of this model we are able to make new numerical tests of the convergence of this fundamental expansion in the modern theory of second order phase transitions.

In the next section we describe the work on stiff walks, while the third section concerns Levy flights and the last section is a discussion including some conjectures about other, related, models.

2. Stiff Self Avoiding Walks.

Our numerical work[1,2] has shown that for an ensemble of walks generated according to the algorithm described in the introduction, the mean square end-to-end distance (or, similarly, radius of gyration) can be described by the following "scaling function" form near the origin in the plane of variables p and $n \equiv N^{-1}$ (N is the "time", the number of steps of the walk or the number of monomeric units along the chain):

$$< R^2 > = N^{1/a_n} f(Np^\Delta) , \tag{1}$$

Here following customary usage, $\Delta = \dfrac{a_n}{a_p}$ is called the crossover exponent and a_p and a_n are called the scaling powers[10] along the $p-$ and n-directions respectively. By matching equation (1) with the one-dimensional behavior which will occur at $p=0$, we conclude that $a_n = 1/2$. Then, various arguments[2] can be used to show that $\Delta = 1$ exactly. The results of Monte Carlo simulations support these conclusions as indicated in Figures 1 and 2 and discussed below. We have also used the numerical results to obtain the entire functional dependence of the scaling function $f(x)$. This is interesting because scaling functions are expected in the theory of second order phase transitions[10] to be the same (up to constant factors) for all systems in the same " universality class " and we wish to explore the conjecture that the stiff self avoiding walk is in the same universality class as the self-avoiding walk. For the stiff *random* walk (no self-avoidance), it is known numerically as well as analytically that the corresponding scaling functions are the same as those of the random walk.

It is convenient to describe the results of this numerical exploration in terms of the following alternative but equivalent form of equation (1):

$$< R^2 > \approx p^{(2y-2)\Delta} \, N^{2y} \, F(\, Np^{\Delta} \,) , \qquad (2)$$

for any real y . If, in the limit $x \equiv Np^{\Delta} >> 1$, we have $<R^2> \propto N^{2z}$, then by choosing $y=z$ in eq.(2) we will have $F(x) \rightarrow constant$ as $x \rightarrow \infty$. The function $F(x)$ obtained using the choice of $y = \nu_{Flory}$, where ν_{Flory} is the Flory exponent for the isotropic self-avoiding walk for the particular system dimensionality,[11] is displayed in Fig.3 for the square lattice and in Fig.4 for the simple cubic lattice.

The choice $y = \nu_{Flory}$ is natural because, for any finite $p > 0$ and for $N \rightarrow \infty$, we expect the ensemble of stiff walks to behave asymptotically like the ensemble of isotropic self-avoiding walks. We do not, however, know exactly what to expect in the "scaling" limit in which we take $x \rightarrow \infty$ *after* letting $p \rightarrow 0$ and $N \rightarrow \infty$, with x fixed. While this scaling limit is not physically attainable for polymer chains, for example, it is still of physical (and possibly even of practical) interest because the rate at which the limit is approached affects physical properties. We note from Figures 1 and 2 that in two dimensions the numerically obtained $F(x)$ appears to approach a constant limiting value as $x \rightarrow \infty$ while in three dimensions $F(x)$ seems to continue falling off towards zero as $x \rightarrow \infty$. This is a strong hint that at least in three dimensions the $x \rightarrow \infty$ limit in the scaling function is not the same as the $p > 0$ and $N \rightarrow \infty$ limit.

It turns out that if the an alternative choice of $y = 1/2$ is made in equation (2) for the stiff self avoiding walk in three dimensions, then the corresponding $F(x)$ approachs a constant as $x \rightarrow \infty$. In order to investigate the differences in two and three dimensions more clearly, let us first note that the scaling of $<R^2>$ as in eq.(2) implies that the following quantity, the local "slope" in the graphical representation of $log <R^2>$ versus log N, also scales:

$$\frac{d(log \, < R^2 >)}{d(log \, N)} \approx 2y + Np^{\Delta} \, \frac{F'(Np^{\Delta})}{F(Np^{\Delta})} = G(Np^{\Delta}) , \qquad (3)$$

where prime indicates the derivative. The scaling function $G(x)$ must satisfy limits $G(x) \rightarrow 2$ as $x \rightarrow 0$ and $G(x) \rightarrow 2z$ as $x \rightarrow \infty$ if the asymptotic behavior for the latter limit is $<R^2> \propto N^{2z}$. Since the way $G(x)$ is defined does not depend on the value of z, we can investigate the $x \rightarrow \infty$ limit without first supposing a trial value for z as in the case of $F(x)$ given in eq.(2).

Our results for $G(x)$ are plotted in Fig.3 and Fig.4 for the square and simple cubic lattices respectively. Also plotted in these figures is the corresponding function $G_o(x)$ exactly obtained for the non-excluded volume, biased *random* walks.[12] This function is calculated to be

$$G_o(x) = \frac{x(1-e^{-x})}{e^{-x}+x-1} , \qquad (4)$$

for both lattices and $G_o(x) \rightarrow 1$ for $x \rightarrow \infty$ implying that the exponent ν is ½ in the $x \rightarrow \infty$ limit for the self-avoiding walk in three dimensions. In other words, we find that in this limit in three dimensions, the exponent ν is the same for the self-avoiding and non-self avoiding cases. In contrast, the two-dimensional results show large deviations between the self avoiding and non-self-avoiding models. In fact it appears that $G(x) \rightarrow 1.5$ for $x \rightarrow \infty$ consistent with the known Flory exponent for two dimensions of approximately 3/4.[11]

The differences in the excluded volume effect for two and three dimensions may be understood physically by the following physical argument, due originally to R. Petschek[13]. Suppose we have a stiff linear chain on a lattice with lattice constant a so that the excluded volume radius is $a/2$. We suppose for the purposes of this argument that this stiff chain can be regarded as if it were a flexible chain of Np steps, each of which is of length $p^{-1}a$. Upon rescaling all lengths by a factor p so that the step length rescales back to a, the excluded volume radius is now $ap/2$. In dimensions $d>2$, this allows a stiffer chain to be more freely penetrable, and in the limit of $p \to 0$, a *random* chain should result. In two dimensions, however, any non-zero excluded volume radius is equally effective in blocking intersections, and thus this limit still yields a fully self-avoiding walk.

In the rest of this section we introduce a lattice spin model which reproduces the statistics of the stiff self-avoiding walk and sketch [14] how we use this spin model to show that $\Delta = 1$ and that self-avoiding effects are expected to be irrelevant for $d > 2$ in the limit $x \to \infty$.. The spin model is defined on the square lattice (the generalization to hypercubic lattices in other dimensions is straightforward). Consider three sets of n-vector spins $\mathbf{U}$, $\mathbf{S}$, $\mathbf{T}$; the spins $\mathbf{U}_i$ and $\mathbf{S}_i$ are defined on the sites i of the square lattice, while the $\mathbf{T}_{i+\delta/2}$ are defined on the sites of the covering lattice of the square lattice. The lengths of all spins are constrained to be $\sqrt{n}$. The model is defined by the Hamiltonian

$$-\beta H = K \sum_{i,\delta} (\mathbf{U}_i \cdot \mathbf{U}_{i+\delta})(\mathbf{T}_{i+\delta/2} \cdot \mathbf{S}_i) \left[(1-p)(\mathbf{T}_{i-\delta/2} \cdot \mathbf{S}_i) + \frac{p}{2} \sum_{\delta_{perp}} (\mathbf{T}_{i+\delta_{perp}/2} \cdot \mathbf{S}_i) \right]. \tag{5}$$

Here β is the inverse temperature ($k_B = 1$), δ is a vector to a nearest neighbor of the site i, K is a coupling constant, and $\sum_{\delta_{perp}}$ is the sum over all nearest-neighbor lattice vectors normal to the lattice vector δ.

This Hamiltonian generates the statistics of the stiff chain just as the single spin n-vector model gives the correct counting [15,16] of an isotropic self avoiding walk: the high-temperature expansion of the spin-spin correlation function takes the form

$$<U_i^\alpha T_{i+\delta/2}^\alpha U_j^\alpha T_{j+\delta'/2}^\alpha>_{n \to 0} = \sum_N K^N \overline{\eta}(i,\delta; j,\delta'; N), \tag{6}$$

where $<...>$ is the usual thermal average taken over all spins, and

$$\overline{\eta}(i,\delta; j,\delta'; N) = \sum_{N_G} (1-p)^{N-N_G} (\frac{p}{2})^{N_G} \eta(i,\delta; j,\delta'; N,N_G). \tag{7}$$

Here, $\eta(i,\delta; j,\delta'; N,N_G)$ is the number of N-step self-avoiding walks with N_G gauche steps starting from site i in the direction δ, and ending at site j in the direction δ'. The equivalence implied by Eq. (6) occurs because the vector spins $\mathbf{U}_i$ assure that the high temperature expansion of the correlation function includes only self-avoiding walks for exactly the same reason as in the case of an unbiased walk [14,15]. That is, terms in the expansion which represent intersecting walks involve four or more $\mathbf{U}$ spins at the intersection site. However, in the $n \to 0$ limit, the only terms which survive have two and only two identical spins at any given site. The spins $\mathbf{T}_{i+\delta/2}$ on the covering lattice then assure in almost every case that these walks are counted with the correct weighting. In a few cases, however, these covering lattice spins alone result in incorrect counting In these cases, the presence of the spins $\mathbf{S}_i$ on the square lattice between each pair of spins on the covering lattice cause the improperly weighted walks to vanish.

In the Hamiltonian (5), the spins $\mathbf{S}_i$ are not coupled to other spins of the same type at lattice sites different from i. This makes it possible to integrate out the $\mathbf{S}_i$ spins in the partition function. Because the correlation function (Eq. (6)) does not involve the $\mathbf{S}$ spins explicitly, they can be integrated over before calculation of the correlation function. Thus we can consider the effective Hamiltonian H' defined by

$$e^{-\beta H'} = n \int \prod_i d^n S_i \delta(S_i^2 - n)e^{-\beta H} \equiv <e^{-\beta H}>_S ,$$ (8)

where n is a suitable normalization constant, chosen such that $n \int \prod_i d^n S_i \delta(S_i^2 - n) = 1.$
$\beta H'$ can be calculated by a cumulant expansion. It turns out that only the first two cumulants contribute to the correlation function. In this way we find [14]

$$-\beta H = \tilde{\kappa_1} + \frac{1}{2}\tilde{\kappa_2},$$ (9)

where

$$\tilde{\kappa_1} = \sum_{(ij)} \lambda_{ij}(\mathbf{U}_{a(ij)} \cdot \mathbf{U}_{b(ij)})(\mathbf{T}_i \cdot \mathbf{T}_j),$$ (10)

and

$$\tilde{\kappa_2} = - \sum_{\substack{(i,j) \neq (k,l) \\ common\ site}} \lambda_{ij}\lambda_{kl}(\mathbf{U}_{a(ij)} \cdot \mathbf{U}_{b(ij)})(\mathbf{U}_{a(kl)} \cdot \mathbf{U}_{c(kl)})(\mathbf{T}_i \cdot \mathbf{T}_j)(\mathbf{T}_k \cdot \mathbf{T}_l).$$ (11)

$\tilde{\kappa_1}$ and $\tilde{\kappa_2}$ include only those parts of κ_1 and κ_2 which ultimately can contribute to the partition or correlation functions . Here, and in what follows, the nearest neighbor bonds (i,j) refer to the *covering lattice,* and the λ_{ij} are the proper weightings for the trans or gauche step on the lattice. $a(ij)$ and $b(ij)$ are the pair of *lattice sites* which correspond to bond (i,j) In other words we have $\lambda_{ij} = 1-p$ if the vector from i to j is parallel to the vector from $a(ij)$ to $b(ij)$ and $\lambda_{ij} = p$ otherwise. We always take $a(ij)$ to be the site which lies between i and j on the covering lattice.

In order to study the crossover induced by the gauche bond probability p, we renormalize [17] the effective Hamiltonian (Eq. (9)) about $p = 0$. Let us first expand κ_1 and κ_2 in powers of p: (We drop tildes from the κ 's in what follows.)

$$\kappa_1 = \kappa_1^{(0)} + p\kappa_1^{(1)},$$ (12)

and

$$\kappa_2 = \kappa_2^{(0)} + p\kappa_2^{(1)} + p^2\kappa_2^{(2)}.$$ (13)

Inserting a coupling constant Λ for the term $\kappa_2^{(1)}$, we then write the Hamiltonian as

$$-\beta H = \kappa_1^{(0)} + \frac{1}{2}\kappa_2^{(0)} + p(\kappa_1^{(1)} + \frac{1}{2}\Lambda\kappa_2^{(1)}) + O(p^2).$$ (14)

The physical meaning of this separation is simply that the $O(p^0)$ terms correspond to an all trans configuration. In studying the crossover at $p = 0$, we need only to look at the linear term in p, neglecting the $O(p^2)$ term.

As we isotropically renormalize the system with a rescaling ratio b, the total Hamiltonian remains invariant while the parameter p must be rescaled with some exponent x:

$$\kappa_i^{(0)\prime} = \kappa_i^{(0)}, \qquad (i=1,2),$$ (15)

$$p' = b^x p.$$ (16)

Since we are renormalizing about the fixed point $p = 0$, the two factors $\mathbf{T}_i \mathbf{U}_{a(ij)}$, and $\mathbf{T}_j \mathbf{U}_{b(ij)}$ in $\kappa_1^{(1)}$ are rescaled independently using the one dimensional rescaling exponent $\eta_0 = 1$:

$$\kappa_1^{(1)\prime} = b^{\eta_\sigma - 2} \kappa_1^{(1)}. \tag{17}$$

Equating the rescaling powers of $\kappa_1^{(0)}$ and $p\kappa_1^{(1)}$, then, we obtain $x = 2 - \eta_0 = 1$, and the crossover exponent $\phi = \Delta^{-1}$ is

$$\phi = \nu_0(2 - \eta_0) = 1, \tag{18}$$

where, as before (Ref. 16), ν is the exponent describing the divergence of the coherence length at the critical point and the subscript 0 denotes one-dimensional values.

There are two terms in $\Lambda\kappa_2^{(1)}$ which may be rewritten by replacing the requirement that the sites $a(ij)$ and $a(kl)$ be the same site in Eq. (11) by a Kronecker delta:

$$\Lambda\kappa_2^{(1)} = \tag{19}$$

$$\Lambda\left[-2p \sum_{ij\,||\,a(ij)-b(ij)kl\,||\,a(kl)-b(kl)} \sum \mathbf{U}_{a(ij)} \cdot \mathbf{U}_{b(ij)} \mathbf{U}_{a(kl)} \cdot \mathbf{U}_{c(kl)} \mathbf{T}_i \cdot \mathbf{T}_j \mathbf{T}_k \cdot \mathbf{T}_l \delta_{a(ij),a(kl)}\right]$$

$$+2nd\ term$$

where the second term is of the same form except that it has the opposite sign and the sum on ij is restricted to cases in which ij is *not* collinear with the vector from $a(ij)$ to $b(ij)$. If we now suppose that the spin factors in this term scale as they did in the first term (as one would expect in the small p limit) and taking into account the scaling behavior associated with the Kronecker delta we obtain

$$\Lambda\kappa_2^{(1)\prime} = b^{y-2+\eta_\sigma+d} \Lambda\kappa_2^{(1)} \tag{20}$$

in which y is the scaling power of Λ. Requiring that the Hamiltonian be scale invariant then gives for the scaling behavior of the coupling constant in this term in the Hamiltonian:

$$\Lambda p \longrightarrow b^{-d+2} \Lambda p \tag{21}$$

using $\eta_0 = 1$. Thus this term $\kappa_2^{(1)}$ is irrelevant near the fixed point $p = 0, 1/N = 0$ if $d > 2$. An essentially identical argument shows that the coupling constant in κ_2^2 scales in the same way and is irrelevant near $p = 0, 1/N = 0$ if $d > 2$. Thus near this fixed point for $d > 2$ we can write eqn. (9) as

$$-\beta H = \tilde{\kappa}_1 \tag{22}$$

The correlation functions associated with the Gibbs ensemble arising from a Hamiltonian of this form can be determined from a Gibbs probability associated with the following Hamiltonian

$$\beta H'' = \sum_{\mu,\nu} \int d\vec{x} \left\{ \frac{1}{2} \nabla\phi^{\mu\nu} : \nabla\phi^{\mu\nu} + \phi^{\mu\nu} \cdot \mathbf{M} \cdot \phi^{\mu\nu} + \frac{u}{2} \phi^2 \cdot \phi^2 \right\} \tag{23}$$

where each component of the tensor field $\phi_{\hat{n}}^{\mu\nu}(\vec{x})$ takes values between $-\infty$ and ∞. In equation (23), the bold faced ϕ denotes a vector whose components refer to different values of the direction $\hat{n}$ of possible nearest neighbor unit vectors pointing away from the point $\vec{x}$ and $\mathbf{M}$ is a square matrix with dimension equal to the number of such unit vectors. $\mathbf{M}$ depends only on the parameter p. ϕ^2 is defined as

$$\phi_{\hat{n}}^2 = \sum_{\mu,\nu} \phi_{\hat{n}}^{\mu\nu} \phi_{\hat{n}}^{\mu\nu} \tag{24}$$

Equation(23) is derived in ref 8 using standard techniques. In eqn (23) we can now explore the scaling power of u for $Np \ll 1$. Using arguments like those used before we then find

$$u \longrightarrow b^{2-d} u \qquad (25)$$

If $<R^2>$ then obeys the scaling relation

$$<R^2> = N^2 F(Np, N^y u) \qquad (26)$$

it then follows that $y = 2 - d$ so that as N becomes $\gg 1/u$ in three dimensions, the walk behaves as if there were no self- avoiding effects at all.

We comment finally that the existence of a regime of random walk behavior in the limit of small p in three dimensions may have some interesting experimental consequences. In considering, for example, the question of how to best form a polymer coating on a two dimensional surface, it may turn out to be best not to bind the polymer too tightly to the surface in some cases, so that a denser coating can be formed. This is discussed in somewhat more detail in section 4.

3. Self-avoiding Levy flights

The biased self-avoiding walks discussed above have a fixed step size (corresponding to a polymer chain with a fixed bond length). However, it is also possible to regard the stiff walk as a problem with a varying step size, taking the separation between two consecutive gauche steps as an effective step. In this case, the step size distribution is exponentially short-ranged since the probability of obtaining an effective step consisting of a series of, say, n trans steps is proportional to $(1-p)^n$ ($\sim e^{-np}$ for small p). While this is sufficient to introduce the various interesting features already discussed, even more drastic step size distributions can be contemplated. One such distribution on which we have performed a preliminary investigation is the power law distribution:

$$P(x) = C \; x^{-\mu-1} , \qquad (26)$$

for some $\mu > 0$.

Such a walk has been introduced and illustrated in detail by Mandelbrot [6] for the case where there is no excluded volume effect. These walks are called Levy flights and, for $0 < \mu < 2$ independent of the spatial dimension d, they have properties strikingly different from those of ordinary random walks. For example [18] the Haussdorff-Besicovitch dimension of the "trail" consisting of the end points of the steps is μ in the limit of a large number of steps (whereas it is two for ordinary random walks). Thus if one associates a mass with the end points of the steps, one has a mass distribution with dimension less than two. It was this feature which motivated the introduction of the model [6,19] , which was originally intended to illustrate how a fractal mass distribution could account for clustering of matter in the universe.

We have studied [8] a self-avoiding extension of the Lévy flight model. Our emphasis here is on the use of this model to study the convergence of the ϵ expansion for the critical exponents of second order phase transitions numerically.

We consider random walks on a lattice which will be taken to be hypercubic for definiteness. We will use the same algorithm for generation of the walks which is used to define Lévy flight on a lattice [18] except for the self-avoiding feature: At each step, a direction for the next step is chosen at random and a step length l is chosen so that the ensemble of step lengths will approach the distribution

$$p(l) = \frac{a-1}{a} \sum_{n=0}^{\infty} a^{-n} \delta_{l,b^n} \; .$$

(27)

If the algorithm allows the resulting step whether that step intersects some part of the path or not, then one has Lévy flight on a lattice as discussed in Ref. 18. Several self-avoiding constraints might be considered. We have studied node-avoiding Lévy flight defined so that the step selected as described above is rejected if the position at the end of the proposed step intersects the position of the end of any previous step in the walk. If a step is disallowed, the entire walk is discarded from the statistical sample, so that the resulting ensemble of walks weights the number of possible Lévy flights subject to the constraint only with the product of step length distribution factors for each step taken from Eq. (27) above. Thus we are considering "equilibrium" ensembles, in the sense in which the self-avoiding walk describes an equilibrium ensemble while the "true" self-avoiding walk does not[3].

We establish a magnetic model yielding the statistics of node-avoiding Lévy flight proceeds as follows: Consider the model described by

$$-\beta H = \sum_{i,j} K_{ij} \mathbf{S}_i \cdot \mathbf{S}_j \; .$$

(28)

Here β is the reciprocal of the temperature, H is the Hamiltonian, and the $\mathbf{S}$'s are n-component vector spins of length $\sqrt{n}$ on the lattice sites. K_{ij} is zero unless i and j lie along one of the d orthogonal directions in the lattice and is $K/r_{ij}^{1+\mu}$ where $r_{ij} = |\vec{r}_i - \vec{r}_j|$ is the distance between sites i and j when they lie along a coordinate axis. As in the Sarma [15,16] derivation of the equivalence of the nearest neighbor n-vector model to a self-avoiding walk in the limit $n \to 0$, we consider the high temperature expansion of the partition function for this model in the $n \to 0$ limit. As in Refs. 15,16,

$$Z = \prod_i \int d\Omega_i [\exp(-\beta H)]/\Omega^N$$

is one in the $n \to 0$ limit and defining

$$<...> = \prod_i \int d\Omega_i [\exp(-\beta H)(...)]/\Omega^N \; ,$$

$$<...>_0 = \prod_i \int d\Omega_i (...)/\Omega^N \; ,$$

one has

$$<S_i^\alpha S_j^\alpha> = \sum_N K^N \bar{\eta}_N(r_{ij})$$

in which

$$\bar{\eta}_N(r_{ij}) = \sum_{\{\vec{r}_{lm}\}} \prod_{lm} p_{lm} \eta_N(r_{ij}, \{\vec{r}_{lm}\}) \; .$$

Here the sum is over all combinations of step lengths given by $\{\vec{r}_{lm}\}$. The factors $p_{lm} = r_{lm}^{-(1+\mu)}$ provide the correct weighting for node-avoiding Lévy flight as explained above. $\eta_N(r_{ij}, \{\vec{r}_{lm}\})$ is the number of node-avoiding walks between $\vec{r}_i$ and $\vec{r}_j$ with N steps of lengths given by $\{\vec{\tau}_{lm}\}$. This relation establishes that the statistics of node-avoiding Lévy flight can be obtained from the statistical mechanical model described by Eq. (28) in exactly the same way that similar equivalences are established for the self-avoiding and biased walk problems[16].

The ϵ -expansion for the model described in Eq. (28) has been studied [9] by Fisher, Ma, and Nickel and by Sak. They showed that if $d/2 < \mu < 2 - \eta_{SAW}$ then the model gives a new kind of critical behavior not seen in models with short range interactions. In the $n \to 0$ limit we call this region the region of node-avoiding Levy flight. Here d is the lattice dimension and η_{SAW} is the exponent characterizing the spin-spin correlation function in the usual notation when $\mu > 2 - \eta_{SAW}$ so that the $n \to 0$ limit of the model describes an ordinary self avoiding walk. For $\mu < d/2$ the model gives critical behavior characteristic of Mandlebrot's Levy flight with no self avoiding constraint. Thus the "upper critical dimension" of this model is 2μ and the ϵ expansion is an expansion in $2\mu - d$ and not in $4 - d$ as it is for the n-vector short range model for which the expansion was orginally invented . The ϵ expansion was carried out for this case by the authors of reference 9. We focus particularly on their result for the exponent ν characterizing the divergence of the correlation length in the model which is as follows:

$$1/\nu = [1 - (2\mu - d)/4\mu - 40/512((2\mu - d)/\mu)^2(3 - \mu^2/4)]\mu + ... \tag{29}$$

It is the fact that the parameter μ is at our disposal in a simulation which provides the unique opportunity to use this model to compare the ϵ expansion for this model with Monte Carlo simulations with $\epsilon \ll 1$. In Fig. 5 we show results for the exponent ν from Monte Carlo simulations of node-avoiding Lévy flight on the square lattice ($d = 2$) where they are compared with the known results for $\mu < 1$ and $\mu > 2 - \eta_{SAW}$ and with first and second order expansions in $\epsilon = 2\mu - d$ (Eqn(29))in the region $1 < \mu < 2 - \eta_{SAW}$. In second order in ϵ we show both the reciprocal of the first two terms in equation (29) and also the reciprocal of equation (29) expanded to second order in ϵ in Figure 5. We have also compared our Monte Carlo results with the '$\Delta\sigma$ expansion ' of reference 9 and find that it agrees less well with the Monte Carlo results than the ϵ expansion results shown. The Monte Carlo calculations are for a minimum of 1,000 walks of 50 steps for a given value of μ and in most cases much more (e.g. 13,000 walks for $\mu=1$) The error bars are somewhat subjective because the occasional large steps make the run to run fluctuations rather large and unpredictable. They represent a conservative estimate of the errors if they are Gaussian distributed but do not take account of any possible systematic errors. We note that both the first order ϵ expansion and the second order expansion of the reciprocal of eqn. (29) agree with our simulation to within the estimated error bars over the whole range from $\mu = 1$ to $\mu = 2 - \eta_{SAW}$. On the other hand , if the reciprocal of equation (29) is used without expansion, then the second order results appear to disagree with the simulations. Over most of the range of μ in both expansions, using only the first order term in ϵ appears to give closer agreement with the simulations than does the second order expansion.

The work of Fisher et al[9] suggests a logarithmic correction with a definite power of logarithm at the upper marginal dimension of $d_c = 2\mu$. There must also be a corresponding logarithmic correction to scaling at the boundary between the long range behavior (the node-avoiding Levy flight regime) and the short range behavior (usual self-avoiding walk regime), for which no previous calculations exist. Our preliminary Monte Carlo calculations strongly suggest the importance of such corrections.

4. Discussion and Conclusions

It is not hard to construct models of stiff chains in which the two and three dimensional situations in which we have discussed here occur as special cases. Such models may have a close relationship to problems of technical interest such as the question of how to produce an impermeable polymer coating on a surface. As an example, consider a biased self-avoiding walk in three dimensions which occurs near a two dimensional surface. The walk is not allowed to have steps below the planar surface so that the surface acts like a hard wall. Above the surface, we suppose that there is a field which causes the walk to be attracted toward the wall. By this we mean that the probability of any step direction (which describes the growth algorithm for the walks) is increased by a factor $1+\delta$ if the step moves the walk toward the surface and is decreased by a factor $1/(1+\delta)$ if the step moves the walk away from the surface. (We require $\delta < 1/p - 1$ and also assume proper normalization will be made for the probabilities describing the other possible step directions.) Then we can think physically of the walk as grown in a field which pulls the walk toward the surface.

In such a case, $\delta = 0$ corresponds to three dimensional growth, while for $\delta > 0$, the walk will behave asymptotically as a two dimensional one in the limit of large N. On the other hand, for small δ, the walk will remain essentially three dimensional. Thus we again have a crossover problem, in which the initial guess is that the crossover from three to two dimensions occurs when $N\delta \sim O(1)$. This picture is complicated further by the interplay of the lengths $1/\delta$ and the persistence length $1/p$: If $1/p \gg 1/\delta$, then the persistence length $1/p$ is much longer than the "width" $1/\delta$ of the surface layer and the walk is intuitively expected to go from essentially one-dimensional behavior when $N \ll 1/p$ to two dimensional behavior when $N \gg 1/p$. On the other hand, if $1/p \ll 1/\delta$, then we expect the walk to crossover from one dimensional to three dimensional behavior when N increases past the value $N=1/p$ and then to cross over from three dimensional to two dimensional behavior when N increases past the value $N=1/\delta$. These conjectures have not yet been tested by analytical or numerical work.

If one grows a "coating" by the rules of the model described in the last two paragraphs, then it is natural to ask questions about the transport of a particle by diffusion through such a "coating". Suppose that a walk is grown as described above in such a way that $N \gg 1/p$ but $N \ll 1/\delta$ so that the walk is essentially flexible and three dimensional. If, as our results so far suggest, the walk (with $p \ll 1$) essentially behaves like a random walk in this case, then the *areal* monomer density of the walk (the number of monomers in the walk per unit area of surface) will be independent of the length N of the walk (because $N/R^2 \sim N^0$). Note that if the walk were not stiff (p not small), then the areal density would decrease with N as $N^{-1/5}$. Suppose on the other hand that the walk is grown under the conditions $N \gg 1/p$ and $N \gg 1/\delta$ so that the walk is essentially two dimensional along the surface. Intuitively, one might expect this essentially two dimensional, tightly bound material to make a better coating. Our results suggest, however, that the self-avoiding character of the walk will be immediately manifest in this case. As a consequence the projected areal monomer density will now decrease as $N^{-1/2}$ (because $N/R^2 \sim N^{1-2\nu}$ where the Flory exponent $\nu=3/4$.) As a consequence, in the case of stiff chains, we can anticipate that tightly bound chains will make worse coverings than weakly bound ones.

To explore these possibilities, one could study the behavior of a diffusing particle on a lattice occupied by chains grown according to the rules discussed. The diffusing particles are not allowed to penetrate the excluded volume occupied by the chains. For definiteness, one could, for example, explore the mean time for the particle to reach the surface as a function of starting height above the surface and as a function of the conditions of growing the chain. We conjecture that this time will have a maximum as a function of δ for fixed starting height, N and p: For small values of δ the chains will be three dimensional (and hence relatively *compact*) but will be essentially unattached to the surface so that the diffusion times will be short. For larger values of δ ($\delta > 1/N$) the chains will be attached to the surface but less compact. At an intermediate value of δ one can conjecture that the diffusion time to the surface will be maximum.

At least for some applications, it would be desirable to have information about a type of self-avoiding Levy flight defined so that the path of the Levy flight avoids all points on the path of the walk ("path-avoiding") rather than just the points at the end of each step ("node-avoiding") as we have assumed here. Almost nothing is presently known about path avoiding Levy flight.

5. Acknowledgements

All the work reported here is a result of collaboration with Hisao Nakanishi. The three dimensional simulations of stiff walks as reported in reference 3. are entirely due to him and S. B. Lee. We are grateful to R. Petschek, both for criticisms and useful insights. The other collaborators in this work were D. Atkatz and R. Sundararajan. The work on this subject began during a program on disordered materials at the Institute for Theoretical Physics at Santa Barbara which is supported by NSF and NASA.

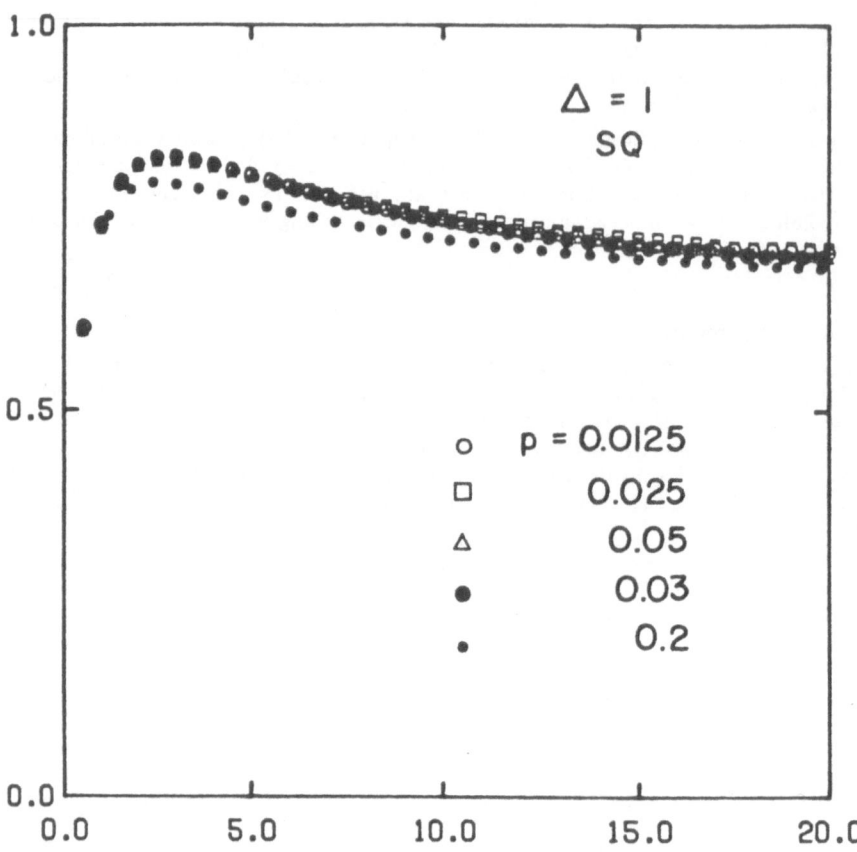

FIG. 1.
The function F of equation (2) by Monte Carlo simulation on a square lattice. Each point represents an average over at least 3000 Monte Carlo walks. Data are shown for p = 0.0125 ( ), .025 ( ) and 0.05 ( ). Δ is taken to be 1 and $y = \nu_{Flory} = .75$. For other values of Δ the data for different values of p would not collapse in this way [1,2].

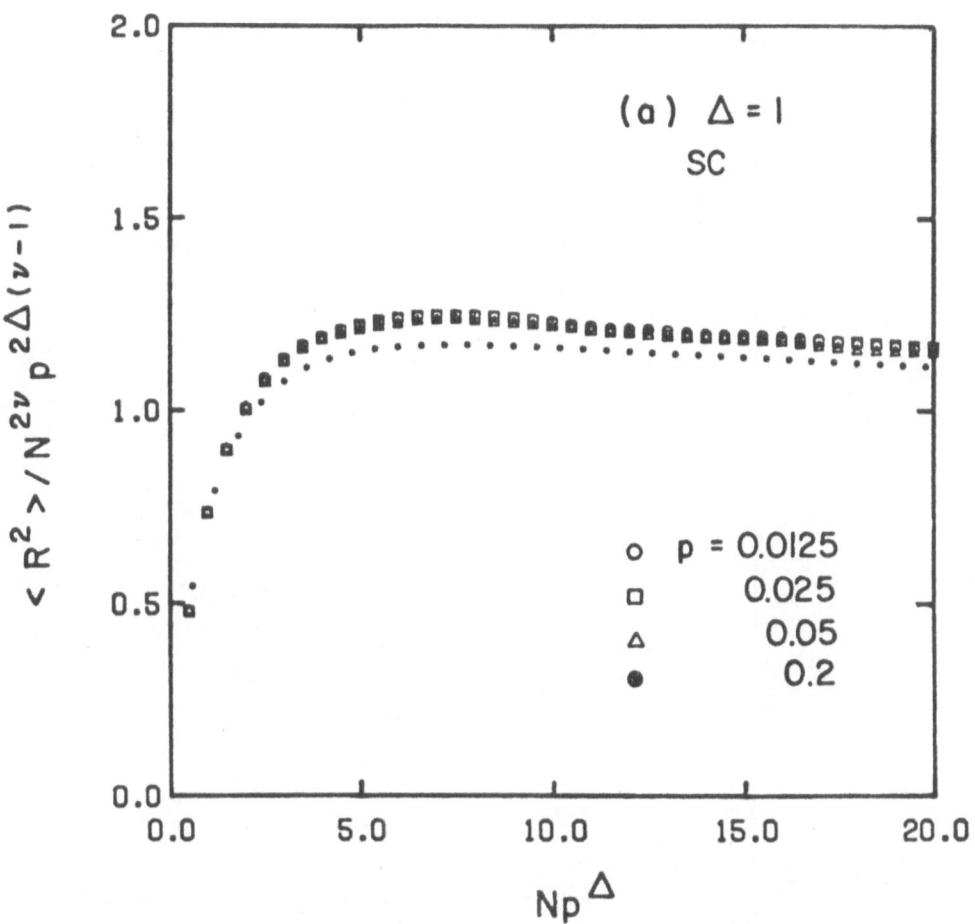

FIG. 2.

Same as Figure 1 but for the simple cubic lattice. Each data point represents an average over at least 7000 walks. Here $\nu_{Flory}=0.588$.

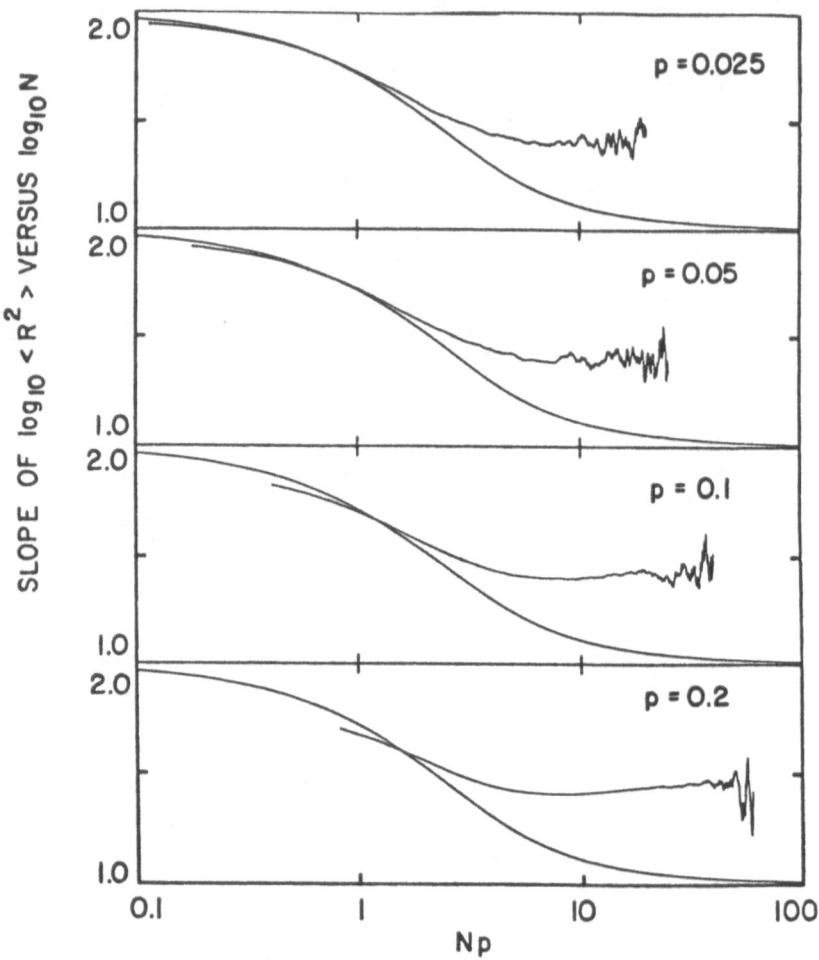

FIG. 3.
    The function $G(x)$ for the square lattice for various values of p as indicated. The
smooth solid line is equation (4).

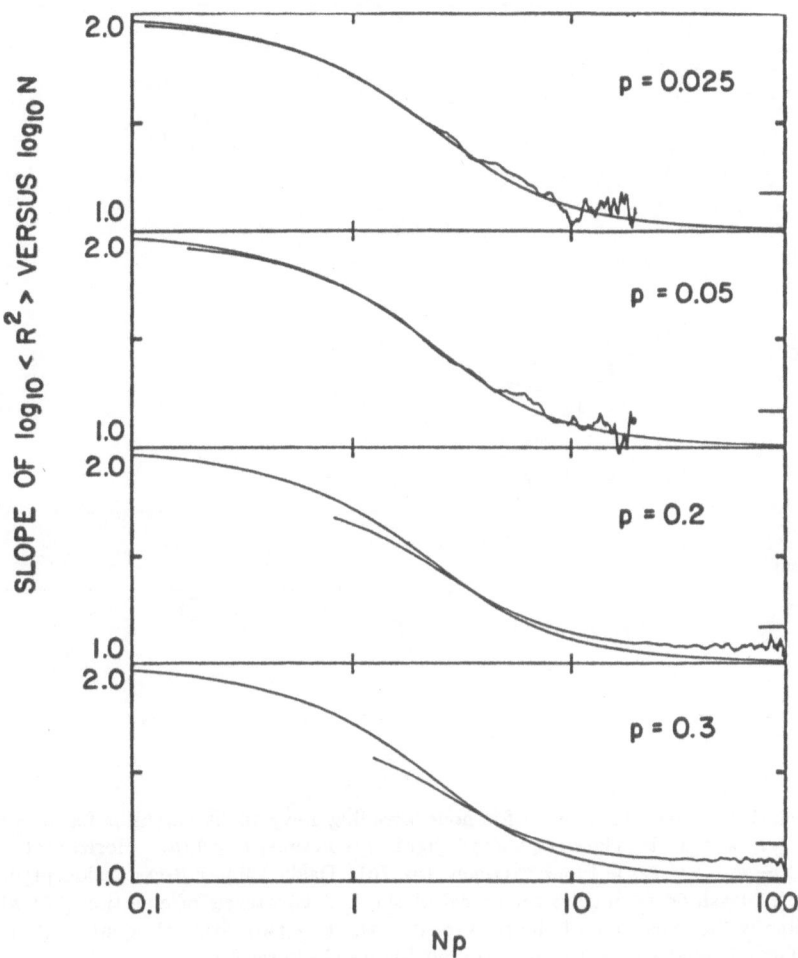

FIG. 4.
    The function $G(x)$ for the simple cubic lattice for values of p as indicated.

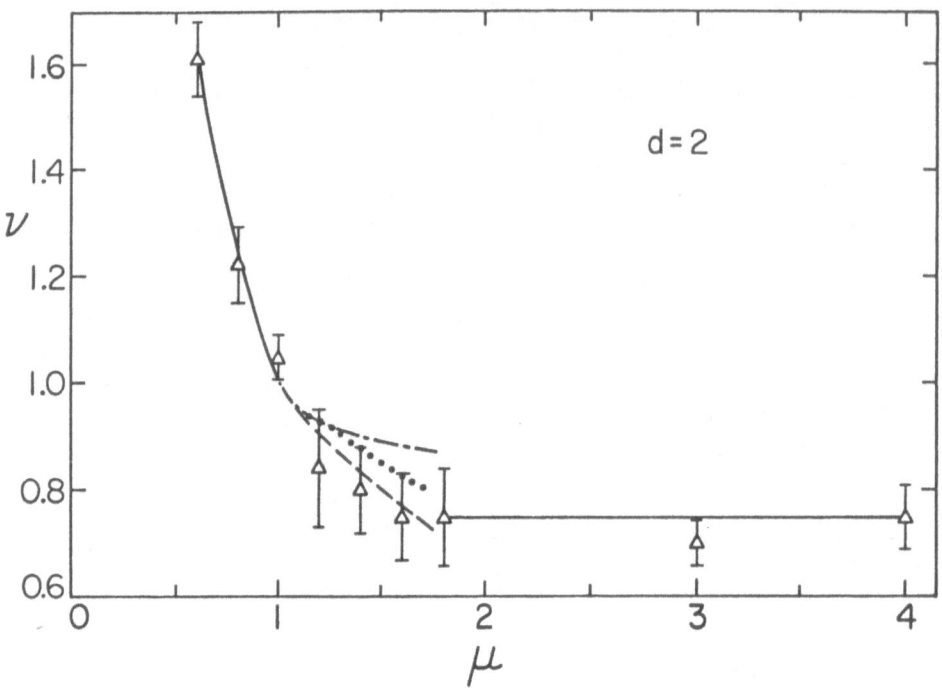

FIG. 5.
Correlation length exponent ν for node avoiding Levy flight versus μ for $d = 2$. Solid line for $\mu < 1$ is classical Lévy flight prediction $\nu = 1/\mu$. Horizontal line for $\mu > 2 - \eta_{SAW} = 1.8$ is Flory exponent $(= .75)$. Dashed line is ϵ-expansion prediction to order ϵ. Dash-dotted line is reciprocal of the first two terms of equation (29) while dotted line is the expansion of the reciprocal of the first two terms of equation (29) to order ϵ^2. The triangles with error bars are our Monte Carlo results.

REFERENCES

1. J.W. Halley, H. Nakanishi, and R. Sundararajan, Phys. Rev. **B31**, 293 (1985).

2. S.B. Lee and H. Nakanishi, Physical Review B (in press)

3. D.J. Amit, G. Parisi, and L. Peliti, Phys. Rev. B **27**, 1635 (1982); S.P. Obukhov and L. Peliti, J. Phys. A **16**, L147 (1983).

4. for example, see K. F. Lim and R. G. Gilbert, J. Chem. Phys. (in press)

5. P. J. Flory, *Statistical Mechanics of Chain Molecules,* John, Wiley and Sons, N. Y. (1969)

6. P.G. de Gennes, *Scaling Concepts in Polymer Physics* (Cornell University Press, Ithaca, 1979).

7. B.B. Mandelbrot, Comptes Rendus de l'Academie des Sciences, Paris **280** A, 1551 (1975).

8. J. W. Halley and H. Nakanishi, Physical Review Letters, **55**, 551(1985)

9. M. Fisher, S.-K. Ma, and B. Nickel, Phys. Rev. Lett. **29**, 917 (1972); J. Sak, Phys. Rev. B **8**, 281 (1973); J. Sak, Phys. Rev. B **15**, 4344 (1977).

10. S.-K. Ma, *Modern Theory of Critical Phenomena* (Benjamin, Reading, Mass, 1976), p. 355.

11. P. J. Flory, *Principles of Polymer Chemistry, Chap. XV.* Cornell University Press, Ithaca, N. Y. (1971)

12. W. K. Schroll, A. B. Walker and M. F. Thorpe, J. Chem. Phys. **76**, 6385(1982)

13. R. Petschek (private communication).

14. For details on the results described here see D. Atkatz, G. ten Brinke, J. W. Halley and H. Nakanishi (unpublished)

15. P. G. de Gennes, Phys. Lett **38A**, 339(1972)

16. M. Daoud *et al.,* Macromolecules **8**, 804 (1975).

17. A. Aharony, in *Phase Transitions and Critical Phenomena,* edited by C. Domb and M.S. Green (Academic, New York, 1976), Vol. 6.

18. B.D. Hughes, M.F. Schlesinger, and E. Montroll, Proc. Natl. Acad. Sci. USA **78**, 3287(1981); J. Stat. Phys. **28**, 111 (1982); B. Hughes and M.F. Schlesinger, J. Math. Phys. **23**, 1688 (1982).

19. P.J.E. Peebles, *The Large-Scale Structure of the Universe* (Princeton University Press, 1980), p. 243. IP 20. M. G. Watts, J. Phys. **A8**, 61(1975), see also B. Nienhuis, Phys. Rev. Lett. **49**, 1062(1982).

ONE DIMENSIONAL STOCHASTIC ISING MODELS

Richard Holley[1]
Department of Mathematics
University of Colorado
Boulder, Colorado 80309

Abstract

We prove that for one dimensional stochastic Ising models with finite range interactions the rate of convergence to equilibrium is at least as fast as $e^{-\text{const } t/\log(t)}$. There is no assumption here that the interaction must be attractive.

0. Introduction

The purpose of this paper is to prove that a one dimensional stochastic Ising model (i.e. a time reversible spin flip system on $\{-1,1\}^Z$) with finite range interactions converges to equilibrium at a rate which is at least as fast as the rate at which $e^{-\text{const } t/\log(t)}$ goes to zero. Moreover this rate of convergence holds for all initial distributions. With the additional assumption that the interaction is attractive an even stronger result is already known, namely that the rate is exponential in that case. However, without the assumption of attractive interactions it was not previously known if the system converged to equilibrium starting from initial states which were not periodic for shifts in space, and even for periodic initial states no rate of convergence to equilibrium was known. The result here does not make use of any hypothesis concerning the attractive nature of the interaction.

The above convergence result is stated as Theorem (3.10), and it is proved in section 3. Section one contains a description of the stochastic Ising model with which we will be dealing together with statements of previous results which we will need. This section also serves to place Theorem (3.10) in

[1]Research supported in part by NSF Grant MCS 8310542.

perspective. Section 2 contains a proof of an infinite collection of inequalities. Each of these inequalities is a logarithmic Sobolev inequality for a Gibbs state on a finite interval. Unfortunately the constant which appears in these logarithmic Sobolev inequalities grows as the logarithm of the length of the interval. (This is the source of the $\log(t)$ in $e^{-\mathrm{const}\ t/\log(t)}$.) I will call this system of inequalities a weak logarithmic Sobolev inequality. Such a system of inequalities holds for all one dimensional Gibbs states with finite range interactions. The weak logarithmic Sobolev inequality is the key to the proof of Theorem (3.10).

The reader will notice that if instead of the weak logarithmic Sobolev inequality proved in section 2, we had the usual type of logarithmic Sobolev inequality for the infinite system (or equivalently if the logarithmic Sobolev constants for the finite systems were bounded in the size of the systems) then by the same argument as is used in the proof of Theorem (3.10), we would obtain an exponential rate of convergence. I have not been able to prove a logarithmic Sobolev inequality for the infinite system at any finite temperature, but I expect it to hold at all finite temperatures in one dimensional systems with finite range interactions.

This work is closely related to the results in [9] done jointly with Dan Stroock. In fact it was his suggestion that logarithmic Sobolev inequalities might be useful in the study of stochastic Ising models which got me started on this project.

1. Background

We begin this section with a description of the Gibbs states and the corresponding stochastic Ising models.

Let $E = \{-1,1\}^Z$ be given the product topology and let B be the associated Borel field. Elements of E will be denoted by $\eta (= \{\eta(k) : k \in Z \})$ and should be thought of as the configuration of an infinite system of spins placed on the integers. Given a non-empty finite subset Λ of Z, let $B_\Lambda = \sigma(\eta(k) : k \in \Lambda)$ (i.e. the smallest σ–algebra of subsets of E with respect to which $\eta(k)$ is measurable for all $k \in \Lambda$). E_Λ will denote $\{-1,1\}^\Lambda$ for $\Lambda \subset Z$. A translation invariant potential with range R, $\{J_F : F$ a finite subset of $Z\}$, is a set of real numbers satisfying.

$$J_{F+k} = J_F, \qquad \text{for all } k \in Z \text{ and all } F,$$

$$\text{and}\quad J_F = 0 \quad\text{if diameter(F)} > R.$$

We adopt the following notation. If $\eta \in E$ and $\sigma \in E_\Lambda$ then $\eta\sigma$ denotes the configuration which is equal to η off of Λ and is equal to σ on Λ.

Given a potential $\{J_F\}$, we say that g is a Gibbs state with potential $\{J_F\}$ if g is a probability measure on E and, for every $\Lambda \subset Z$ with $|\Lambda| < \infty$,

$$(1.1)\qquad g_{\Lambda,\eta}(\sigma) = \exp\Big(-\sum_{F\cap\Lambda\neq\varnothing} J_F \prod_{k\in F} \eta\sigma(k)\Big)/Z(\Lambda,\eta)$$

is the regular conditional probability distribution of g on B_Λ given B_{Λ^c}. Here $Z(\Lambda,\eta)$ is the normalizing constant needed to make $g_{\Lambda,\eta}$ have total mass one.

Since we are in one dimension and the range of the interaction is finite, there is only one Gibbs state for each potential, and that Gibbs state is an R-Markov chain on $\{-1,1\}$. (cf. [12])

A spin flip system on E with range R is a Markov processes on E with a Feller continuous semi-group $\{P_t : t \geq 0\}$ on $C(E)$ (the continuous real valued functions on E with sup norm denoted by $\|\cdot\|_u$), having an infinitesimal generator, Ω, which is given as follows. First denote those real valued functions on E which are B_Λ measurable by C_Λ. Then let D be the union over all finite subsets $\Lambda \subset Z$ of C_Λ. (i.e. D is the set of cylinder functions or local observables on E.) For $f \in C(E)$, let $\Delta_k f(\eta) = f(\eta^k) - f(\eta)$. Here η^k is the configuration which is equal to η except at k, where it is equal to $-\eta$. Finally let $\{c_k : k \in Z\}$ be a collection of non-negative functions with $c_k \in C_{[k-R,k+R]}$. We assume in addition that the c_k's are shift invariant. That is, $c_0(\eta) = c_k(\sigma)$ whenever η and σ satisfy $\eta(j) = \sigma(j+k)$ for all $j,k \in Z$. The generator of $\{P_t : t \geq 0\}$, when restricted to D, is given by

$$\Omega f(\eta) = \sum_{k\in Z} c_k(\eta)\Delta_k f(\eta).$$

So far we have not related the spin flip system to any potential, but it is still possible to say something about the convergence of $P_t f$ for $f \in D$. The proof of the following theorem may be found in [10] Theorem (1.3.9).

(1.2) Theorem: Set

$$\epsilon = \inf\{c_0(\eta_1) + c_0(\eta_2) : \eta_1(k) = \eta_2(k) \text{ if } k \neq 0 \text{ and } \eta_1(0) \neq \eta_2(0)\}$$

$$\text{and}\qquad M = \sum_{u\neq 0} \sup\{|c_0(\eta_1) - c_0(\eta_2)| : \eta_1(y) = \eta_2(y),\ y \neq u\}$$

Then if $\epsilon > M$, P_t has a unique stationary measure, μ, and for every $f \in D$ there is a constant A_f such that

$$\sup_{\eta} |P_t f(\eta) - \int f d\mu| \leq A_f \, e^{(M-\epsilon)t}.$$

The above theorem does not make any use of the time reversibility, which plays an important role in what follows. Time reversibility is what connects the spin flip system to the Gibbs states. (cf. Theorem 4.2.13 in [10].)

In particular, if a finite range spin flip system with strictly positive flip rates is time reversible for some measure ν, then ν is a Gibbs state for some translation invariant finite range potential $\{J_F\}$ and the flip rates $\{c_k : k \in Z\}$ must satisfy

$$(1.3) \qquad c_k(\eta) e^{-\sum_F J_F \prod_{j \in F} \eta(j)} = c_k(\eta^k) e^{\sum_F J_F \prod_{j \in F} \eta(j)}.$$

(cf. Theorem 4.2.13 in [10].) Conversely if the c_k's satisfy (1.3) then the spin flip process which they determine is time reversible with respect to the Gibbs state, g, determined by $\{J_F\}$. We will assume from now on that the c_k's are uniformly positive and satisfy (1.3). Note that for any given potential there are an infinite number of ways to choose the c_k's to satisfy (1.3).

The most useful expression of time reversibility is the statement that for a time reversible process with generator Ω and Gibbs state g, Ω is self-adjoint on $L^2(g)$. A straight forward calculation (see [6]) leads to the following equality.

$$(1.4) \qquad \int \phi \Omega \psi dg = -\frac{1}{2} \sum_k \int c_k (\Delta_k \phi)(\Delta_k \psi) dg.$$

If we represent Ω in terms of its spectral resolution (i.e. $\Omega = -\int_0^{\infty} \lambda dE_\lambda$ with $\{E_\lambda : \lambda \geq 0\}$ a resolution of the identity in $L^2(g)$) then $E_0 =$ Projection on the constants (see [6]), and thus

$$\|P_t \phi - \int \phi dg\|^2_{L^2(g)} = \int_{0^+}^{\infty} e^{-2\lambda t} d_\lambda(E_\lambda \phi, \phi) \leq e^{-2\epsilon_0 t} \|\phi - \int \phi dg\|^2_{L^2(g)},$$

where ϵ_0 is the distance between 0 and the rest of the spectrum of Ω. The following theorem will be important for us (see [4]).

(1.5) Theorem: $\epsilon_0 > 0$ for all one dimensional finite range stochastic Ising models.

That is, there is an $\epsilon\ (=2\epsilon_0\min_\eta c_0(\eta)) > 0$ such that for all $\phi \in L^2(g)$

(1.6) $$\sum_k \int (\Delta_k\phi(\eta))^2 g(d\eta) \geq \epsilon \int (\phi(\eta) - \int \phi(\sigma)g(d\sigma))^2 g(d\eta).$$

From Theorem (1.5) we can conclude that if $\mu \ll g$ and $\dfrac{d\mu}{dg} \in L^2(g)$ then for all $\phi \in L^2(g)$,

$$\left| \int P_t\phi(\eta)\mu(d\eta) - \int \phi(\eta)g(d\eta) \right| \leq \|\phi - \int \phi dg\|_{L^2(g)} \|\frac{d\mu}{dg}\|_{L^2(g)} e^{-\epsilon t},$$

however; we cannot draw any conclusion about $P_t\phi(\eta)$ for any given $\eta \in E$.

By adding the hypothesis that the c_k's are attractive (i.e. that $c_0(\eta)$ is an increasing function of η on $\{\eta : \eta(0) = -1\}$ and a decreasing function of η on $\{\eta : \eta(0) = 1\}$ see [2]) we can conclude from Theorem (1.5) that there is a $\gamma > 0$ such that for all $f \in D$ there is a constant A_f such that

(1.7) $$\|P_t f - \int f dg\|_u \leq A_f e^{-\gamma t}, \quad t \geq 0$$

(see [4]). Thus the convergence to equilibrium of one dimensional stochastic Ising models with attractive finite range interactions is fairly well understood. Since we are trying to avoid the assumption of attractive interactions in this paper, we will not pursue this further here.

Aside from attractiveness the most successful tool for studying ergodic properties of the stochastic Ising model has been the Helmholtz free energy (see [1], [7], or [11]). By studying the specific free energy, or free energy per lattice site, it has been possible to prove theorems such as:

(1.8) **Theorem:** If μ_0 is periodic for shifts in space and $\mu_t(\phi) = \int P_t\phi(\eta)\mu_0(d\eta)$, then for all $\phi \in C(E)$, $\lim_{t\to\infty} \mu_t(\phi) = \int \phi(\eta)g(d\eta)$.

The proof of Theorem (1.8) fails if μ_0 is not periodic, and even if $\phi \in D$ the proof of Theorem (1.8) does not yield a rate of convergence.

We are going to work with free energy below, but in this paper we will be concerned with free energy in finite intervals rather than for the infinite system. This permits us to get away from the assumption that μ_0 is periodic. It also eliminates the need to deal with the boundary terms which show up when dealing with the infinite system, but it necessitates the introduction of stochastic Ising models on finite intervals. We pause to introduce these finite systems before defining the free energy.

Given a finite interval $\Lambda \subset Z$, and $\eta \in E$, let $\Omega_{\Lambda,\eta}$ be the operator on C_Λ defined by

$$\Omega_{\Lambda,\eta}\phi(\sigma) = \sum_{k \in \Lambda} c_k(\eta\sigma)\Delta_k\phi(\sigma), \quad \sigma \in E_\Lambda.$$

Let $T_t^{\Lambda,\eta} = e^{t\Omega_{\Lambda,\eta}}$. Then $\{T_t^{\Lambda,\eta} : t \geq 0\}$ is a semigroup on C_Λ which is time reversible with respect to $g_{\Lambda,\eta}$. Moreover if $\Lambda_0 \subset \Lambda$ and $\phi \in C_{\Lambda_0}$ then

$$(1.9) \qquad \sup_{\sigma \in E_\Lambda} |T_t^{\Lambda,\eta}\phi(\sigma) - P_t\phi(\eta\sigma)| \leq A_\phi e^{Ct}\frac{(Ct)^N}{N!},$$

where A_ϕ depends only on ϕ, $C = \max_\eta c_0(\eta)$ and $N = 2 + 2\text{dist}(\Lambda_0,\Lambda^c)/R$ (see Lemma (4.15) in [5]).

Since g is R-Markov, one easily sees from the Frobenius theorem that there is a $\delta > 0$ such that if $\phi \in C_{\Lambda_0}$ and $\Lambda_0 \subset \Lambda$ then

$$(1.10) \qquad |\int \phi(\sigma)g_{\Lambda,\eta}(d\sigma) - \int \phi(\eta)g(d\eta)| \leq 2\|\phi\|_u e^{-\delta M},$$

where $M = \text{dist}(\Lambda_0,\Lambda^c)$. Thus if we want to prove that $\|P_t\phi - \int \phi dg\|_u$ is small, a first step will be to get control of the size of

$$(1.11) \qquad \|T_t^{\Lambda,\eta}\phi - \int \phi(\sigma)g_{\Lambda,\eta}(d\sigma)\|_u.$$

This is where the free energy enters the picture.

If Λ is a finite interval in Z and $\eta \in E$ and $\mu^{(\Lambda)}$ is a probability measure on E_Λ, then the free energy of $\mu^{(\Lambda)}$ in Λ with boundary conditions η, denoted $h_{\Lambda,\eta}(\mu^{(\Lambda)})$ is given by

$$h_{\Lambda,\eta}(\mu^{(\Lambda)}) = \sum_{\sigma \in E_\Lambda} \mu^{(\Lambda)}(\sigma)\log\left(\frac{\mu^{(\Lambda)}(\sigma)}{g_{\Lambda,\eta}(\sigma)}\right).$$

If μ_0 is a fixed probability measure on E and $\eta \in E$ is fixed we let $\mu_t^{(\Lambda,\eta)}$ be defined by

$$\sum_{\sigma \in E_\Lambda} \phi(\sigma)\mu_t^{(\Lambda,\eta)} = \sum_{\sigma \in E_\Lambda} T_t^{\Lambda,\eta}\phi(\sigma)\mu_0^{(\Lambda)},$$

where $\mu_0^{(\Lambda)}$ is the marginal distribution of μ_0 on E_Λ. Then for fixed μ_0 and $\eta \in E$ we denote $f_t^{(\Lambda)}(\sigma) = \mu_t^{(\Lambda,\eta)}(\sigma)/g_{\Lambda,\eta}(\sigma)$ and write

$$h_\Lambda(t) \equiv h_{\Lambda,\eta}(\mu_t^{(\Lambda,\eta)}) = \sum_{\sigma \in E_\Lambda} f_t^{(\Lambda)}(\sigma)\log(f_{t_{(\Lambda)}}(\sigma))g_{\Lambda,\eta}(\sigma) = \int f_t^\Lambda \log(f_t^\Lambda)dg_{\Lambda,\eta}.$$

In section 3 we will show how to bound the expression in (1.11) by using $h_\Lambda(t)$. Thus what we need to do is to show that $h_\Lambda(t)$ is small. Begin by differentiating $h_\Lambda(t)$, and using the equality

$$\frac{d}{dt}\int \phi d\mu_t^{(\Lambda,\eta)} = \int \Omega_{\Lambda,\eta}\phi d\mu_t^{(\Lambda,\eta)}$$

to conclude that

$$(1.12) \qquad \frac{d}{dt}h_\Lambda(t) = \int f_t^{(\Lambda)} \Omega_{\Lambda,\eta} \log(f_t^{(\Lambda)}) dg_{\Lambda,\eta}.$$

From (1.4) it is easy to see that the right side of (1.12) is negative. What we want to do is to bound it away from 0 from above by a constant multiple of $h_\Lambda(t)$. That is the point of the next section. We will return to free energy considerations in section 3.

2. A Weak Logarithmic Sobolev Inequality

Our goal in this section is to prove a collection of inequalities which we call a weak logarithmic Sobolev inequality. Each of the inequalities is a logarithmic Sobolev inequality for a $g_{\Lambda,\eta}$ for some finite $\Lambda \subset Z$ and $\eta \in E$. We say that a measure μ on E_Λ satisfies a logarithmic Sobolev inequality with logarithmic Sobolev constant $\gamma(\mu)$ if there is a constant $\gamma(\mu) < \infty$ such that for all $f \in C_\Lambda$

$$(\text{L.S.}) \qquad \int f^2(\xi) \log(f^2(\xi)) \mu(d\xi) \leq \gamma(\mu) \sum_{k \in \Lambda} \int |\Delta_k f(\xi)|^2 \mu(d\xi) + \int f^2(\xi) \mu(d\xi) \log(\int f^2(\xi) \mu(d\xi)).$$

It is not immediately obvious that (L.S.) holds for any $g_{\Lambda,\eta}$. Thus our first objective is to prove that for each finite Λ there is a constant $\gamma(|\Lambda|)$, which depends only on $|\Lambda|$ and for which $\gamma(g_{\Lambda,\eta}) \leq \gamma(|\Lambda|)$ for all $g_{\Lambda,\eta}$, $\eta \in E$.

We start with a special case.

(2.2) Lemma: If μ on $E_{\{0\}}$ is given by $\mu(\sigma) = \frac{1}{2}$ for each $\sigma \in E_{\{0\}}$, then $\gamma(\mu) = 1$.

Proof. Since $C_{\{0\}}$ is only two dimensional, this is just a tedious but straight forward calculus exercise, which is left to the reader.

The next lemma is well known in other contexts (cf. Lemma (9.17) in [13]).

(2.3) Lemma: If $\Lambda_1 \cap \Lambda_2 = \emptyset$ and μ,ν are probability measures on $E_{\Lambda_1}, E_{\Lambda_2}$ respectively and both μ and ν satisfy (L.S.) then $\mu \times \nu$ satisfies (L.S.) and $\gamma(\mu \times \nu) \leq \gamma(\mu) \vee \gamma(\nu)$.

Proof.

$$\int \int f^2(\xi,\sigma) \log(f^2(\xi,\sigma) \mu(d\xi) \nu(d\sigma)$$

$$\leq \int [\gamma(\mu) \sum_{k\in\Lambda_1} \int |\Delta_k f(\xi,\sigma)|^2 \mu(d\xi) + \int f^2(\xi,\sigma)\mu(d\xi)\log(\int f^2(\xi,\sigma)\mu(d\xi))]\nu(d\sigma)$$

$$\leq \gamma(\mu) \sum_{k\in\Lambda_1} \int |\Delta_k f(\xi,\sigma)|^2 \mu(d\xi)\nu(d\sigma) + \gamma(\nu) \sum_{j\in\Lambda_2} \int |\Delta_j (\int f^2(\xi,\sigma)\mu(d\xi))^{1/2}|^2 \nu(d\sigma)$$

$$+ \int\int f^2(\xi,\sigma)\mu(d\xi)\nu(d\sigma) \log(\int\int f^2(\xi,\sigma))\mu(d\xi)\nu(d\sigma).$$

The proof is completed by observing that the Cauchy Schwartz inequality yields

$$|\Delta_j (\int f^2(\xi,\sigma)\mu(d\xi))^{1/2}|^2 \leq \int |\Delta_j f(\xi,\sigma)|^2 \mu(d\xi), \quad j \in \Lambda_2.$$

Q.E.D.

Now let $\rho_\Lambda(\sigma) = 2^{-|\Lambda|}$ for all $\sigma \in E_\Lambda$. The following lemma is immediate from Lemmas (2.2) and (2.3).

(2.4) Lemma: ρ_Λ satisfies (L.S.) with $\gamma(\rho_\Lambda) = 1$.

(2.5) Lemma: Let μ and ν be probability measures on E_Λ. Assume that there is a constant $\alpha > 1$ such that $\alpha^{-1} \leq \frac{d\mu}{d\nu} \leq \alpha$ and that ν satisfies (L.S.). Then μ satisfies (L.S.) with $\gamma(\mu) \leq \alpha^2 \gamma(\nu)$.

Proof.

Rewrite (L.S.) as

$$\int f^2(\xi)\log(\frac{f^2(\xi)}{\|f\|^2_{L^2(\mu)}})\mu(d\xi) \leq \gamma(\mu) \sum_{k\in\Lambda} \int |\Delta_k f(\xi)|^2 \mu(d\xi).$$

It is clear that

$$\sum_{k\in\Lambda} \int |\Delta_k f(\xi)|^2 \nu(d\xi) \leq \alpha \sum_{k\in\Lambda} \int |\Delta_k f(\xi)|^2 \mu(d\xi).$$

Thus we need only show that

(2.6) $$\int f^2(\xi)\log(\frac{f^2(\xi)}{\|f\|^2_{L^2(\mu)}})\mu(d\xi) \leq \alpha \int f^2(\xi)\log(\frac{f^2(\xi)}{\|f\|^2_{L^2(\nu)}})\nu(d\xi).$$

But for any probability measure m on E_Λ and any $f \in C_\Lambda$

$$\int f^2(\xi)\log(\frac{f^2(\xi)}{\|f\|^2_{L^2(m)}})m(d\xi) = \inf_{x>0} \int (f^2(\xi)\log(f^2(\xi)) - f^2(\xi)\log(x) - f^2(\xi) + x)m(d\xi),$$

and for each $x > 0$ the integrand on the right side of the above equation is non-negative. Also the infimum on the right side is achieved when $x = \int f^2(\xi)m(d\xi)$. Hence one easily checks that (2.6) holds.

Q.E.D.

(2.7) Theorem: There is a sequence of constants, $\{\gamma(l) : l \geq 1\}$ such that for every finite interval, $\Lambda \subset Z$, and every $\eta \in E$, $g_{\Lambda,\eta}$ satisfies (L.S.), with $\gamma(g_{\Lambda,\eta}) \leq \gamma(|\Lambda|)$ for all $\eta \in E$.

Proof. First observe that for every finite Λ there is a constant $\alpha_\Lambda < \infty$ depending only on $|\Lambda|$ such that $\alpha_\Lambda^{-1} \leq \dfrac{dg_{\Lambda,\eta}}{d\rho_\Lambda} \leq \alpha_\Lambda$. The theorem now follows immediately from Lemmas (2.4) and (2.5).

Q.E.D.

We assume that each $\gamma(l)$ in Theorem (2.7) has been chosen to be as small as possible. The rest of this section is devoted to proving that $\gamma(l) \leq \gamma \log(l)$ for some finite γ. We begin with two lemmas, the first of which follows easily form Theorem (1.5) (cf. Lemma (2.7) in [4]).

(2.8) Lemma: There is an $\epsilon > 0$ such that for all finite intervals $\Lambda \subset Z$, all $\eta \in E$, and all $\phi \in C_\Lambda$,

$$(2.9) \qquad \sum_{k \in \Lambda} \int (\Delta_k \phi(\eta))^2 g_{\Lambda,\eta}(d\eta) \geq \epsilon \int (\phi(\eta) - \int \phi(\sigma) g_{\Lambda,\eta}(d\sigma))^2 g_{\Lambda,\eta}(d\eta).$$

(2.10) Lemma: There is a constant $K_0 < \infty$ such that for all $l_1, l_2 \geq 1$, $\gamma(l_1 + l_2 + R) \leq \gamma(l_1) \vee \gamma(l_2) + K_0$.

Proof. Let Λ_1, Λ_2, and Λ_3 be adjacent intervals of Z with $|\Lambda_1| = l_1$, $|\Lambda_2| = R$, and $|\Lambda_3| = l_2$, and set $\Lambda = \Lambda_1 \cup \Lambda_2 \cup \Lambda_3$. If $\sigma \in E_\Lambda$ and $\omega_i \in E_{\Lambda_i}$, $i = 1,2,3$, we write $\sigma = \omega_1 \omega_2 \omega_3$ to mean $\sigma(k) = \omega_i(k)$ if $k \in \Lambda_i$. We denote the conditional distribution of g given $B_{\Lambda^c \cup \Lambda_2}$ by $\bar{g}_{\eta \omega_2}(\cdot)$ and note that since $|\Lambda_2| = R$, $\bar{g}_{\eta \omega_2} = g_{\Lambda_1,\eta \omega_2} \times g_{\Lambda_3,\eta \omega_2}$. If $A \in B_{\Lambda_2}$ we denote $g_{\Lambda,\eta}(A)$ by $g_\Lambda^{(\Lambda_2)}(A|\eta)$.

Let $f \in C_\Lambda$. Note that the left side of (L.S.) remains unchanged, and the right side of (L.S.) decreases if we replace f with $|f|$. Thus we may as well assume that $f \geq 0$.

By conditioning on B_{Λ_2} we have

$$(2.11) \qquad \int f^2(\sigma) \log f^2(\sigma) g_{\Lambda,\eta}(d\sigma)$$

$$= \int \int f^2(\omega_1 \omega_2 \omega_3) \log f^2(\omega_1 \omega_2 \omega_3) \bar{g}_{\eta \omega_2}(d\omega_1 d\omega_3) g_\Lambda^{(\Lambda_2)}(d\omega_2 | \eta)$$

Thus by first factoring $\bar{g}_{\eta \omega_2}(\cdot)$ and then applying Lemma (2.3) we bound the right side of (2.11) above by

$$(2.12) \quad \int \{|\gamma(l_1) \vee \gamma(l_2)| \sum_{k \in \Lambda_1 \cup \Lambda_3} \int \int |\Delta_k f(\omega_1 \omega_2 \omega_3)|^2 g_{\Lambda_1, \eta \omega_2}(d\omega_1) g_{\Lambda_3 \eta \omega_2}(d\omega_3)$$

$$+ F^2(\eta \omega_2) \log F^2(\eta \omega_2)\} g_\Lambda^{(\Lambda_2)}(d\omega_2 | \eta) ,$$

where

$$F^2(\eta \omega_2) = \int \int f^2(\omega_1 \omega_2 \omega_3) g_{\Lambda_1, \eta \omega_2}(d\omega_1) g_{\Lambda_3 \eta \omega_2}(d\omega_3) = \int\int f^2(\omega_1 \omega_2 \omega_3) \overline{g}_{\eta \omega_2}(d\omega_1 d\omega_2) .$$

Note that $g_\Lambda^{(\Lambda_2)}(\sigma | \eta)$ is bounded away from 0 and ∞ uniformly in Λ and η. Thus just as in Theorem (2.7) we may use Lemma (2.5) to conclude that (L.S.) holds for $g_\Lambda^{(\Lambda_2)}(\cdot | \eta)$ with a constant, K_1 which is independent of Λ, and η. By applying (L.S.) to the part of (2.12) which involves F^2 we may bound (2.12) above by

$$(2.13) \quad [\gamma(l_1) \vee \gamma(l_2)] \sum_{k \in \Lambda_1 \cup \Lambda_3} \int |\Delta_k f(\sigma)|^2 g_{\Lambda, \eta}(d\sigma)$$

$$+ K_1 \sum_{k \in \Lambda_2} \int |\Delta_k F(\eta \omega_2)|^2 g_\Lambda^{(\Lambda_2)}(d\omega_2)$$

$$+ \int f^2(\sigma) g_{\Lambda, \eta}(d\sigma) \log \left(\int f^2(\sigma) g_{\Lambda, \eta}(d\sigma) \right) .$$

Concentrate on the second term in (2.13). For any $k \in \Lambda_2$

$$(2.14) \quad |\Delta_k F(\eta \omega_2)|^2 = ((\int \int f^2(\omega_1 \omega_2^k \omega_3) \overline{g}_{\eta \omega_2^k}(d\omega_1 d\omega_3))^{1/2} - (\int\int f^2(\omega_1 \omega_2 \omega_3) \overline{g}_{\eta \omega_2}(d\omega_1 d\omega_2))^{1/2})^2$$

$$\leq 4((\int \int f^2(\omega_1 \omega_2^k \omega_3) \overline{g}_{\eta \omega_2}(d\omega_1 d\omega_2))^{1/2} - (\int \int f^2(\omega_1 \omega_2 \omega_3) \overline{g}_{\eta \omega_2}(d\omega_1 d\omega_2))^{1/2})^2$$

$$+ 4((\int \int f^2(\omega_1 \omega_2^k \omega_3) \overline{g}_{\eta \omega_2}(d\omega_1 d\omega_2))^{1/2} - \int \int f(\omega_1 \omega_2^k \omega_3) \overline{g}_{\eta \omega_2}(d\omega_1 d\omega_2))^2$$

$$+ 4(\int \int f(\omega_1 \omega_2^k \omega_3) \overline{g}_{\eta \omega_2}(d\omega_1 d\omega_2) - \int \int f(\omega_1 \omega_2^k \omega_3) \overline{g}_{\eta \omega_2^k}(d\omega_1 d\omega_2))^2$$

$$+ 4(\int \int f(\omega_1 \omega_2^k \omega_3) \overline{g}_{\eta \omega_2^k}(d\omega_1 d\omega_2) - (\int \int f^2(\omega_1 \omega_2^k \omega_3) \overline{g}_{\eta \omega_2^k}(d\omega_1 d\omega_2))^{1/2})^2$$

$$\leq 4 \int \int |\Delta_k f(\omega_1 \omega_2 \omega_3)|^2 \overline{g}_{\eta \omega_2}(d\omega_1 d\omega_2)$$

$$+ 4 \int \int (f^2(\omega_1 \omega_2^k \omega_3) - \int \int f(\omega_1 \omega_2^k \omega_3) \overline{g}_{\eta \omega_2}(d\omega_1 d\omega_3))^2 \overline{g}_{\eta \omega_2}(d\omega_1 d\omega_3)$$

$$+ 4 (\int \int f(\omega_1 \omega_2^k \omega_3) \overline{g}_{\eta \omega_2^k}(d\omega_1 d\omega_3) - \int \int f(\omega_1 \omega_2^k \omega_3) \overline{g}_{\eta \omega_2}(d\omega_1 \omega_3))^2$$

$$+ 4 \int \int (f^2(\omega_1 \omega_2^k \omega_3) - \int \int f(\omega_1 \omega_2^k \omega_3) \overline{g}_{\eta \omega_2^k}(d\omega_1 \omega_3))^2 \overline{g}_{\eta \omega_2^k}(d\omega_1 d\omega_3).$$

In the last inequality in (2.14) we have used the fact that $f \geq 0$ in the second and fourth terms.

The first term on the right side of (2.14) is the type of term which we want. Using Lemma (2.8) and the fact that $\bar{g}_{\eta\omega_2} = g_{\Lambda_1,\eta\omega_2} \times g_{\Lambda_3\eta\omega_2}$, we may bound the second and fourth terms by

(2.15)
$$4\epsilon^{-1} \sum_{j\in\Lambda_1\cup\Lambda_3} \int\int |\Delta_j f(\omega_1\omega_2^k\omega_3)|^2 \bar{g}_{\eta\omega_2}(d\omega_1 d\omega_3)$$

and

(2.16)
$$4\epsilon^{-1} \sum_{j\in\Lambda_1\cup\Lambda_3} \int\int |\Delta_j f(\omega_1\omega_2^k\omega_3)|^2 \frac{d\bar{g}_{\eta\omega_2^k}}{d\bar{g}_{\eta\omega_2}}(\omega_1\omega_3)\bar{g}_{\eta\omega_2}(d\omega_1 d\omega_3)$$

$$\leq 4\epsilon^{-1}A_1 \sum_{j\in\Lambda_1\cup\Lambda_3} \int\int |\Delta_j f(\omega_1\omega_2^k\omega_3)|^2 \bar{g}_{\eta\omega_2}(d\omega_1 d\omega_3) \ ,$$

where $A_1 < \infty$ is such that $\|\frac{d\bar{g}_{\eta\omega_2^k}}{d\bar{g}_{\eta\omega_2}}\|_u \leq A_1$ for all η, ω_2, Λ, and $k\in\Lambda_2$.

Now consider the third term.

(2.17)
$$(\int\int f(\omega_1\omega_2^k\omega_3)\bar{g}_{\eta\omega_2^k}(d\omega_1 d\omega_3) - \int\int f(\omega_1\omega_2^k\omega_3)\bar{g}_{\eta\omega_2}(d\omega_1 d\omega_3))^2$$

$$= (\int\int (f(\omega_1\omega_2^k\omega_3) - \int\int f(\omega_1\omega_2^k\omega_3)\bar{g}_{\eta\omega_2}(d\omega_1 d\omega_3))(\frac{d\bar{g}_{\eta\omega_2^k}}{d\bar{g}_{\eta\omega_2}}(\omega_1\omega_3) - 1)\bar{g}_{\eta\omega_2}(d\omega_1 d\omega_3))^2$$

$$\leq A_1^2 \int\int (f(\omega_1\omega_2^k\omega_3) - \int\int f(\omega_1\omega_2^k\omega_3)\bar{g}_{\eta\omega_2}(d\omega_1 d\omega_3))^2 \bar{g}_{\eta\omega_2}(d\omega_1 d\omega_3) \ .$$

Thus by again applying Lemma (2.8) we may bound the third term by

(2.18)
$$4A_1^2\epsilon^{-1} \sum_{j\in\Lambda_1\cup\Lambda_3} \int\int |\Delta_j f(\omega_1\omega_2^k\omega_3)|^2 \bar{g}_{\eta\omega_2}(d\omega_1 d\omega_3) \ .$$

On substituting (2.15), (2.16), and (2.18) into the right side of (2.14) we see that

(2.19)
$$\int |\Delta_k F(\eta\omega_2)|^2 g_\Lambda^{(\Lambda_2)}(d\omega_2|\eta)$$

$$\leq 4\int |\Delta_k f(\sigma)|^2 g_{\Lambda,\eta}(d\sigma) + 4(1 + A_1 + A_1^2)\epsilon^{-1} \sum_{j\in\Lambda_1\cup\Lambda_3} \int |\Delta_j f(\sigma^k)|^2 g_{\Lambda,\eta}(d\sigma) \ .$$

But,

(2.20)
$$\int |\Delta_j f(\sigma^k)|^2 g_{\Lambda,\eta}(d\sigma) = \sum_{\sigma\in E_\Lambda} |\Delta_j f(\sigma^k)|^2 \frac{g_{\Lambda,\eta}(\sigma)}{g_{\Lambda,\eta}(\sigma^k)} g_{\Lambda,\eta}(\sigma^k)$$

$$\leq A_2 \sum_{\sigma\in E_\Lambda} |\Delta_j f(\sigma^k)|^2 g_{\Lambda,\eta}(\sigma^k) = A_2 \int |\Delta_j f(\sigma)|^2 g_{\Lambda,\eta}(d\sigma),$$

where $A_2 < \infty$ is such that $g_{\Lambda,\eta}(\sigma)/g_{\Lambda,\eta}(\sigma^k) \leq A_2$ for all η, σ, k, and Λ.

Finally substituting (2.20) into (2.19) and the resulting inequality into (2.13) we see that the lemma is proved with $K_0 = 4K_1[1 + A_2(1 + A_1 + A_1^2)R\epsilon^{-1}]$.

QED.

(2.21) Theorem: There is a constant $\gamma < \infty$ such that $\gamma(l) \le \gamma \log(l)$ for all $l \ge 2$.

Proof. By induction on i it is easily seen from Lemma (2.10) that if $(2^i-1)R < m \le (2^{i+1}-1)R$, then

$$(2.22) \qquad \gamma(m) \le \bar{\gamma} + iK_0,$$

where $\bar{\gamma} = \max_{1 \le i \le R} \gamma(i)$. Also if $(2^i-1)R < m \le (2^{i+1}-1)R$, then $\log R + (i-1)\log 2 \le \log m$. Thus

$$\varlimsup_{m \to \infty} \frac{\gamma(m)}{\log m} \le \varlimsup_{m \to \infty} \frac{\bar{\gamma}+iK_0}{\log R+(i-1)\log 2} = K_0 / \log 2,$$

and hence there is a constant $\gamma < \infty$ such that

$$\gamma(m) \le \gamma \log m \quad \text{for all} \quad m \ge 2.$$

Q.E.D.

3. Rapid Convergence

We now return to considerations of free energy and the rate of convergence to equilibrium. The first two lemmas are concerned with controlling the variation from equilibrium in terms of the free energy.

(3.1) Lemma: Let (Ω, F, μ) be a probability space and let ϕ be a bounded measurable real valued function on Ω such that $\int_\Omega \phi(x)\mu(dx) = 0$. Define

$$\Phi(a) = \int e^{a\phi(x)} \mu(dx).$$

Then for all $\epsilon > 0$,

$$\inf\{\int f(x)\log(f(x))\mu(dx) \colon f \ge 0, \ \int f(x)\mu(dx) = 1, \quad \text{and}$$
$$\int \phi(x)f(x)\mu(dx) \ge \epsilon\} \ge \sup_a(a\epsilon - \log(\Phi(a))).$$

Proof. By a theorem of Sanov (see Lemma (3.38) in [13]), for each $f \ge 0$ such that $\int f(x)\mu(dx) = 1$, we have

(3.2) $\int f(x)\log(f(x))\mu(dx) = \sup_{\psi} \{\int \psi(x)f(x)\mu(dx)-\log(\int e^{\psi(x)} \mu(dx))\},$

where the supremum over ψ is over all bounded measurable functions ψ. Letting ψ be of the form $\psi(x) = a\phi(x)$ we see that

(3.3) $\int f(x)\log(f(x))\mu(dx) \geq \sup_{a}\{\int a\phi(x)f(x)\mu(dx)-\log(\int e^{a\phi(x)} \mu(dx))\}.$

Note that $\int \phi(x)\mu(dx) = 0$ implies that $\log(\int e^{a\phi(x)} \mu(dx)) \geq 0$ for all a. Thus if in addition $\int f(x)\phi(x)\mu(dx) \geq \epsilon$, then we have

(3.4) $\qquad \int f(x)\log(f(x))\mu(dx) \geq \sup_{a}\{a\epsilon-\log(\int e^{a\phi(x)} \mu(dx))\}$

$\qquad$ Q.E.D.

(3.5) Lemma: Let ϕ be a bounded measurable function such that

$\qquad \int \phi(x)\mu(dx) = 0.$ Then for all $f \geq 0$ such that $\int f(x)\mu(dx) = 1,$

(3.6) $\qquad \int \phi(x)f(x)\mu(dx) \leq 2^{3/2}\|\phi\|(\int f(x)\log f(x)\mu(dx))^{1/2}$

Proof. Let $\epsilon = \int \phi(x)f(x)g(dx).$ If $\epsilon \leq 0,$ then (3.6) is immediate. Thus we may assume that $\epsilon > 0.$ We denote $\log(\int e^{a\phi(x)} \mu(dx))$ by $F(a).$

Define $K(\epsilon) = \sup_{a}\{a\epsilon-F(a)\}.$ Since $F(0) = 0$ and $F'(0) = 0$ and $F(a) \geq 0$ for all a we have $K(0) = 0$ and $K(\epsilon) > 0$ for all $\epsilon > 0.$ Note that if $G(x) \geq F(x)$ for all $x \geq 0,$ then

(3.7) $\qquad K(\epsilon) = \sup_{a \geq 0} (\epsilon a-F(a)) \geq \sup_{a \geq 0} (\epsilon a-G(a)).$

Since $F(0) = F'(0) = 0$ and $F''(a) \leq 4\|\phi\|_u^2$ for all $a,$ we have $F(a) \leq 2a^2\|\phi\|_u^2$ for all $a.$ Thus by (3.7), $K(\epsilon) \geq \dfrac{\epsilon^2}{8\|\phi\|_u^2}$ for all $\epsilon > 0.$ From Lemma (3.1) we have $\int f(x)\log f(x)\mu(dx) \geq K(\epsilon) \geq \epsilon^2/8\|\phi\|_u^2.$

$\qquad$ Q.E.D.

(3.8) Lemma: Let $\Lambda \subset Z$ be a finite interval and let $f \in C_\Lambda$ with $f \geq 0$ and $\int f(\sigma)g_{\Lambda,\eta}(d\sigma) = 1.$ Set $\alpha = \inf_{\eta} c_0(\eta) > 0.$ Then

$$\int f(\sigma)\Omega_{\Lambda,\eta}(\log f(\sigma))g_{\Lambda,\eta}(d\sigma) \leq -\frac{2\alpha}{\gamma\log(|\Lambda|)}\int f(\sigma)\log(f(\sigma))g_{\Lambda,\eta}(d\sigma).$$

Proof. It is enough to show that

(3.9) $$\int f(\sigma)\Omega_{\Lambda,\eta}\log(f(\sigma))g_{\Lambda,\eta}(d\sigma) \leq 4\int f^{1/2}(\sigma)\Omega_{\Lambda,\eta}f^{1/2}(\sigma)g_{\Lambda,\eta}(d\sigma),$$

and then apply (1.4) (for $\Omega_{\Lambda,\eta}$) and Theorem (2.21).

To prove (3.9) first let $m_t(d\omega,d\sigma)$ be defined by

$$\int h(\omega)k(\sigma)m_t(d\omega,d\sigma) = \int k(\sigma)T_t^{\Lambda,\eta}h(\sigma)g_{\Lambda,\eta}(d\sigma),$$

and then note that by the time reversibility of $T_t^{\Lambda,\eta}$ with respect to $g_{\Lambda,\eta}$,

$$\int h(\sigma)\Omega_{\Lambda,\eta}k(\sigma)g_{\Lambda,\eta}(d\sigma) = \lim_{t\to 0}-\frac{1}{2t}\int\int(h(\omega)-h(\sigma))(k(\omega)-k(\sigma))m_t(d\omega,d\sigma).$$

Thus

$$4\int f^{1/2}(\sigma)\Omega_{\Lambda,\eta}f^{1/2}(\sigma)g_{\Lambda,\eta}(d\sigma) - \int f(\sigma)\Omega_{\Lambda,\eta}\log(f(\sigma))g_{\Lambda,\eta}(d\sigma)$$

$$= \lim_{t\to 0}t^{-1}[-2\int\int(f^{1/2}(\omega)-f^{1/2}(\sigma))^2m_t(d\omega,d\sigma) + \frac{1}{2}\int\int(f(\omega)-f(\sigma))(\log(f(\omega))-\log(f(\sigma)))m_t(d\omega,d\sigma)]$$

$$= \lim_{t\to 0}\frac{1}{2t}[\int\int\{(f(\omega)-f(\sigma))(\log(f(\omega))-\log(f(\sigma))) - 4(f^{1/2}(\omega)-f^{1/2}(\sigma))^2\}m_t(d\omega,d\sigma).$$

Thus it suffices to show that for all $a,b > 0$ $(a-b)(\log(a)-\log(b)) - 4(a^{1/2}-b^{1/2})^2 \geq 0$. Equivalently we must show that $(x-1)\log(x) \geq 4(x^{1/2}-1)^2$ for all $x > 0$. But $(x-1)\log(x) - 4(x^{1/2}-1)^2$ is a convex function on $(0,\infty)$ and its minimum occurs at $x = 1$.

Q.E.D.

(3.10) Theorem: For every finite range one dimensional stochastic Ising model there is an $\epsilon > 0$ such that for all initial distributions μ_0 and all $\phi \in D$:

(4.8) $$|\int P_t\phi(\xi)\mu_0(d\xi) - \int \phi(\xi)g(d\xi)| \leq B(\phi)e^{-\epsilon\frac{t}{(\log t)}}, \quad t \geq 2,$$

where $B(\phi)\in(0,\infty)$.

Proof. It suffices to prove the theorem for μ_0 concentrated on a point, η. Let $\phi \in C_{\Lambda_0}$. If Λ_0 has side length l let $\Lambda(t)$ be the interval with side length $l + 8CRt$ centered at the center of Λ_0. Here C is as in (1.9). Then

(3.11) $\quad |P_t\phi(\eta) - \int \phi(\xi)g(d\xi)| \leq |P_t\phi(\eta) - T_t^{\Lambda(t),\eta} \phi(\eta)|$

$$+ |T_t^{\Lambda(t),\eta} \phi(\eta) - \int \phi(\xi)g_{\Lambda,\eta}(d\xi)|$$

$$+ |\int \phi(\xi)g_{\Lambda,\eta}(d\xi) - \int \phi(\xi)g(d\xi)| .$$

By (1.9) the first term on the right side of (3.11) is bounded by $A_\phi e^{Ct} \dfrac{(Ct)^{|4Ct|+2}}{(|4Ct|+2)!} \leq A_\phi(e(\dfrac{e}{4})^4)^{Ct} \leq$

$A_\phi e^{-\frac{Ct}{2}}$. By (1.10) the third term on the right side of (3.11) is bounded by $2\|\phi\|_u e^{-\delta 4RCt}$. Thus we need only bound the second term. To do that we use Lemma (3.5). By that Lemma,

(3.12) $\quad\quad\quad\quad |T_t^{\Lambda(t),\eta} \phi(\eta) - \int \phi(\xi)g_{\Lambda(t),\eta}(d\xi)|$

$$\leq 2^{5/2}\|\phi\|_u(\int f_t^{\Lambda(t)}(\xi)\log f_t^{\Lambda(t)}(\xi)g_{\Lambda(t),\eta}(d\xi))^{1/2} ,$$

where $f_s^{\Lambda(t)}(\cdot) = \dfrac{d\mu_s^{\Lambda(t)}(\cdot)}{dg_{\Lambda(t),\eta}(\cdot)}$ and $\mu_s^{\Lambda(t)} = (T_s^{\Lambda(t),\eta})^*\delta_\eta(\cdot)$. Now it is easily seen that there is a constant Γ

such that for any μ_0

(3.13) $\quad\quad\quad \int f_0^{\Lambda(t)}(\xi)\log f_0^{\Lambda(t)}(\xi)g_{\Lambda(t),\eta}(d\xi) \leq \Gamma|\Lambda(t)| .$

Also by (1.12), and Lemma (3.8)

(3.14) $\quad\quad\quad \dfrac{d}{ds} \int f_s^{\Lambda(t)}(\xi)\log f_s^{\Lambda(t)}(\xi)g_{\Lambda(t),\eta}(d\xi)$

$$= \int f_s^{\Lambda(t)}(\xi)\, \Omega^{\Lambda(t),\eta} (\log f_s^{\Lambda(t)}(\xi))g_{\Lambda(t),\eta}(d\xi)$$

$$\leq - \dfrac{2\alpha}{\gamma\log(|\Lambda(t)|)} \int f_s^{\Lambda(t)}(\xi)\log(f_s^{\Lambda(t)}(\xi))g_{\Lambda(t),\eta}(d\xi) .$$

Thus

(3.15) $\quad\quad\quad \int f_t^{\Lambda(t)}(\xi)\log(f_t^{\Lambda(t)}(\xi))g_{\Lambda(t),\eta}(d\xi) \leq \Gamma|\Lambda(t)|e^{-\frac{2\alpha t}{\gamma|\Lambda(t)|}}$

$$\leq \Gamma(l + 8CRt)e^{-2\alpha t/\gamma\log(l+8CRt)}$$

$$\leq B_0 e^{-\epsilon t/(\log t)}$$

for some $B_0 < \infty$ which depends on ϕ only through l , and some $\epsilon > 0$ which does not depend on l , and all $t \geq 2$.

Q.E.D.

References

[1]   Holley, R., Free energy in a Markovian model of a lattice spin system, Comm. Math. Phys. 23 (1971), 87-99.

[2]   Holley, R., An ergodic theorem for interacting systems with attractive interactions, Z. Wahr. verw. Geb. 24 (1972), 325-334.

[3]   Holley, R., Convergence in L^2 of stochastic Ising models: jump processes and diffusions, Proceedings of the Tanaguchi International Symposium on Stochastic Analysis at Katata, 1982, North Holland Math. Lib. 32 (1984), 149-167.

[4]   Holley R., Rapid convergence to equilibrium in one-dimensional stochastic Ising models, Ann. Prob. 13 (1985), 72-89.

[5]   Holley R., and Stroock D. W., Applications of the Stochastic Ising model to the Gibbs states, Comm. Math. Phys. 48 (1976), 246-265.

[6]   Holley R., and Stroock D. W., L_2 theory of the stochastic Ising model, Z. Wahr, verw. Geb. 35 (1976), 87-101.

[7]   Holley, R., and Stroock, D.W., In one and two dimensions every stationary measure for a stochastic Ising model is a Gibbs state, Comm. Math. Phys. 55 (1977), 37-45.

[8]   Holley R., and Stroock D. W., Diffusions on an infinite dimensional Torus, J. Funct. Anal. 42 (1981), 29-63.

[9]   Holley R., and Stroock D. W., Logarithmic Sobolev inequalities and stochastic Ising models, to appear.

[10] Liggett, T.M., Interacting Particle Systems, Grundlehren der mathematischen Wissenschaften, 276 (1985) Berlin-Heidelberg-New York: Springer.

[11] Moulin Ollagnier, J., and Pinchon, D. Free energy in a spin-flip process is non-increasing, Comm. Math. Phys. 55 (1977), 29-35.

[12] Ruelle, D., Statistical Mechanics, New York, Benjamin 1969.

[13] Stroock D., An introduction to the theory of large deviations, Universitext, Springer-Verlag, New York, 1984.

A Scaling relation at criticality for 2D-Percolation

Harry Kesten[1]

Department of Mathematics, Cornell University, Ithaca, NY 14853

Abstract. We prove a relation between the radius and volume of a
two-dimensional percolation cluster. This implies for 2D
percolation that the critical exponents δ and η satisfy
$\eta = 4/(\delta + 1)$ (provided η exists).

1. Relation Between Radius and Volume of a Percolation Cluster.

In two recent papers, [1] and [2], we studied the size of a 2D
percolation cluster, given that its "radius" is at least n , as
n → ∞ . The purpose of these papers was to give a rigorous
definition of the incipient infinite cluster and to study the
asymptotic behavior of a random walk on this cluster. We wish to
explain here that the results of these references also imply a
strong relationship between the distributions of the radius and
size of a 2D percolation cluster at criticality. For simplicity
we restrict ourselves here to bond percolation on z^2 . As
explained in [1] the results hold more generally for any 2D
(Benoulli) percolation model which satisfies condition (1) of [1].
All results here deal with the system at criticality. Thus our
basic probability measure is always P_{cr}, according to which all
bonds of z^2 are independently open (or closed) with probability
1/2. (Recall that 1/2 is the critical probability for bond
percolation on z^2). We use the following additional notation.

[1] Research supported by the NSF through a grant to Cornell
University.

W is the open cluster of $\underline{O}$ (= the origin),

#W = number of sites in W ,

R = radius of W = max{$|x|$: x $\in$ W} , where

$\quad$ $|x|$ = max($|x_1|$,$|x_2|$) when x = (x_1,x_2) ,

A $\to$ B means that there exists an open path on Z^2 from some

$\quad$ vertex in A to some vertex in B ,

$\tau(x,y)$ = P_{cr}{x $\to$ y} ,

S(n) = [-n,n] $\times$ [-n,n] = {x: $|x| \leq$ n} and

∂S(n) = boundary of S(n) ,

π(n) = P_{cr}{$\underline{O} \to \partial$S(n)} [2] .

We showed in [1] that π is decreasing but that[3]

$$\pi(2n) \geq C_1\pi(n) \quad , \quad C_2 \sum_1^n \pi(k) \leq n\pi(n) \leq \sum_1^n \pi(k) \quad . \tag{1}$$

It is also easy to see (cf. (32) of [1] that

$$\pi(n) \leq P_{cr}\{R \geq n\} \leq 4\pi(n) \quad . \tag{2}$$

We further know (cf. Cor. 3.15 in [3] that

$$\pi(n) \geq C_3 n^{-1/2} \quad . \tag{3}$$

---

[2] This definition of π differs from that of [1] and [2].
However, by virtue of eq. (32) in [1] the change is insignificant.

[3] C_i always stands for a strictly positive but finite constant
whose value is of no importance for our considerations. Its value
may vary from appearance to appearance.

(In fact for bond percolation on z^2 this can be improved to $\pi(n) \geq C_3 n^{-1/3}$, but this will not be used here.)

__Lemma.__ __For suitable constants__ C_i __and all vertices__ x __of__ z^2

$$C_1 [\pi(|x|)]^2 \leq \tau(0,x) \leq C_2 [\pi(|x|)]^2 \quad . \tag{4}$$

__Theorem.__ __For each__ $\epsilon > 0$ __there exists a__ $\lambda = \lambda(\epsilon) < \infty$ __such that__

$$P_{cr}\left\{\frac{\#W}{R^2 \pi(R)} \leq \frac{1}{\lambda} \; \middle| \; n \leq R \leq 2n\right\} \leq \epsilon \quad \text{for all large} \quad n \; . \tag{5}$$

__Also, for__ $\lambda \geq 1, n \geq 1$ __and__ $t \geq 1$ __there is some__ C_t, __depending__ __on__ t __only such that__

$$P_{cr}\left\{\frac{\#W}{R^2 \pi(R)} \geq \lambda \,\middle|\, n \leq R \leq 2n\right\} \leq C_t \lambda^{-t} \quad . \tag{6}$$

__Corollary.__ __If one of the two limits__

$$-\frac{1}{\delta_r} = \lim_{n \to \infty} \frac{\log \pi(n)}{\log n} \quad \text{or} \quad -\eta = \lim_{n \to \infty} \frac{\log \tau(0,(n,0))}{\log n} \tag{7}$$

__exist, then both these limits as well as__

$$-\frac{1}{\delta} = \lim_{n \to \infty} \frac{\log P_{cr}\{\#W \geq n\}}{\log n} \tag{8}$$

__exist, and__

$$\eta = \frac{2}{\delta_r} \quad , \quad \delta = 2\delta_r - 1 = \frac{4}{\eta} - 1 \quad . \tag{9}$$

Remarks. (i) Using simple Tanberian arguments and $P_{cr}\{\#W \geq n\} \geq$
$\pi(n) \geq c_3 n^{-1/2}$ (cf. (3)) one can show that the limit in (8)
equals the usual definition of $-1/\delta$, namely

$$\lim_{h \downarrow 0} \left[\log \frac{1}{h}\right]^{-1} \log \left[\sum_{n=0}^{\infty} P_{cr}\{\#W = n\}(1 - e^{-hn})\right]$$

provided at least one of these limits exists. We also note that
the Hausdorff dimension of the incipient infinite cluster as
defined in [1] is $2-\eta$ (see Theorem 8 in [1]). The second
relation in (9) therefore corresponds to one of the Buckingham-
Gunton or Fisher-Stell relations in the case of 2D percolation.
(Buckingham and Gunton [4] proposed only an _inequality_; their
argument was made more rigorous by Fisher [5]. The equality seems
to have been proposed first by Stell [6] and put on a solid
footing by Fisher [7], eg. (9.3.14). These references mainly deal
with the Ising model.) Note that $\pi(n) \geq c_3 n^{-1/3}$ says $\delta_r \geq 3$,
and thus by (9) $\delta \geq 5$. The conjectured value for δ in two
dimensions is 91/5 (see [8]).   The value of δ for percolation
on a Bethe tree is 2.

ii) Stronger hypotheses on the existence of critical exponents
yield stronger conclusions. E.g. if one strengthens assumption
(7) to

$$c_1 n^{-1/\delta_r} \leq \pi(n) \leq c_2 n^{-1/\delta_r} \quad \text{or} \quad c_1 n^{-\eta} \leq \tau(\underline{0},(n,0)) \leq c_2 n^{-\eta}$$

for all n , then one obtains

$$c_3 n^{-1/6} \leq P_{cr}(\#W \geq n) \leq c_4 n^{-1/6}$$

for suitable c_3, c_4 . We also remark that the method of Steps
(i) and (ii) of the proof of (3.24) in [2] allows us to give some
estimates for $\lambda(\epsilon)$ in (5), in terms of ϵ .

2. Proofs. (4) is easy. For the right hand inequality observe
that $\underline{0} \to x$ implies that $\underline{0} \to \partial S(\frac{1}{2}|x|)$ and that x is connected
by an open path to the boundary of the square of edge size $|x|$
and centered at x . The interiors of the latter square and of
$S(\frac{1}{2}|x|)$ are disjoint, so that

$$\tau(\underline{0}, x) = P_{cr}(\underline{0} \to x) \leq [\pi(\frac{1}{2}|x|)]^2$$

Now apply (1). For the left hand inequality of (4) observe that
$\underline{0}$ will be connected to x if there exist open paths from $\underline{0}$ to
$\partial S(2|x|)$ and from x to $\partial S(2|x|)$ and an open circuit in
$S(2|x|)\backslash S(|x|)$ (see Figure 1).

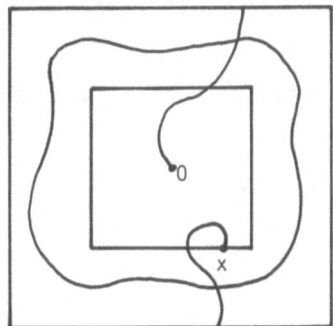

Figure 1. The inner square is $S(|x|)$ and the outer square is
$S(2|x|)$.

Therefore, by the Harris-FKG inequality,

$$\tau(0,x) \geq \pi(2|x|)\pi(|x|)P_{cr} \{\exists \text{ open circuit in } S(2|x|)\backslash S(|x|)\} \quad .$$

Since the last factor is bounded away from 0 (see [9],[10] or eq.
(28) in [1]), (4) follows, again by an appeal to (1).

For the Theorem we observe that

$$P_{cr}\{n \leq R \leq 2n\} \geq P_{cr}\{\Omega \to \partial S(n)\} \cdot P_{cr}\{\exists \text{ closed circuit on the}$$
$$\text{dual of } \mathbf{Z}^2 \text{ in } S(2n)\backslash S(n)\} \geq C_1\pi(n) \quad \text{(again by [9], [10] or}$$
$$\text{eq. (28) in [1])} \tag{10}$$

In view of (1) it therefore suffices for (5) to prove

$$\frac{1}{\pi(n)}P_{cr}\left\{\#W \leq \frac{1}{\lambda} n^2\pi(n) \quad \text{and} \quad R \geq n\right\}$$

$$= \frac{1}{\pi(n)}P_{cr}\left\{\#W \leq \frac{1}{\lambda} n^2\pi(n) \quad \text{and} \quad \Omega \to \partial S(n)\right\} \leq \epsilon$$

for sufficiently large λ . For this one can use the proof of
(54) in the second part of Theorem 8 in [1] without any
significant changes. Similarly, for (6) it suffices to prove

$$\frac{1}{C_1\pi(n)} P_{cr}\{\#[W \cap S(2n)] \geq \lambda n^2\pi(n) \quad \text{and} \quad \Omega \to \partial S(n)\}$$

$$\leq c_t\lambda^{-t} \quad .$$

But the same argument as used to prove $\pi(2n) \geq C_1\pi(n)$ (see proof of (6) in [1]) shows that

$$P_{cr}\{\#[W \cap S(2n)] \geq \lambda n^2\pi(n) \quad \text{and} \quad \underline{0} \to \partial S(n)\}$$

$$\leq C_2 P_{cr}\{\#[W \cap S(2n)] \geq \lambda n^2\pi(n) \quad \text{and} \quad \underline{0} \to \partial S(2n)\}$$

and the last probability is (by Markov's inequality) at most

$$[\lambda n^2\pi(n)]^{-t}\pi(2n)E_{cr}\{\#[W \cap S(2n)]^t|\underline{0} \to \partial S(2n)\} \qquad .$$

Finally, the proof of the first part of Theorem 8 in [1] gives without essential changes that (E_{cr} denotes expectation with respect to P_{cr})

$$E_{cr}\{[\#[W \cap S(2n)]^t]|\underline{0} \to \partial S(2n)\} \leq C_t[n^2\pi(n)]^t \qquad .$$

As for the <u>Cor.</u> it is clear from (4) that if either one of the limits in (7) exists, then so does the other, and $\eta = 2/\delta_r$. Next for fixed $\epsilon > 0$ and large n choose

$$m = n^{\delta_r/((2-\epsilon)\delta_r-1)} \qquad .$$

Then for large n , under (7),

$$n = m^{2-\epsilon-1/\delta_r} \leq [\lambda(\tfrac{1}{2})]^{-1}m^2\pi(m)$$

and

$$P_{cr}\{\#W \geq n\} \geq P_{cr}\{\#W \geq [\lambda(\tfrac{1}{2})]^{-1}m^2\pi(m)]$$

$$\geq \tfrac{1}{2} P_{cr}\{m \leq R \leq 2m\} \text{ (by (5))} \geq \tfrac{1}{2} C_1\pi(m) \text{ (by 10))}$$

$$\geq C_2 m^{-1/(\delta_r-\epsilon)} \text{ (by (7))} .$$

Therefore (7) implies, for any $\epsilon > 0$

$$\liminf_{n\to\infty} \frac{\log P_{cr}\{\#W \geq n\}}{\log n} \geq - \frac{\delta_r}{\delta_r-\epsilon} \frac{1}{(2-\epsilon)\delta_r-1} \tag{11}$$

For an inequality in the other direction we note that for any k_0, by (6),

$$P_{cr}\{\#W \geq n\} \leq \sum_{k<k_0} \pi(2^k)P_{cr}\{\#W \geq n | 2^k \leq R \leq 2^{k+1}\}$$

$$+ P_{cr}\{R \geq 2^{k_0}\} \leq \sum_{k<k_0} \pi(2^k)C_t\left[\frac{2^{2k}\pi(2^k)}{n}\right]^t$$

$$+ \pi(2^{k_0}) = \pi(2^{k_0})\left\{1 + C_t \sum_{k<k_0} \frac{\pi(2^k)}{\pi(2^{k_0})}\left[\frac{2^{2k}\pi(2^k)}{n}\right]^t\right\} .$$

By (1) $n\pi(n)$ is practically increasing. Therefore

$$\sum_{k<k_0} \frac{\pi(2^k)}{\pi(2^{k_0})}\left[\frac{2^{2k}\pi(2^k)}{2^{2k_0}\pi(2^{k_0})}\right]^t$$

$$= \sum_{k<k_0} 2^{(k-k_0)(t-1)}\left[\frac{2^k\pi(2^k)}{2^{k_0}\pi(2^{k_0})}\right]^{t+1}$$

$$\leq c_2^{-t-1}c_3 \quad ,$$

provided $t > 1$. Consequently

$$P_{cr}\{\#W \geq n\} \leq C_4\pi(2^{k_0})\left\{1 + \left[\frac{2^{2k_0}\pi(2^{k_0})}{n}\right]^t\right\} \quad ,$$

and if we choose $k_0 = (1-\epsilon)\log n[(\log 2)(2 - 1/\delta_r)]^{-1}$ we obtain from (7) that

$$\liminf_{n\to\infty} \frac{\log P_{cr}\{\#W \geq n\}}{\log n} \leq -(1-\epsilon)\frac{1}{\delta_r}\left[2 - \frac{1}{\delta_r}\right]^{-1} \quad .$$

The existence of the limit in (8) and the second relation of (9) now follow from the first relation in (9) and (11).

REFERENCES

[1] H. Kesten, The incipient infinite cluster in two-dimensional percolation, submitted to Theor. Probab. Rel. Fields.

[2] H. Kesten, Subdiffusive behavior of random walk on a random cluster, submitted to Ann. Inst. H. Poincaré.

[3] J. van den Berg and H. Kesten, Inequalities with applications to percolation and reliability, J. Appl. Prob. 22, 556-569 (1985).

[4] M.J. Buckingham and J.D. Gunton, Correlations at the critical point of the Ising model, Phys. Rev. 178, 848-853 (1969).

[5] M.E. Fisher, Rigorous inequalities for critical-point correlation exponents, Phys. Rev. 180, 594-600 (1969).

[6] G. Stell, Extension of the Ornstein-Zernike theory of the critical region, Phys. Rev. Lett. 20, 533-536 (1968).

[7] M.E. Fisher, The theory of equilibrium statistical phenomenon, Rep. Prog. Phys. 30, 615-730 (1967).

[8] D. Stauffer, Scaling properties of percolation clusters, pp. 9-25 in Disordered Systems and Localization, C. Castellani, C. Di Castro and L. Peliti eds., Lecture notes in physics, vol. 149, Springer, 1981.

[9] P.A. Seymour and D.J.A. Welsh, Percolation probabilities on the square lattice, Ann. Discrete Math. 3, 227-245 (1978).

[10] L. Russo, On the critical percolation probabilities, Z. Wahrsch. verw. Geb. 56, 229-237 (1981).

REVERSIBLE GROWTH MODELS ON Z^d: SOME EXAMPLES

Thomas M. Liggett[1]

Department of Mathematics
University of California, Los Angeles
Los Angeles, California   90024

1. Introduction

In a recent paper [3], a class of reversible growth models on a fairly
general set of sites was introduced and studied. These models are generalizations
of the finite reversible nearest particle systems on the integers, which have been
considered in several papers in recent years (see [1] and [2], for example). The
focus of attention in these growth models is the probability of survival of the
system. Typically there are natural one parameter families of models, and one
wishes to determine the critical value for that parameter, which is the point at
which survival with positive probability begins to occur. Once this is done, it
is of interest to determine the manner in which the survival probability
approaches its limit (which is usually zero) as the parameter approaches the critical value from above.

The purpose of this paper is to apply the results of [3] to various examples
of models on the d-dimensional integer lattice Z^d. We begin in Section 2 by
reviewing the main results of [3], as they appear when specialized to Z^d. The
reader is referred to [3] for the proofs of these results. The later sections
deal with various special classes of models in which the birth rates have a specified form. Among these, we will find examples in which the critical value λ_c
is zero, and others in which the critical value is positive. One typical feature
of the results is that the survival probability decays like a power of $\lambda - \lambda_c$ as
$\lambda \downarrow \lambda_c$ if $\lambda_c > 0$, while it decays much faster than any power of λ if $\lambda_c = 0$.

[1] Research supported in part by NSF Grant MCS 83-00836. This work was carried
out at the Institute for Mathematics and its Applications at the University of
Minnesota. Its hospitality is gratefully acknowledged.

2. The General Model on Z^d

The birth rates of the growth model are expressed in terms of a collection of weights $\pi(A) > 0$ defined for nonempty finite subsets A of Z^d. We will assume that these weights satisfy the following properties:

(a) $\pi(A) = \pi(B)$ whenever A and B are translates of each other,

(b) $A \subset B$ and $\pi(A) = 0$ imply $\pi(B) = 0$, and

(c) there are sets of arbitrarily large cardinality with $\pi(A) > 0$.

A consequence of these assumptions is that $\pi(\{x\})$ is positive and independent of $x \in Z^d$. Since only ratios of π's will be relevant, we adopt the normalization $\pi(\{x\}) \equiv 1$.

The growth model is a continuous time Markov chain on

$$\mathcal{S} = \{\phi\} \cup \{A \subset Z^d : A \text{ is finite and } \pi(A) > 0\}.$$

Its transition rates are determined as follows:

(a) The empty set ϕ is a trap.

(b) $A \to A \setminus \{x\}$ at rate 1 for each $x \in A$.

(c) $A \to A \cup \{x\}$ at rate $\pi(A \cup \{x\})/\pi(A)$ for each $x \notin A$ if $A \neq \phi$ and $\pi(A) > 0$.

In order to guarantee that all states of this Markov chain are stable, we will impose the following additional assumption:

(2.1) $$\sum_x \pi(A \cup \{x\}) < \infty \quad \text{for each } A \neq \phi.$$

The survival probability ρ is defined by

(2.2) $$\rho = P^{\{x\}}(A_t \neq \phi \text{ for all } t),$$

which does not depend on x because the transition mechanism for A_t is translation invariant. Our primary tool is the following variational characterization of ρ which is provided by the Dirichlet principle for reversible Markov chains. See [1] for a general statement of this principle.

__Theorem 2.3__ Let H be the collection of all functions $h: \mathcal{A} \to [0,1]$ which satisfy

(a) $h(\phi) = 0$,

(b) $h(A) = h(B)$ if A and B are translates of each other, and

(c) $\lim\limits_{n \to \infty} \inf\limits_{|A|=n} h(A) = 1$.

For $h \in H$, define

$$\phi(h) = \sum_{A \ni 0} \frac{\pi(A)}{|A|} \sum_{x \in A} [h(A) - h(A \setminus \{x\})]^2.$$

Here $|A|$ denotes the cardinality of A. Then

$$\rho = \inf_{h \in H} \phi(h).$$

An immediate consequence of this theorem is the following monotonicity statement.

__Corollary 2.4.__ Let ρ and ρ' be the survival probabilities for the systems corresponding to the weights $\pi(\cdot)$ and $\pi'(\cdot)$ respectively. If

$$\pi(A) \leq \pi'(A) \quad \text{for all} \quad A,$$

then $\rho \leq \rho'$.

Upper bounds for the survival probability can be easily deduced from Theorem 2.3 by computing $\phi(h)$ for a carefully chosen $h \in H$. The following result is obtained in this way.

__Corollary 2.5.__ For $n > 1$, let

(2.6)
$$\pi_n = \sum_{\substack{|A|=n \\ A \ni 0}} \pi(A).$$

Then

$$\rho \leq [\sum_{n=1}^{\infty} \pi_n^{-1}]^{-1},$$

where the upper bound is interpreted as zero if the series diverges.

To obtain a lower bound of a similar type, it is necessary to make an additional assumption. The following definition is appropriate for this purpose.

Definition 2.7. The weights $\pi(\cdot)$ are said to be monotone if for each $n > 2$ there is a nonnegative function $m_n(A,B)$ which is translation invariant, is zero unless $|A| = n$, $|B| = n - 1$ and $B \subset A$, and satisfies

$$(2.8) \qquad \sum_B m_n(A,B) = \pi(A)/\pi_n \qquad \text{if} \quad |A| = n, \text{ and}$$

$$(2.9) \qquad \sum_A m_n(A,B) = (n - 1)\pi(B)/n\pi_{n-1} \quad \text{if} \quad |B| = n - 1.$$

Corollary 2.10. Suppose the weights $\pi(\cdot)$ are monotone. Then

$$\rho \geq [\sum_{n=1}^{\infty} n\pi_n^{-1}]^{-1}.$$

This corollary is proved by using the Schwarz inequality twice. For its proof, as well as that of Theorem 2.3 and Corollary 2.5, see [3].

In order to see that the monotonicity assumption is needed, consider the following example.

Example 2.11. Take $d = 1$. For $A = \{x_1,\ldots,x_n\} \subset Z^1$ with $x_1 < \ldots < x_n$, let $D(A) = x_n - x_1$. Choose a sequence A_n of subsets of Z^1 so that $|A_n| = n$ and A_n is not a subset of any translate of A_{n+1} if $n > 2$. Define weights $\pi(A)$ by

$$\pi(A) = \begin{cases} \frac{1}{2}\gamma^{D(A)} + \frac{1}{2}n^2 & \text{if} \quad A \text{ is a translate of } A_n \\ \\ \frac{1}{2}\gamma^{D(A)} & \text{if} \quad A \text{ is not a translate of any } A_n. \end{cases}$$

Then

$$\pi_n = \sum_{\substack{A \ni 0 \\ |A|=n}} \pi(A) \geq \frac{1}{2}n^3,$$

since there are n translates of A_n which contain zero. Therefore the lower bound for ρ in Corollary 2.10 is strictly positive. However, we will show now that $\rho = 0$ if $\gamma < \frac{1}{2}$. To do so, let $h(n)$ be defined on $\{0,1,\ldots\}$ in such a way that $h(0) = 0$, $0 \le h(n) \le 1$, and $h(n) \to 1$ as $n \to \infty$. Define $h \in H$ by $h(\phi) = 0$, and if $|A| = n > 1$,

$$h(A) = \begin{cases} h(n + 1) & \text{if } A \text{ is contained in a translate of } A_{n+1}, \text{ and} \\ h(n) & \text{otherwise.} \end{cases}$$

Then compute

$$\phi(h) = \sum_{A \ni 0} \frac{\pi(A)}{|A|} \sum_{x \in A} [h(A) - h(A \setminus \{x\})]^2$$

$$< \frac{1}{2} \sum_{n=1}^{\infty} [h(n + 1) - h(n - 1)]^2 \sum_{\substack{A \ni 0 \\ |A|=n}} \gamma^{D(A)}$$

$$+ \frac{1}{2} \sum_{n=1}^{\infty} n^2 \sum_{x \in A_n} [h(A_n) - h(A_n \setminus \{x\})]^2$$

$$= \frac{1}{2} \sum_{n=1}^{\infty} [h(n + 1) - h(n - 1)]^2 n \left(\frac{\gamma}{1 - \gamma} \right)^{n-1} + \frac{1}{2} h^2(2).$$

By Theorem 2.3, this provides an upper bound to ρ . Now evaluate this bound for

$$h(n) = \min\{ \frac{n}{N} , 1 \},$$

and then let $N \to \infty$ to conclude that $\rho = 0$ if $\gamma < \frac{1}{2}$. Therefore the monotonicity assumption is needed in Corollary 2.10.

While the monotonicity assumption is automatically satisfied for reversible nearest particle systems on Z^1 , this assumption is often either not satisfied or difficult to check for higher dimensional models. Therefore it is useful to have some classes of examples for which the weights are automatically monotone. Even if these are not themselves of great interest, they can sometimes be used in conjunction with Corollary 2.4 to obtain lower bounds for the survival probability of more interesting examples. Let $\gamma_A(x)$ be defined for finite $A \subset Z^d$ and

$x \in Z^d$ so that

(2.12) $\qquad \gamma_A(x) \geq 0,\ \gamma_A(x) = 0$ if $x \in A$, and $\sum\limits_x \gamma_A(x) = 1$.

Assume further that $\gamma_A(x)$ is translation invariant in the sense that $\gamma_{A+y}(x + y) = \gamma_A(x)$ for all $y \in Z^d$. Define weights $\pi(A)$ recursively via

(2.13)
$$\pi(A) = 1 \quad \text{if} \quad |A| = 1, \text{ and}$$
$$\pi(A) = c_n \sum_{x \in A} \pi(A \setminus \{x\}) \gamma_{A \setminus \{x\}}(x) \quad \text{if} \quad |A| = n > 2,$$

where the c_n's are positive constants. Assume that the $\gamma_A(x)$ are such that $A \subset B$ and $\pi(A) = 0$ imply $\pi(B) = 0$.

Theorem 2.14. Suppose $\pi(\cdot)$ is constructed as in (2.13). Then these weights are monotone, and

(2.15)
$$\pi_n = nc_n c_{n-1} \cdots c_2.$$

This result is proved in [3]. In checking the monotonicity, use

$$m_n(A, A \setminus \{x\}) = \gamma_{A \setminus \{x\}}(x)\ \pi(A \setminus \{x\}) \frac{c_n}{\pi_n}$$

for $|A| = n$ and $x \in A$.

As mentioned earlier, it is often not easy to determine whether or not the monotonicity assumption in Corollary 2.10 is satisfied. The following result is useful in showing that it is not satisfied in some cases.

Theorem 2.16. Suppose that the weights $\pi(\cdot)$ are monotone, and let $f(\cdot)$ be a translation invariant bounded function on the collection of all finite subsets of Z^d such that $f(B) < f(A)$ whenever $B \subset A$. For $n > 1$, let

(2.17)
$$\phi(n) = \sum_{\substack{A \ni 0 \\ |A| = n}} \frac{f(A) \pi(A)}{\pi_n}.$$

Then $\phi(n)$ is a nondecreasing function of n.

Proof: Fix $n > 2$. For $u, v \in Z^d$, let

$$F(u,v) = \sum_{\substack{A \ni u \\ B \ni v}} m_n(A,B) f(B),$$

where $m_n(A,B)$ is the function appearing in Definition 2.7. Then

$$\sum_u F(u,v) = \sum_{\substack{A \\ B \ni v}} |A| m_n(A,B) f(B)$$

$$= \frac{(n-1)}{\pi_{n-1}} \sum_{\substack{B \ni v \\ |B| = n-1}} f(B) \, \pi(B)$$

$$= (n-1) \phi(n-1), \text{ while}$$

$$\sum_v F(u,v) = \sum_{\substack{A \ni u \\ B}} |B| \, m_n(A,B) f(B)$$

$$< (n-1) \sum_{\substack{A \ni u \\ B}} m_n(A,B) f(A)$$

$$= \frac{(n-1)}{\pi_n} \sum_{\substack{A \ni u \\ |A| = n}} f(A) \, \pi(A)$$

$$= (n-1) \phi(n).$$

However, $F(u,v) = F(0, v - u)$, so it follows that

$$\phi(n-1) \leq \phi(n) \quad \text{for} \quad n > 2.$$

3. First Example

Define the projection $R_i: Z^d \to Z^1$ by $R_i(x^1, \ldots, x^d) = x^i$. Suppose $\lambda > 0$, $\beta(n) > 0$ for $n > 1$, and $B = \sum_{n=1}^{\infty} \beta(n) < \infty$. In this section, we will consider the growth model corresponding to the following weights:

$$(3.1) \qquad \pi(A) = \lambda^{|A|-1} \prod_{i=1}^{d} \prod_{k=1}^{n_i - 1} \beta(x_{k+1}^i - x_k^i),$$

where $x_1^i < \ldots < x_{n_i}^i$ and

$$\{x_k^i: 1 \leq k \leq n_i\} = \{R_i x: x \in A\} = R_i(A).$$

Note that when $d = 1$, this is simply a reversible nearest particle system.

Our first job is to compute the π_n's, which are defined in (2.6), for this model. Given positive integers $n_1,\ldots,n_d$, first carry out the sum in (2.6) over those sets A for which $|R_i(A)| = n_i$ for $1 \leq i \leq d$. This gives

$$(3.2) \qquad \pi_n = n\lambda^{n-1} \sum_{n_1,\ldots,n_d} N_n(n_1,\ldots,n_d)B^{n_1+\ldots+n_d-d},$$

where $N_n(n_1,\ldots,n_d)$ is the number of subsets A of

$$\prod_{i=1}^d \{1,\ldots,n_i\}$$

for which $R_i(A) = \{1,\ldots,n_i\}$ for $1 \leq i \leq d$ and $|A| = n$. It is somewhat easier to compute $M_n(n_1,\ldots,n_d)$, which is the number of subsets A of

$$\prod_{i=1}^d \{1,\ldots,n_i\}$$

with $|A| = n$. These quantities are related by

$$(3.3) \qquad M_n(n_1,\ldots,n_d) = \sum_{\substack{1 \leq k_1 \leq n_1 \\ 1 \leq k_d \leq n_d}} \binom{n_1}{k_1} \cdots \binom{n_d}{k_d} N_n(k_1,\ldots,k_d).$$

Since

$$(3.4) \qquad \sum_{n=k}^{\infty} \binom{n}{k} z^n = \frac{z^k}{(1-z)^{k+1}}$$

for $|z| < 1$, we can deduce from (3.3) that

$$\sum_{n_1,\ldots,n_d} z^{n_1+\ldots+n_d} M_n(n_1,\ldots,n_d) =$$

$$= \sum_{k_1,\ldots,k_d} N_n(k_1,\ldots,k_d)\left(\frac{z}{1-z}\right)^{k_1+\ldots+k_d} (1-z)^{-d}.$$

Therefore, using (3.2), it follows that

$$(3.5) \qquad \pi_n = n\lambda^{n-1} \sum_{n_1,\ldots,n_d} M_n(n_1,\ldots,n_d) \left(\frac{B}{B+1}\right)^{n_1+\ldots+n_d} (B^2 + B)^{-d}.$$

However,

$$M_n(n_1,\ldots,n_d) = \binom{n_1 n_2 \cdots n_d}{n},$$

so

$$(3.6) \qquad \pi_n = n\lambda^{n-1} \sum_{n_1,\ldots,n_d \geq 1} \binom{n_1 n_2 \cdots n_d}{n} \left(\frac{B}{B+1}\right)^{n_1+\ldots+n_d} (B^2 + B)^{-d}.$$

Using the inequality

$$\binom{n_1 \cdots n_d}{n} \leq \binom{n_1+n-1}{n} \cdots \binom{n_d+n-1}{n} (n!)^{d-1}$$

and (3.4), we have

$$(3.7) \qquad \pi_n \leq n\lambda^{n-1} (n!)^{d-1} (B + 1)^{d(n-1)}.$$

Therefore Corollary 2.5 gives the following upper bound:

$$(3.8) \qquad \rho \leq \left[\sum_{n=1}^{\infty} \frac{1}{n(n!)^{d-1}[\lambda(B+1)^d]^{n-1}}\right]^{-1}_0.$$

This bound is strictly positive for all $\lambda > 0$ if $d > 2$, and tends to zero as $\lambda \to 0$. To see how rapidly it tends to zero, take $c > (B + 1)^d$ and use Holder's inequality if $d \geq 3$ and a simple comparison if $d = 2$ to obtain

$$\sum_{n=1}^{\infty} \frac{1}{n(n!)^{d-1}[\lambda(B+1)^d]^{n-1}} \geq \varepsilon \left[\sum_{n=1}^{\infty} \frac{1}{(n-1)!(\lambda c)^{(n-1)/(d-1)}}\right]^{d-1}$$

for some positive ε. Therefore

$$(3.9) \qquad \rho \leq \varepsilon^{-1} \exp\left\{-\frac{d-1}{(\lambda c)^{1/(d-1)}}\right\}.$$

Unfortunately, the weights in this class of examples are not necessarily monotone, so that we cannot apply Corollary 2.10 to obtain an analogous lower bound for ρ. To see that monotonicity may not hold, take $d = 2$, and let $f(A)$

be the function defined by $f(A) = 1$ if A contains three points of the form $(x - 1,y)$, (x,y), $(x + 1,y)$ for some x and y, and $f(A) = 0$ it if contains no such triple. Let $\phi(n)$ be the corresponding sequence defined in (2.17). Then $\phi(1) = \phi(2) = 0$,

$$\phi(3) = \frac{3\lambda^2 B^2(1)}{\pi_3} \ , \quad \text{and}$$

$$\phi(4) = \frac{16\lambda^3 B^2(1)B(B + 2)}{\pi_4} \ .$$

If the weights were monotone, it would follow from Theorem 2.16 that

$$3\pi_4 \leq 16\lambda B(B + 1)\pi_3.$$

Using (3.6), it would then follow that

$$\sum_{k,\ell \geq 1} \binom{k\ell}{4}\left(\frac{B}{B + 1}\right)^{k+\ell} \leq 4B(B + 1) \sum_{k,\ell \geq 1} \binom{k\ell}{3}\left(\frac{B}{B + 1}\right)^{k+\ell}.$$

However, this is not correct for small B.

In view of these remarks, it is necessary to use a different approach to obtain a lower bound for ρ. Let $\pi'(A)$ be the weights constructed as in (2.13) for some sequence c_n and the following choice of $\gamma_A(x)$: Given A and $x \notin A$, let

$$I = \{i: \ R_i(x) \ \epsilon \ R_i(A)\}$$
$$J = \{i: \ R_i(x) > R_i(A)\}$$
$$K = \{i: \ R_i(x) < R_i(A)\},$$

where the inequality in the definition of J, for example, means that $R_i(x)$ lies strictly to the right of the set $R_i(A)$. For $i \ \epsilon \ J$, let j_i be the distance from $R_i(x)$ to the rightmost point in $R_i(A)$, and for $i \ \epsilon \ K$, let k_i be the distance from $R_i(x)$ to the leftmost point in $R_i(A)$. Note that in general $I \cup J \cup K$ does not exhaust all of $\{1,\ldots,d\}$. Now set $\gamma_A(x) = 0$ unless $J \cup K \neq \phi$ and $I \cup J \cup K = \{1,\ldots,d\}$. When these two conditions are satisfied, let

$$\gamma_A(x) = \frac{\prod_{i \in J} \beta(j_i) \prod_{i \in K} \beta(k_i)}{\sum_{L \subsetneq \{1,\ldots,d\}} (2B)^{d-|L|} \prod_{i \in L} |R_i(A)|} .$$

We will show by induction on the cardinality of A that for an appropriate choice of the sequence $\{c_n\}$, $\pi'(A) \leq \pi(A)$ for all A. This is true when $|A| = 1$ by our normalization. So, suppose it is true for all sets A with $|A| < n$, and take an A with $|A| = n$. Then by the induction hypothesis and the definition of π' in terms of the $\gamma_A(x)$'s,

$$\pi'(A) \leq c_n \sum_{x \in A} \pi(A \setminus \{x\}) \gamma_{A \setminus \{x\}}(x)$$

$$\leq c_n \lambda^{-1} \pi(A) \sum_x \frac{1}{\prod_{i=1}^{d} [2B + |R_i(A \setminus \{x\})|] - \prod_{i=1}^{d} |R_i(A \setminus \{x\})|} ,$$

where the sum is over those $x \in A$ for which $\gamma_{A \setminus \{x\}}(x) \neq 0$. There are at most $2d$ such x's, and for each of them,

$$n - 1 \leq \prod_{i=1}^{d} |R_i(A \setminus \{x\})|.$$

Therefore

$$\pi'(A) \leq \frac{c_n \lambda^{-1} \pi(A)}{B(n-1)^{(d-1)/d}} .$$

So, we can complete the induction step provided that we take

$$c_n = \lambda B(n - 1)^{(d-1)/d}.$$

Combining Corollaries 2.4 and 2.10 with Theorem 2.14, it then follows that the following lower bound is valid for the survival probability of the system with weights given by (3.1):

$$\rho \geq \left[\sum_{n=0}^{\infty} \frac{1}{(\lambda B)^n (n!)^{(d-1)/d}} \right]^{-1} .$$

Note again that this lower bound is strictly positive for all positive λ. Using

Hölder's inequality as before, it follows that if $c < B$ and $d > 1$, then there is a positive ε so that

(3.10)
$$\rho \geq \varepsilon \exp\left[-\left[\frac{d-1}{d}\frac{1}{(\lambda c)^{d/(d-1)}}\right]\right].$$

It would be interesting to determine the following limit for $d > 2$, if it exists:

$$\lim_{\lambda \to 0} \frac{\log|\log \rho(\lambda)|}{\log \lambda^{-1}}.$$

In particular, does this limit depend on d? Bounds (3.9) and (3.10) imply that this limit must lie between $\frac{1}{d-1}$ and $\frac{d}{d-1}$.

4. Second Example

This example is not of great interest in its own right, since it is essentially one dimensional in nature. However, it is sometimes useful as a comparison system when one is analyzing more natural higher dimensional models. Let $\leq$ be a partial order on Z^d. A natural choice, for example, is the lexicographical ordering in which $(x^1,\ldots,x^d) < (y^1,\ldots,y^d)$ if $(x^1,\ldots,x^d) = (y^1,\ldots,y^d)$ or if for some $1 \leq k \leq d$, $x^i = y^i$ for all $i < k$ and $x^k < y^k$.

Let $B(x)$ be a strictly positive summable function on

$$\{x \in Z^d: x > 0, x \neq 0\}.$$

Define weights $\pi(A)$ by

(4.1)
$$\pi(A) = \prod_{k=1}^{n-1} B(x_{k+1} - x_k)$$

if $A = \{x_1,\ldots,x_n\}$, $|A| = n$, and $x_1 \leq x_2 \leq \cdots \leq x_n$, and $\pi(A) = 0$ if A is not totally ordered by $\leq$. These weights are monotone, as can be seen by letting

$$m_n(A,B) = \pi(A)/\pi_n$$

if $A = \{x_1,\ldots,x_n\}$ and $B = \{x_1,\ldots,x_{n-1}\}$ if $x_1 \leq x_2 \leq \cdots \leq x_n$. Also,

225

$$\pi_n = n[\sum_x \beta(x)]^{n-1}.$$

Therefore Corollaries 2.5 and 2.10 imply that $\rho = 0$ if $\lambda \leq 1$ and

$$\frac{\lambda - 1}{\lambda} < \rho < |\lambda \log \frac{\lambda - 1}{\lambda}|^{-1}$$

for $\lambda > 1$, where

$$\lambda = \sum_x \beta(x).$$

It is instructive to see what happens when this example is used as a comparison system for the example of Section 3. To do so, let λ and $\beta(n)$ now be as in Section 3, and let $\pi'(A)$ be the weights defined in (4.1) in terms of the lexicographical partial order and the function

$$\beta'(x) = \lambda \prod_{i=1}^{d} \beta(R_i(x)),$$

where $\beta(0)$ is taken to be 1. Then

$$\pi(A) > \pi'(A)$$

for each A, so $\rho \geq \rho'$ by Corollary 2.4. Since

$$\sum_{x \in Z^d} \beta'(x) = \lambda \sum_{i=1}^{d} [\sum_{n=1}^{\infty} \beta(n)]^i,$$

we conclude that the example of Section 3 survives provided that

$$\lambda \sum_{i=1}^{d} [\sum_{n=1}^{\infty} \beta(n)]^i > 1.$$

In particular, we cannot conclude from this comparison that it survives for all $\lambda > 0$, as we saw was the case in Section 3. Hence, the more involved analysis given there is needed.

5. Third Example

Now let $\beta(x)$ be a strictly positive summable function on $Z^d \setminus \{0\}$ such that $\beta(x) = \beta(-x)$ for every $x \in Z^d \setminus \{0\}$. Define weights via

$$\pi(A) = \max_{k=1}^{n-1} \prod \beta(x_{k+1} - x_k), \quad |A| = n,$$

where the max is taken over all orderings of the n points $\{x_1, \ldots, x_n\}$ in A. Note that if $d = 1$ and $\beta(x)$ is decreasing for $x > 0$, then these weights correspond exactly to a reversible nearest particle system. In general, however they give rise to a genuinely d-dimensional model.

Note that this example can be compared with the example in Section 4 (with the lexicographical ordering) using Corollary 2.4 to yield

(5.1)
$$\rho \geq 1 - (\tfrac{1}{2} \sum_x \beta(x))^{-1},$$

so that the process survives if $\sum_x \beta(x) > 2$.

To obtain an upper bound on ρ, we will establish an upper bound on π_n and then use Corollary 2.5. If $A \ni 0$ and $|A| = n$, then there exists $1 \leq k \leq n$ and $y_1, \ldots, y_{n-1} \in Z^d$ so that

$$A = \{-y_1 - y_2 - \cdots - y_{k-1}, y_2 - \cdots - y_{k-1}, \ldots, 0, y_k, y_k + y_{k+1}, \ldots, y_k + \cdots + y_{n-1}\}$$

and $\pi(A) = \beta(y_1)\beta(y_2) \cdots \beta(y_{n-1})$. Since the k and $y_1, \ldots, y_{n-1}$ determine A, it follows that

$$\pi_n = \sum_{\substack{|A|=n \\ A \ni 0}} \pi(A) \leq n \sum_{y_1, \ldots, y_{n-1}} \beta(y_1) \cdots \beta(y_{n-1})$$

$$= n \left[\sum_y \beta(y)\right]^{n-1}.$$

Therefore by Corollary 2.5,

(5.2)
$$\rho \leq |\lambda \log \tfrac{\lambda - 1}{\lambda}|^{-1},$$

where

$$\lambda = \sum_y \beta(y).$$

So, the process dies out if $\sum_y \beta(y) \leq 1$.

This result can be improved somewhat by noting that the y_i's used in the above expression for A must be such that the points in that expression are all distinct. This yields a smaller bound on π_n which then implies that the process dies out whenever

$$\sum_{y+z \neq 0} \beta(y)\beta(z) = [\sum_y \beta(y)]^2 - \sum_y \beta^2(y) \leq 1.$$

Further improvements of this type are possible.

In general, it appears that when

$$1 < \sum_y \beta(y) < 2,$$

the survival of the system depends very much on the particular choice of $\beta(\cdot)$ with the given sum. For $\beta(\cdot)$ with a sum lying between 1 and 2, it seems to be difficult to determine exactly when survival occurs, except in certain degenerate cases. There is a similarity between the computations required to do this, and those which would be required to determine the rate of exponential growth in n of the number of self-avoiding random walks of length n. The connection can most easily be seen by considering the degenerate case in which $\beta(x) = \lambda$ if x is a neighbor of 0 in Z^d, and $\beta(x) = 0$ otherwise. Then, in terms of rate of exponential growth, π_n is essentially λ^n times the number of sets of cardinality n which are the sets a self-avoiding random walk starting at 0 visits.

References

1. Griffeath, D. and Liggett, T.M. (1982). Critical phenomena for Spitzer's reversible nearest particle systems. The Annals of Probability 10, 881-895.

2. Liggett, T.M. (1986). Applications of the Dirichlet principle to finite reversible nearest particle systems. Probability Theory and Related Fields. To appear.

3. Liggett, T.M. (1986). Reversible growth models on symmetric sets. Proceedings of the 1985 Taniguchi Symposium on Probabilistic Methods in Mathematical Physics.

INEQUALITIES FOR γ AND RELATED CRITICAL EXPONENTS
IN SHORT AND LONG RANGE PERCOLATION

C. M. Newman

Department of Mathematics, University of Arizona

Tucson, Arizona 85721

Abstract

We relate the cluster size distribution $P_n(p)$ at the percolation critical point, $p = p_c$, to the critical exponent γ $\left(\sum n P_n(p) \approx |p_c - p|^{-\gamma} \text{ as } p \uparrow p_c \right)$. If $P_\infty(p_c) > 0$ (i.e., if $P_\infty(p)$ is discontinuous at $p = p_c$), then $\gamma \geq 2$. If $P_n(p_c) \approx n^{-1-1/\delta}$ as $n \to \infty$, then $\gamma \geq 2(1 - 1/\delta)$. Related inequalities are valid for γ_r $\left(\sum n^r P_n(p) \approx |p_c - p|^{-\gamma_r} \text{ as } p \uparrow p_c \right)$ and γ'_r (defined analogously as $p \downarrow p_c$) when $r > 1/\delta$: $\gamma_r, \gamma'_r \geq 2(r - 1/\delta)$. These results are valid for Bernoulli site or bond percolation on d-dimensional lattices for any d with p the site or bond occupation probability. They are also valid for long range translation invariant Bernoulli bond percolation with p the occupation probability for bonds of some given length.

1. Introduction and Main Results

The purpose of this paper is to present a number of results about the expected cluster size critical exponent γ which are valid for a large class of independent translation invariant percolation models in any dimension- site or bond, short or long range. Complete proofs of these results are given in this and later sections; for less detail, see [N2].

For simplicity, we will begin by considering standard nearest neighbor site percolation on the hypercubic lattice Z^d. In this context, p denotes the probability of any site being occupied (the site occupation density), P_n (for $n = 0, 1, 2, \ldots, \infty$) denotes the probality that $C(0)$, the cluster (or connected component) of the origin, contains exactly n (occupied) sites and χ denotes the expected cluster size,

$$\chi = \sum_{n \leq \infty} n P_n = \begin{cases} \sum_{n < \infty} n P_n & \text{if } P_\infty = 0 \\ \infty & \text{if } P_\infty > 0. \end{cases} \tag{1.1}$$

χ and the P_n's are functions of p in $[0,1]$. The critical point p_c is defined as $\sup\{p : P_\infty(p) = 0\}$. It was recently proved by Aizenman and Barsky [AB] that p_c coincides with $\sup\{p : \chi(p) < \infty\}$; this was previously proved only for d = 2 [K1].

Our first result, Theorem 1.1, relates the rate of divergence of $\chi(p)$ as $p \uparrow p_c$ to the continuity of $P_\infty(p)$ at p_c. $P_\infty(p)$ is always right continuous and was recently shown to be continuous for all $p > p_c$ [BK; AKN]. It is at present only proved to be continuous at p_c for $d = 2$ [R]. Theorem 1.1 may be regarded as a step toward proving continuity at p_c for $d > 2$ in the sense that it "reduces" the problem to showing that $\chi(p) = 0\left(|p_c - p|^{-\gamma}\right)$ as $p \uparrow p_c$ with $\gamma < 2$. We note that numerical results show that γ is approximately 1.7 for $d = 3$ (see e.g., [St] and the references given there). The proof of Theorem 1.1 is sufficiently pleasant that we give it immediately and then continue with the presentation of our other results.

Theorem 1.1. For standard nearest neighbor Bernoulli site percolation on Z^d with $d > 1$: if

$$\int_0^{p_c-0} (\chi(p))^{1/2} dp < \infty, \tag{1.2}$$

then $P_\infty(p_c) = 0$.

Proof. The standard lattice animal expansion for P_n is

$$P_n(p) = \sum_\ell a_{n\ell} p^n (1 - p)^\ell \tag{1.3}$$

where $a_{n\ell}$ is the number of lattice animals with n occupied sites and ℓ vacant boundary sites, and the sum in (1.3) is clearly finite. Now

$$P_\infty(p) = 1 - \sum_{n<\infty} P_n(p) \tag{1.4}$$

and $P_\infty(p) = 0$ for $p < p_c$. We may use (1.3) and differentiate (1.4) once to obtain

$$\sum_{\substack{\ell \\ n<\infty}} [n/p - \ell/(1 - p)]a_{n\ell} p^n (1 - p)^\ell = 0 \quad \text{for} \quad 0 < p < p_c \tag{1.5}$$

and twice to obtain

$$\sum_{\substack{\ell \\ n<\infty}} [n/p - \ell/(1 - p)]^2 a_{n\ell} p^n (1 - p)^\ell$$

$$= \sum_{\substack{\ell \\ n<\infty}} [n/p^2 + \ell/(1 - p)^2]a_{n\ell} p^n (1 - p)^\ell$$

$$= p^{-2}(1 - p)^{-1}\chi(p) \quad \text{for} \quad 0 < p < p_c, \tag{1.6}$$

where the last equality of (1.6) uses (1.5). The term-by-term differentiations used to obtain (1.5)-(1.6) are justified by the exponential falloff of P_n [H] for $p < p_c$ [AB]. Now for small $\varepsilon > 0$,

$$P_\infty(p_c) = P_\infty(p_c) - P_\infty(p_c - \varepsilon) = \lim_{N\to\infty} \int_{p_c-\varepsilon}^{p_c-0} \left(\sum_{n<N} (-d/dp)P_n(p) \right) dp$$

$$= \lim_{N\to\infty} \int_{p_c-\varepsilon}^{p_c-0} \left(\sum_{\substack{\ell \\ n<N}} - [n/p - \ell/(1 - p)]a_{n\ell}p^n(1 - p)^\ell \right) dp$$

$$\leq \lim_{N\to\infty} \int_{p_c-\varepsilon}^{p_c-0} \left(\sum_{\substack{\ell \\ n<\infty}} [n/p - \ell/(1 - p)]^2 a_{n\ell}p^n(1 - p)^\ell \right)^{1/2} dp$$

$$= \int_{p_c-\varepsilon}^{p_c-0} [p^{-2}(1 - p)^{-1}\chi(p)]^{1/2} dp, \tag{1.7}$$

where the inequality in (1.7) is a consequence of the Cauchy-Schwarz inequality on the sequence space, $\{f_{n\ell}: \sum_{\substack{\ell \\ n<\infty}} |f_{n\ell}|^2 a_{n\ell}p^n(1 - p)^\ell < \infty\}$.

Since for $d > 1$, p_c is in $(0,1)$ and hence $p^{-1}(1 - p)^{-1/2}$ does not diverge on the region of integration in (1.7), we see that if (1.2) is valid, then letting $\varepsilon \to 0$ in (1.7) yields $P_\infty(p_c) = 0$. □

Remark. The proof of Theorem 1.1 is related to arguments used in [G], [K2; p. 252] and [AKN, Sec. 3]. Here we bound $\sum |dP_n/dp|$ to obtain continuity of $\sum P_n$ on $[0,p_c]$; there $\sum n^{-1}|dP_n/dp|$ is bounded to obtain continuous differentiability of $\sum n^{-1}P_n$.

Clearly, Theorem 1.1 extends to site percolation on lattices other than the hypercubic. It also extends to bond percolation on d-dimensional lattices and perhaps more interestingly it extends also to long range bond percolation, as we now explain.

In a translation invariant independent bond percolation model on Z^d, bonds $\{x,y\}$ between pairs of sites in Z^d are independently occupied with probability p_{y-x}. There are two standard ways of choosing a one-parameter family of such models. We may pick a finite set Γ of nonzero z's in Z^d which is invariant under reflection $(z \to -z)$ and consider the models in which $p_z = p$ for each z in Γ where p varies in $[0,1]$ while all other p_x's are held fixed. For example, in the standard model, Γ consists of the nearest neighbors of the origin and all other p_x's are zero. An alternative parametrization is obtained by setting

$$p_x = 1 - \exp(-\beta J_x) \qquad (1.8)$$

where $0 \leq J_x < \infty$ and allowing β to vary in $[0,\infty]$.

In these models, the cluster of the origin $C(0)$ consists of those sites connected to the origin by paths of occupied bonds. P_n, χ and p_c are then all defined exactly as in site percolation; i.e., n refers to the number of <u>sites</u> connected to the origin. We consider χ or P_n as functions either of p or of β; in the latter case β_c is of course defined as $\sup\{\beta : P_\infty(\beta) = 0\}$.

The next theorem is the extension of Theorem 1.1 to bond percolation. It can similarly be regarded as a step toward proving continuity of P_∞ at the critical point for those models in which continuity should hold. On the other hand, there is one class models, namely the one-dimensional models with $J_x = 1/|x|^2$, where $\beta_c < \infty$ [NS] and P_∞ is discontinuous at the critical point [AN2]. For these models, the next theorem implies a new result on the critical exponent γ (assuming it exists), namely $\gamma \geq 2$, which improves the previous result [AN1] that $\gamma \geq 1$. We state the theorem in a form most suitable for that application; its proof, which mimics that of Theorem 1.1, will be given in Section 2 below.

<u>Theorem 1.2</u>.　<u>For a translation invariant independent bond percolation model on</u> Z^d:

<u>if</u> p_c <u>is in</u> $(0,1)$ <u>and</u> $P_\infty(p_c) > 0$, <u>then</u> $\displaystyle\int_0^{p_c-0} (\chi(p))^{1/2} dp = \infty,$ $\quad(1.9)$

<u>if</u> β_c <u>is in</u> $(0,\infty)$ <u>and</u> $P_\infty(\beta_c) > 0$, <u>then</u> $\displaystyle\int_0^{\beta_c-0} (\chi(\beta))^{1/2} d\beta = \infty.$ $\quad(1.10)$

<u>Remark</u>. $\beta_c > 0$ if and only if $\sum\limits_{\text{all } x} J_x < \infty$; p_c may be 0 even if $\sum\limits_{\text{all } x} p_x < \infty$, but $p_c > 0$ if e.g., $\sum\limits_{x \in \Gamma} p_x < 1$ (see e.g., [S] or [AN1]). If the model is not essentially one-dimensional, i.e., if $d > 1$ and the x's for which $p_x > 0$ do not all lie on a one-dimensional line, then $\beta_c < \infty$ and $p_c < 1$. For $d = 1$: if $\lim \inf x^2 J_x > 0$, then $\beta_c < \infty$ and if $\lim \inf x^2 p_x > 1$, then $p_c < 1$ [NS]; if in addition $\lim \sup x^2 J_x < \infty$, then P_∞ is strictly positive at the critical point [AN2].

Theorems 1.1 an 1.2 relate the rate of divergence of $\chi(p)$ as $p \uparrow p_c$ to the positivity of P_∞ at p_c; our next results relate this divergence to the

decay rate of P_n as $n \to \infty$ at p_c. They also relate the same decay rate to the divergence of

$$\chi'(p) = \sum_{n < \infty} nP_n(p) \tag{1.11}$$

as $p \to p_c$. Note that there is at present no analogue of the result of [AB] which would insure, even for nearest neighbor models, that $\chi'(p) < \infty$ for all $p > p_c$, except for $d = 2$ [K2] (but see [CCN] for a partial result for $d > 2$). Indeed, such a result is not true for general long range models since it has recently been shown to fail for (some) one dimensional $1/|x - y|^2$ models [ACCIN].

Nevertheless, if we suppose that a model of interest has $P_n(p_c) \approx n^{-(1+1/\delta)}$ as $n \to \infty$, $\chi(p) \approx |p_c - p|^{-\gamma}$ as $p \to p_c$ and $\chi'(p) \approx |p - p_c|^{-\gamma'}$ as $p \to p_c$, then the next theorem implies that

$$\gamma \geq 2(1 - 1/\delta), \tag{1.12}$$

$$\gamma' \geq 2(1 - 1/\delta). \tag{1.13}$$

Since, it is already known [AB] that $\delta \geq 2$, (1.12) is an improvement over the previous result [AN1] that $\gamma \geq 1$. For $d = 2$ nearest neighbor models, much better results have recently been obtained by Kesten which give equalities among various critical exponents [K3; K4]. There also exist nonrigorous but probably exact values for these $d = 2$ exponents: $\delta = 91/5$ and $\gamma = 43/18$ (see [St] and the references listed there). Other known rigorous inequalities, which concern the exponent b (usually denoted β) "defined" by $P_\infty(p) \approx |p - p_c|^b$ are that $b \leq 1$ [CC] and that $b(\delta - 1) \geq 1$, The latter inequality was derived for Ising models in [N1] as a consequence of a differential inequality related to Burgers' equation; the differential inequality and hence the exponent inequality were extended to percolation models in [AB]. There are also inequalities for the "specific heat" exponent α [AKN].

Our analysis of γ and γ' automatically yields inequalities for the critical exponents of other moments of the finite cluster size distribution. Let us define for real $r \geq 0$,

$$\chi_r = \sum_{n \leq \infty} n^r P_n, \tag{1.14}$$

$$\chi_r' = \sum_{n < \infty} n^r P_n. \tag{1.15}$$

If we suppose that $\chi_r \approx |p_c - p|^{-\gamma_r}$ as $p \to p_c$ and $\chi_r' \approx |p - p_c|^{-\gamma_r'}$ as $p \to p_c$, then the next theorem extends (1.12)-(1.13) to

$$\gamma_r, \quad \gamma'_r \geq 2(r - 1/\delta) \quad \text{for} \quad r > 1/\delta. \tag{1.16}$$

We will use the fact that $\chi_r = \chi'_r$ for $p < p_c$ (since $P_\infty = 0$) and that $\chi' = \chi'_1$ to abbreviate the statement of the theorem. The proof will be given below in Section 3.

Theorem 1.3. For standard site percolation or for any independent translation invariant bond percolation model on z^d: if $0 < p_c < 1$ and for some $\delta > 1$ and $B_1 > 0$,

$$P_n(p_c) \geq B_1 n^{-(1+1/\delta)} \quad \text{for} \quad n \geq 1, \tag{1.17}$$

then for any $r > 1/\delta$ there is some $B_2 > 0$ so that

$$\chi'_r(p) \geq B_2 \cdot |(p - p_c)^2 \cdot \ln(|p - p_c|)|^{-(r-1/\delta)} \tag{1.18}$$

as $p \uparrow p_c$ or as $p \downarrow p_c$.

Remarks. i) A technically different version of (1.12) than the one which follows from (1.18) can be obtained by combining some of the arguments used for Theorem 1.3 with some of those used for Theorems 1.1 and 1.2. For example, we can define critical exponents by

$$\bar{\delta} = \sup\{\delta: \sum_{n<\infty} n^{1/\delta} P_n(p_c) = \infty\}, \tag{1.19}$$

$$\bar{\gamma} = \sup\{\gamma: \int_0^{p_c-0} [\chi(p)]^{1/\gamma} dp = \infty\}, \tag{1.20}$$

and then show that

$$\bar{\gamma} \geq 2(1 - 1/\bar{\delta}). \tag{1.21}$$

Since the distinctions between various versions of (1.12) (or of (1.16)) currently seem to be of limited technical interest, we will not pursue the issue.

ii) Improvements of (1.16) can be obtained by using the fact that $\ln(\chi'_r)$ is convex in r. For example, suppose in addition to (1.17), one assumes that $P_n(p_c) = 0(n^{-(1+1/\delta+\varepsilon)})$ for every $\varepsilon > 0$ so that $\gamma_r = 0$ for $r < 1/\delta$. It follows that

$$\gamma_r \geq \frac{\gamma_1}{(1-1/\delta)} \cdot (r - 1/\delta) \quad \text{for} \quad r > 1. \tag{1.22}$$

iii) For a variant of (1.18) which eliminates the logarithmic factor, see the remarks at the end of Section 3.

2. Proof that $\gamma \geq 2$ when $P_\infty > 0$ at the Critical Point

In this section we mimic the proof of the site model Theorem 1.1 to obtain a proof of the bond model Theorem 1.2. As was noted in the case of Theorem 1.1, the general strategy is similar to that used in [G], [K2; p. 252] and [AKN, Sec. 3]. We will show that if the integral of $\chi^{1/2}$ in (1.9) (or (1.10)) is convergent, then we can obtain a bound on $\sum |dP_n/dp|$ (or $\sum |dP_n/d\beta|$) which implies continuity of $P_\infty = 1 - \sum P_n$ on $[0, p_c]$. We first prove (1.9) and then (1.10).

Proof of (1.9). The analogue of the lattice (site) animal representation (1.3) for these bond models is

$$P_n(p) = \sum_{m, \ell} a_{nm\ell} p^m (1 - p)^\ell \tag{2.1}$$

where $a_{nm\ell} \geq 0$ is independent of p, but depends on the p_z's for z not in Γ. This formula my be obtained by summing over all lattice (bond) animals which connect the origin with $n - 1$ other sites and which contain exactly m occupied Γ-bonds and ℓ vacant (boundary) Γ-bonds (see [AKN] for more details). A Γ-bond is a bond $\{x, x + z\}$ with z in Γ. The sum in (2.1) is finite because $a_{nm\ell} = 0$ unless $n|\Gamma|/2 \leq m + \ell \leq n|\Gamma|$ where $|\Gamma|$ denotes the (finite) number of sites in Γ.

Differentiating (2.1) yields

$$dP_n/dp = \sum_{m, \ell} [m/p - \ell/(1 - p)] a_{nm\ell} p^m (1 - p)^\ell, \tag{2.2a}$$

$$d^2 P_n/dp^2 = \sum_{m, \ell} \{[m/p - \ell/(1 - p)]^2 - [m/p^2 + \ell/(1 - p)^2]\} a_{nm\ell} p^m (1 - p)^\ell. \tag{2.2b}$$

The proof now proceeds essentially as in Theorem 1.1 except that the double sums over n and ℓ become triple sums over n, m and ℓ and the Cauchy-Schwarz inequality is on a triply indexed sequence space. The resulting analogue of the basic site model estimate (1.7) is

$$P_\infty(p_c) \leq \int_{p_c - \epsilon}^{p_c - 0} (p^{-2}(1 - p)^{-1} \sum_{n, m, \ell} m a_{nm\ell} p^m (1 - p)^\ell)^{1/2} dp$$

$$\leq |\Gamma|^{1/2} p^{-1} (1 - p)^{-1/2} \int_{p_c - \epsilon}^{p_c - 0} (\chi(p))^{1/2} dp, \tag{2.3}$$

where the last inequality uses the fact that $a_{nm\ell} = 0$ unless $m \leq n|\Gamma|/2$. (The last inequality in this argument will need to be improved in the proof below of (1.10)). This yields (1.9).　　　　　　　　　　　　　　　　　　　　　□

Proof of (1.10). According to the remark following Theorem 1.2, we may assume that

$$\sum_{\text{all } x} J_x < \infty. \tag{2.4}$$

(J_0 may be set to zero by convention.) Let us denote by N the number of sites in the cluster of the origin, by M_x the number of occupied bonds of the form $\{y, y + x\}$ in the (bond) cluster of the origin and by L_x the analogous number of vacant (boundary) bonds. We write $I(A)$ to denote the indicator function of an event A. The analogoues of (2.2a) and (2.2b) are, by (1.8) and the chain rule,

$$dP_n/d\beta = (d/d\beta)E_\beta(I(N = n)) = E_\beta(I(N = n) \cdot Q), \tag{2.5}$$

$$d^2 P_n/d\beta^2 = E_\beta\left(I(N = n) \cdot Q^2\right) - E_\beta(I(N = n)W), \tag{2.6}$$

where E_β denotes expected value in the model with parameter β,

$$Q = (1/2)\sum_x J_x(1 - p_x)[M_x/p_x - L_x/(1 - p_x)], \tag{2.7}$$

$$W = (1/2)\sum_x (J_x)^2(1 - p_x)p_x^{-2}M_x, \tag{2.8}$$

and the sums in (2.7)-(2.8) are over those x in Z^d with $J_x \neq 0$. The finiteness of the expectations in (2.5)-(2.6) follows from Lemma 2.1. We may also use Lemma 2.1 (and the uniformity for $\beta \leq \beta_0 < \beta_c$ implicit in it) to sum (2.6) over n and exchange the order of summation and differentiation to obtain

$$E(Q^2) = E_\beta(W) \quad \text{for} \quad \beta < \beta_c. \tag{2.9}$$

Proceeding as in the site model estimate (1.7) and the parameter-p bond model estimate (2.3), we obtain

$$P_\infty(\beta_c) \leq \lim_{n \to \infty} \int_{\beta_c - \varepsilon}^{\beta_c - 0} [E_\beta(I(N < n)Q^2]^{1/2} d\beta = \int_{\beta_c - \varepsilon}^{\beta_c - 0} [E_\beta(W)]^{1/2} d\beta. \tag{2.10}$$

Next we define $M = \sum_x M_x$, the total number of occupied bonds in the cluster of the origin, and note that

$$E_\beta(W) \leq (2\beta^2)^{-1}(\sup_{J>0}[(\beta J)^2 e^{-\beta J}/(1 - e^{-\beta J})^2])E_\beta(M) = (2\beta^2)^{-1}E_\beta(M),$$

so that our estimate for $P_\infty(\beta_c)$ becomes

$$P_\infty(\beta_c) \leq \int_{\beta_c-\epsilon}^{\beta_c-0} (2\beta^2)^{-1/2}[E_\beta(M)]^{1/2}d\beta. \tag{2.11}$$

Combining this with (2.12) of the next lemma yields (1.10). □

Lemma 2.1. For any $\beta < \beta_c$, N, M, Q and W have finite moments of all orders. For any $\delta > 0$, there is some $B(\beta)$, which is bounded on compact subsets of β in $(0,\infty)$, so that

$$E_\beta(M) \leq (\delta + \sum_x p_x)E_\beta(N) + B(\beta) \leq (\delta + \beta\sum_x J_x)\chi(\beta) + B(\beta). \tag{2.12}$$

Proof. Since $\chi = E(N) < \infty$ for $\beta < \beta_c$ by [AB], it follows [AN1] that $P_n \doteq \text{Prob}(N = n)$ is exponentially decreasing in n so that

$$E_\beta(e^{rN}) < \infty \quad \text{for} \quad r < r_0(\beta) \quad \text{in} \quad (0,\infty), \quad \beta < \beta_c, \tag{2.13}$$

and hence N has finite moments of all orders. Note that this is uniform for $\beta \leq \beta_0 < \beta_c$ since r_0 is decreasing in β. We will next show that M also has a finite moment generating function (for small r); the same will follow for Q and W by elementary considerations (including use of the bound $L_x \leq 2N$).

Let us now consider a sequence of i.i.d. random variables U_1, U_2, U_3, ... with each U_j having the same distribution as the number of occupied bonds touching the origin (or any other site) so that

$$E(\exp(rU_j)) = \phi(r) \doteq \prod_x (1 - p_x + p_x e^r), \tag{2.14}$$

$$E(U_j) = \mu \doteq \sum_x p_x. \tag{2.15}$$

Now by a variant of the algorithmic construction of [PS; KS] (see also [AKN, Sec. 3]), in which the occupation status of the bonds touching the origin are first determined, followed (according to some initially prescribed ordering on z^d) by those undetermined bonds touching the sites connected to the origin by a single occupied bond, followed by ..., one can easily obtain the bound

$$M \leq \sum_{j=1}^{N} U_j. \tag{2.16}$$

(The algorithmic construction has M, N and the U_j's all defined on the same probability space.)

Even though the U_j's are <u>not</u> independent of N, (2.16) still yields the following bound, where we denote $U_1 + \ldots + U_n$ by S_n:

$$E_\beta(e^{rM}) \le E_\beta(e^{r(\mu+\delta)N}) + \sum_{n=1}^{\infty} E_\beta(e^{rS_n} I(S_n \ge (\mu + \delta)n)). \qquad (2.17)$$

The first term on the RHS of (2.17) is finite for small r $(r < r_1(\beta))$ by (2.13). The second term may be bounded, using the Cauchy-Schwarz inequality, by

$$\sum_{n=1}^{\infty} [E_\beta(e^{2rS_n})]^{1/2}[P(S_n \ge (\mu + \delta)n)]^{1/2} \le \sum_{n=1}^{\infty} (\phi(2r))^{n/2}e^{-nG/2}, \qquad (2.18)$$

where the bound on $P(S_n \ge (\mu + \delta)n)$ is based on a standard large deviation estimate which yields $G = G_\beta(\delta) > 0$ by means of a Legendre transform of $\ln(\phi(r))$. For fixed $\delta > 0$, the RHS of (2.18) will be finite for r sufficiently small - indeed for $r < r_2(\beta,\delta)$ where r_2 is bounded away from zero on compact subsets of β in $(0,\infty)$.

Finally to obtain (2.12), we may obtain from (2.16), analogously to (2.17),

$$E_\beta(M) \le (\mu + \delta)E_\beta(N) + \sum_{n=1}^{\infty} E_\beta(S_n I(S_n \ge (\mu + \delta)n))$$

$$\le (\mu + \delta)E_\beta(N) + \sum_{n=1}^{\infty} [E_\beta(S_n^2)]^{1/2}e^{-nG/2}. \qquad (2.19)$$

Since G is bounded away from zero on compact subsets of β in $(0,\infty)$ and since

$$E_\beta(S_n^2) = (n\mu)^2 + nE((U_j - \mu)^2) = (n\mu)^2 + n\sum_x p_x(1 - p_x),$$

we easily obtain (2.12) from (2.19).

3. Derivation of γ_r, $\gamma_r' \ge 2(r - 1/\delta)$

In this section we prove Theorem 1.3. The proof, like that of Theorems 1.1 and 1.2, is based on the lattice animal expansions (1.3) and (2.1). We first give the proof for site percolation, then the proof for bond percolation which has some small extra complications.

<u>Proof of Theorem 1.3 for Site Percolation</u>. We define $P_{n\ell}(p)$ $= a_{n\ell}p^n(1 - p)^\ell$. Then, letting $\varepsilon = p - p_c$, one has

$$P_{n\ell}(p) = (1 + \varepsilon/p_c)^n(1 - \varepsilon/(1 - p_c))^\ell P_{n\ell}(p_c) \doteq \psi_{n\ell}(\varepsilon)P_{n\ell}(p_c). \qquad (3.1)$$

For some constant K_2 (to be determined) we define, for $n > 1$,

$$\Sigma_a = \Sigma_{\ell: |n/p_c - \ell/(1-p_c)| \leq K_2[n \ln(n)]^{1/2}/(1-p_c)} \tag{3.2}$$

and Σ_b to be the complementary sum over the remaining values of ℓ.
We also define

$$\psi(n,\epsilon) = \begin{cases} (1 + \epsilon/p_c)^n (1 - \epsilon/(1 - p_c))^{(1-p_c)n/p_c + K_2[n \ln(n)]^{1/2}} & , \text{ for } \epsilon > 0 \\ \\ (1 + \epsilon/p_c)^n (1 - \epsilon/(1 - p_c))^{(1-p_c)n/p_c - K_2[n \ln(n)]^{1/2}} & , \text{ for } \epsilon < 0 \end{cases}.$$

Then

$$P_n(p) = \sum_\ell P_{n\ell}(p) \geq \sum_a P_{n\ell}(p) = \sum_a \psi_{n\ell}(\epsilon) P_{n\ell}(p_c)$$

$$\geq \psi(n,\epsilon) \sum_a P_{n\ell}(p_c) = \psi(n,\epsilon) P_n(p_c) - \psi(n,\epsilon) \sum_b P_{n\ell}(p_c). \tag{3.3}$$

Given any $K_1 < \infty$, K_2 can be chosen sufficiently large so that (for $n > 1$)

$$\sum_b P_{n\ell}(p_c) \leq n^{-K_1}. \tag{3.4}$$

(For more details on this large deviation result, see [K2, Lemma 5.1] or [AKN].)
In particular, we may take $K_1 = r + 2$ and the corresponding K_2, so that (3.3)
yields

$$\chi'_r(p) \geq \sum_{1 < n < \infty} \psi(n,\epsilon) n^r P_n(p_c) - \sum_{1 < n < \infty} \psi(n,\epsilon) n^{-2}$$

$$\geq \sum_{1 < n < \infty} \psi(n,\epsilon) (n^r P_n(p_c) - n^{-2}). \tag{3.5}$$

In order to obtain the conclusion (1.18) of the theorem from (3.5) it suffices to
show that

$$\lim_{\epsilon \to 0} \sum_{1 < n < \infty} \psi(n,\epsilon) n^{\theta-1} = \sum_{1 < n < \infty} n^{\theta-1} < \infty \text{ for } \theta < 0, \tag{3.6}$$

while

$$\sum_{1 < n < \infty} \psi(n,\epsilon) n^r P_n(p_c) \geq B|\epsilon^2 \ln(|\epsilon|)|^{-\theta} \text{ as } \epsilon \to 0 \text{ for } \theta > 0. \tag{3.7}$$

By considering the Taylor series in powers of ε for $\ln(1 + \varepsilon/p_c)$ and $\ln(1 - \varepsilon/(1 - p_c))$, one finds that

$$\psi(n,\varepsilon) = \exp\left[-\left(2p_c^2(1 - p_c)\right)^{-1}\varepsilon^2 n - K_2(1 - p_c)^{-1}|\varepsilon|[n \ln(n)]^{1/2}\right.$$

$$\left. + O\left(|\varepsilon|^3 n + \varepsilon^2[n \ln(n)]^{1/2}\right)\right]. \tag{3.8}$$

We see that for small ε, $\psi(n,\varepsilon) \leq 1$ for all n greater than some n_0 (independent of ε). Since $\psi(n,\varepsilon) \to 1$ as $\varepsilon \to 0$ for fixed n, (3.6) follows by dominated convergence. To prove (3.7), we use (3.8) to see that as $\varepsilon \to 0$,

$$\sum_{1 < n < \infty} \psi(n,\varepsilon)n^r P_n(p_c) \geq \sum_{1 < n < [\varepsilon^2 \ln(1/\varepsilon)]^{-1}} \psi(n,\varepsilon)n^r P_n(p_c)$$

$$\geq B' \sum_{1 < n < [\varepsilon^2 \ln(1/\varepsilon)]^{-1}} n^r P_n(p_c). \tag{3.9}$$

The hypothesis (1.17) about $P_n(p_c)$ then easily yields (3.7). $\qquad\square$

Remark. It is clear from the last portion of the proof (which was suggested by Harry Kesten), that the hypothesis (1.17) can be weakened to:

$$\sum_{n \leq N} n^r P_n(p_c) \geq B_1' N^{r-1/\delta} \quad \text{as} \quad N \to \infty. \tag{3.10}$$

Proof of Theorem 1.3 for Bond Percolation. The proof for bond percolation is essentially the same as for site percolation, but with some differences.

Defining $P_{nm\ell}(p) = a_{nm\ell} p^m (1 - p)^\ell$, we have

$$P_{nm\ell}(p) = \psi_{m\ell}(\varepsilon) P_{nm\ell}(p_c) \tag{3.11}$$

as a replacement for the site model identity (3.1). We replace (3.2) with

$$\sum_a^* = \sum_{m,\ell : |m/p_c - \ell/(1-p_c)| \leq K_2[n \ln(n)]^{1/2}/(1-p_c)} \tag{3.12}$$

and let $\sum_b^*$ be the complementary sum. The basic inequality between $\psi_{n\ell}(\varepsilon)$ and $\psi(n,\varepsilon)$ used in (3.3) is replaced here by the following, which is valid for m, ℓ in the summation region of $\sum_a^*$:

$$\psi_{m\ell}(\varepsilon) \geq \psi^*(n,\varepsilon) \doteq \left[(1 + \varepsilon/p_c)(1 - \varepsilon/(1 - p_c))\right]^{(1-p_c)/p_c]|r|n/2}$$

$$\cdot (1 - \varepsilon/(1 - p_c))^{\text{sgn}(\varepsilon) \cdot K_2(n \ln(n))^{1/2}} \tag{3.13}$$

where $\text{sgn}(\epsilon) = +1$ for $\epsilon > 0$ and -1 for $\epsilon < 0$. This inequality is easily obtained by using the facts that the quantity in square brackets on the RHS of (3.13) is not greatger than one (which can be verified by elementary arguments), while $m \leq |\Gamma| n/2$.

As a consequence of (3.13), (3.3) is replaced by

$$P_n(p) \geq \psi^*(n,\epsilon) P_n(p_c) - \psi^*(n,\epsilon) \sum_b^* P_{nm\ell}(p_c). \qquad (3.14)$$

The analogue of (3.4) is (see [AKN, Lemma 3.2])

$$\sum_b^* P_{nm\ell}(p_c) \leq n^{-K_1}. \qquad (3.15)$$

Since $\psi^*(n,\epsilon)$ and $\psi(n,\epsilon)$ differ only by scale factors in the exponent, the remainder of the proof is basically unchanged from that already given for site percolation. □

Remarks. i) A variant of the above proof of Theorem 1.3 can be used to eliminate the logarithmic factor in the critical exponent inequality (1.18) at the cost of combining the $p < p_c$ and $p > p_c$ asymptotics. In the case of site percolation, for example, the lattice animal expansion (see (1.3) and (3.1)) yields

$$P_n(p_c + \epsilon) = (1 + \epsilon/p_c)^n \phi_n(\ln[1 - \epsilon/(1 - p_c)]) P_n(p_c)$$

where $\phi_n(r)$ is the moment generating function at $p = p_c$ for the surface size of clusters of size n:

$$\phi_n(r) = \sum_\ell e^{r\ell} P_{n\ell}(p_c)/P_n(p_c).$$

By the log-convexity of such a generating function,

$$\phi_n(r) \geq \exp(\phi_n'(0) \cdot r)$$

where $\phi_n'(0)$ is the mean surface size

$$\phi_n'(0) = \sum_\ell \ell P_{n\ell}(p_c)/P_n(p_c) \doteq (1 - p_c)n/p_c + b_n.$$

Combining these expressions yields

$$P_n(p_c + \epsilon) \geq \exp[-(2p_c^2(1 - p_c))^{-1}\epsilon^2 n + O(|\epsilon|^3 n)] \cdot [1 - \epsilon/(1 - p_c)]^{b_n} P_n(p_c).$$

Without knowing the sign of b_n or using any estimates on its magnitude, we may still write

$$P_n(p_c + \epsilon) + P_n(p_c - \epsilon)$$
$$\geq \exp[-(2p_c^2(1 - p_c))^{-1}\epsilon^2 n + o(|\epsilon|^3 n)]\, P_n(p_c). \tag{3.16}$$

This easily implies, under the assumptions of Theorem 1.3, that

$$\chi_r'(p_c - \epsilon) + \chi_r'(p_c + \epsilon) \geq B_2|\epsilon|^{-2(r-1/\delta)} \tag{3.17}$$

without the logarithmic factor of (1.18).

ii) A second variant of Theorem 1.3 can be obtained by altering the argument in the preceding remark to use the log-convexity inequality,

$$[\phi_n(-r)\phi_n(r)]^{1/2} \geq \phi_n(0) = 1.$$

This yields

$$[P_n(p_c - \epsilon)\, P_n(p_c + \tfrac{\epsilon}{1+\epsilon/(1-p_c)})]^{1/2} \geq \exp[-(2p_c^2(1 - p_c))^{-1}\epsilon^2 n + o(|\epsilon|^3 n)]\, P_n(p_c)$$

which implies (after multiplying by n^r, summing over n and using the Cauchy-Schwartz inequality) that

$$[\chi_r'(p_c - \epsilon) \cdot \chi_r'(p_c + \epsilon + o(\epsilon^2))]^{1/2} \geq B_2|\epsilon|^{-2(r-1/\delta)}. \tag{3.18}$$

Note that to obtain (3.17) or (3.18), say for $r = 1$, one may weaken the hypothesis (1.17) on $P_n(p_c)$ to

$$\sum_{n<\infty} e^{-nh} n P_n(p_c) \geq B_1' h^{1/\delta-1} \quad \text{as} \quad h \to 0. \tag{3.19}$$

Acknowledgments

The author thanks the Institute for Mathematics and its Applications for its hospitality. He also thanks Harry Kesten for his comments on an earlier draft of this paper. The research reported here was supported in part by NSF Grants MCS-8019384 and DMS-8514834.

References

[AB]   Aizenman, M. and Barsky, D. J., Proof of the sharpness of the phase transition in translation invariant percolation models, Rutgers University preprint, in preparation.

[ACCIN] Aizenman, M., Chayes, J. T., Chayes, L., Imbrie, J. and Newman, C. M., An intermediate phase with slow decay of correlations in one-dimensional $1/|x - y|^2$ percolation, Ising and Potts models, in preparation.

[AKN]   Aizenman, M., Kesten, H. and Newman, C. M., Uniqueness of the infinite cluster and continuity of connectivity functions for short and long range percolation, Rutgers University/Cornell University/University of Arizona preprint, in preparation.

[AN1]   Aizenman, M. and Newman, C. M., Tree graph inequalities and critical behavior in percolation models, J. Stat. Phys. 36 (1984), 107-143.

[AN2]   Aizenman, M. and Newman, C. M., Discontinuity of the percolation density in one-dimensional $1/|x - y|^2$ percolation models, Rutgers University/University of Arizona preprint, 1986.

[BK]   van den Berg, J. and Keane, M., On the continuity of the percolation probability function, Contemporary Mathematics 26 (1984), 61-65.

[CC]   Chayes, J. T. and Chayes, L., An inequality for the infinite cluster density in Bernoulli percolation, Phys. Rev. Lett. 56 (1986), 1619-1622.

[CCN]   Chayes, J. T., Chayes, L. and Newman, C. M., Bernoulli percolation above threshold: an invasion percolation analysis, Ann. Prob., to appear.

[G]   Grimmett, G. R., On the differentiability of the number of clusters per vertex in the percolation model, J. London Math. Soc. (2) 23 (1981), 372-384.

[H]   Hammersley, J. M., Percolation processes. Lower bounds for the critical probability, Ann Math. Statist. 28 (1957), 790-795.

[K1]   Kesten, H., The critical probability of bond percolation on the square lattice equals 1/2, Comm. Math. Phys. 74 (1980), 41-59.

[K2]   Kesten, H., Percolation Theory for Mathematicians, Birkhäuser, 1982.

[K3]   Kesten, H., A scaling relation at criticality for 2D-percolation, in Proceedings of the IMA workshop on percolation theory and ergodic theory of infinite particle systems (H. Kesten, Ed.), IMA Volumes in Mathematics and its Applications, Springer-Verlag, to appear.

[K4]   Kesten, H., Scaling relations for 2D-percolation, Institute for Mathematics and its Applications (Minneapolis) preprint, 1986.

[KS]   Klein, S. T. and Shamir, E., An algorithmic method for studying percolation clusters, Stanford Univ. Dept. of Computer Science, Report No. STAN-CS-82-933 (1982).

[N1]   Newman, C. M. Shock waves and mean field bounds. Concavity and analyticity of the magnetization at low temperature, University of Arizona preprint (1981), published as an appendix to Percolation theory: a selective survey of rigorous results, to appear in Proceedings of the SIAM Workshop on Multiphase Flow (G. Papanicolaou, Ed.).

[N2]   Newman, C. M., Some critical exponent inequalities for percolation, J. Stat. Phys., to appear.

[NS]   Newman, C. M. and Schulman, L. S., One-dimensional $1/|j - i|^s$ percolation models: the existence of a transition for $s \leq 2$, Comm. Math. Phys. 104 (1986), 547-571.

[PS]   Pike, R. and Stanley, H. E., Order propagation near the percolation threshold, J. Phys. A 14 (1981), L169-L177.

[R]   Russo, L., On the critical percolation probabilities, Z. Wahrsch. verw. Geb. 56 (1981), 229-237.

[S]   Schulman, L. S., Long range percolation in one dimension, J. Phys. A Lett. 16 (1983), L639-L641.

[St]   Stauffer, D., Scaling properties of percolation clusters, in Disordered Systems and Localization (C. Castellani, C. DiCastro and L. Peliti, Eds.), Springer, 1981, 9-25.

A New Look at Contact Processes in Several Dimensions

Roberto H. Schonmann
Rutgers University
and
São Paulo University - Brazil

Summary

We prove that if the complete convergence theorem holds for the basic contact process in dimension d with infection parameter λ larger than the critical value in this dimension, then the same theorem holds for this process in any dimension $d' \geqslant d$ for any $\lambda' \geqslant \lambda$.

Contact processes were first studied by Harris (1974). They can be considered as very idealized models for the spread of an infection and are also closely related to oriented percolation (see for instance [2] or [7]).

One can think of individuals which are located at all the sites of a lattice $\mathbb{Z}^d$ (one individual at each site) and each one can be either infected or healthy. The infected individuals recover at a constant rate, which can be chosen as 1; and the healthy individuals become infected at a rate which is proportional to the number of infected neighbors.

More precisely the basic contact process in d dimensions with infection parameter $\lambda > 0$ is the continuous time Markov and Feller process on $P(\mathbb{Z}^d)$ = set of subsets of $\mathbb{Z}^d$ (the state of the process is the set of infected individuals) whose evolution is defined by the rates

$$\xi \to \xi \setminus \{x\} \quad \text{if } x \in \xi \text{ at rate } 1$$
$$\xi \to \xi \cup \{x\} \quad \text{if } x \notin \xi \text{ at rate } \lambda \cdot |\xi \cap \{y \in \mathbb{Z}^d : ||y-x|| = 1\}|$$

where $|\eta|$ is the cardinality of η and $||\cdot||$ is the Euclidean norm of $\mathbb{Z}^d$. For proofs of the existence and uniqueness of the process defined by these rates see for instance [6] or [11].

We will use the notation $(\xi_t^\eta, t \geqslant 0)$ for the process starting from the configuration $\eta \subset \mathbb{Z}^d$. The corresponding semigroup acting on measures on $P(\mathbb{Z}^d)$ will be denoted by $S(t)$.

One of the basic problems for interacting particle systems like the contact processes is the determination of the set of invariant measures and their domains of attraction. Before presenting our contribution let us review some fundamental facts about the model. The reader is referred to [6], [7] and [11] for the proofs and more information.

(1) ϕ is a trap. Therefore δ_ϕ, the unit mass on ϕ, is invariant for the process.

(2) $\delta_{\mathbb{Z}^d} S(t)$ converges weakly to a measure $\nu_{d,\lambda}$ as $t \to \infty$.

(3) The process is ergodic if and only if $\delta_\phi = \nu_{d,\lambda}$.

(4) For any dimension d there exists a critical value $\lambda_d \in (0,\infty)$ such that:

if $\lambda < \lambda_d$, the system is ergodic,

if $\lambda > \lambda_d$, the system is not ergodic; δ_ϕ and $\nu_{d,\lambda}$ are extremal invariant measures.

λ_d does not increase as d increases.

Remark: It is still an open problem for any dimension whether the system is ergodic or not for $\lambda = \lambda_d$.

For $\lambda > \lambda_d$ one knows therefore two extremal invariant measures and the next question is whether there are others. In $d=1$ the answer is negative. The first proof of this fact (for every λ) was given by Liggett (1978). His theorem applies to a large class of one dimensional systems, but unfortunately it does not give information about domains of attraction and it seems difficult to generalize this technique to larger dimensions. Another approach is due to Griffeath (1978); it relies on proving that for any measure μ on $P(\mathbb{Z}^d)$,

$$\mu S(t) \to \gamma_\mu \, \delta_\phi + (1-\gamma_\mu) \, \nu_{d,\lambda} \qquad (5a)$$

weakly as $t \to \infty$, where

$$\gamma_\mu = \int P(\xi_t^\eta = \phi \text{ for some } t) \ d\mu(\eta) \tag{5b}.$$

Remark: (5) is known as complete convergence theorem.

Griffeath (1978) proved (5) for $d = 1$ and $\lambda > \lambda_1^*$, where λ_1^* is a constant larger than λ_1 (in fact it is the critical value for the one dimensional contact process with infection only in one direction). Durrett (1980) extended this result for $\lambda > \lambda_1$. One should observe that (5) is trivially true (and non informative) if $\lambda < \lambda_1$. Finally for $\lambda = \lambda_1$ (5) is also true, as observed in [3], by a remark on p. 1013 of [2]. From (5) it follows not only that the extremal invariant measures are δ_ϕ and $\nu_{d,\lambda}$, but also that for any initial measure μ, $\mu S(t)$ converges weakly. This is false for contact processes in one dimension for which the infection has not the right-left symmetry [6], [13]. The coefficient γ_μ can also be computed in some cases; for instance if μ is concentrated on configurations with infinite cardinality, then $\gamma_\mu = 0$.

In $d \geq 2$ much less is known. Durrett and Griffeath (1982) proved that for $\lambda > \lambda_1$ the basic contact process in any dimension d satisfies (5).

In Schonmann (1986a) a simpler proof of the same statement was given. Here we slightly simplify our argument and verify that in fact it can be used to prove a more general statement:

Theorem: If (5) holds in dimension d for some $\lambda > \lambda_d$, then it holds also in dimension $d' \geq d$ for any $\lambda' \geq \lambda$.

We will prove this statement in two steps:

a. (5) holds in dimension d for any $\lambda' > \lambda$.

b. (5) holds in dimension $d' = d + 1$ for infection parameter λ.

The main tool that we use is a lemma in p. 383 of [5].

Lemma (Griffeath) - Let (ξ_t^A) and $(\bar\xi_t^B)$ be two independent versions of the basic contact process in dimension d with the same infection parameter λ, starting from A and B respectively.

Define

$$\tau^A = \inf \{t > 0 \; : \xi^A_t = \phi\},$$

$$\overline{\tau}^B = \inf \{t > 0 \; : \overline{\xi}^B_t = \phi\}$$

$$(\inf \; \phi = \infty).$$

Then (5) is equivalent to

$$P(\tau^A > t, \; \overline{\tau}^B > t, \; \xi^A_t \; \overline{\xi}^B_t = \phi) \to 0 \tag{6}$$

as $t \to \infty$, for any $A, B \subset Z^d$ such that $|A| < \infty$ and $|B| < \infty$.
Another tool is a consequence of lemma 9.14 in [9]:

$$\text{if} \quad \lambda > \lambda_d, \quad \sup_{|A| = n} \; P(\tau^A < \infty) \to 0 \quad \text{as} \quad n \to \infty \tag{7}$$

The proofs of (a) and (b) are similar to the proof of theorem 1 in [12] so
we only sketch them.

Sketch of the proof of (a):

For fixed n define the stopping time

$$\theta_n = \inf \{t > 0 \; : \; |\xi^A_t| > n, \; |\overline{\xi}^B_t| > n\}$$

Then, by elementary arguments, $P(\tau^A > t, \; \overline{\tau}^B > t, \theta_n > t) \to 0$ as $t \to \infty$, so
that θ_n probably occurs before a fixed time t_o suitably chosen.
From θ_n on modify the processes (ξ^A_t) and $(\overline{\xi}^B_t)$ by reducing the
infection parameter from λ' to λ. Since $\lambda > \lambda_d$ it follows from (7) that if
n is large enough the modified processes will probably survive. And from
Griffeath's lemma, if we wait long enough after θ_n the modified processes will
probably intercept each other (here we used the fact that (6) is not only sufficient but also necessary for (5) to hold) and this is clearly also true then for
the original processes.

Sketch of the proof of b:

Define $Q : \mathbb{R}^{d+1} \to \mathbb{R}$ by $Q(x_1,\ldots,x_{d+1}) = x_1$. The next definition is different from the corresponding one in $\lceil 12 \rceil$ in order to simplify the proof. For fixed n let

$$K_n = \{(C,D) \subset \mathbb{Z}^{d+1} \times \mathbb{Z}^{d+1} : \exists (c_1,\ldots,c_n) \subset C, \exists (d_1,\ldots,d_n) \subset D \text{ s.t.}$$

$$Q(c_1) = \ldots = Q(c_n) < Q(d_1) = \ldots = Q(d_n)\} .$$

Given $c_1,\ldots, c_n, d_1,\ldots, d_n$ in the conditions above, and $\ell \in \mathbb{N}$, define the following d dimensional structure imbedded in $\mathbb{R}^{d+1}$

$$H = \{(x_1,\ldots,x_d) \in \mathbb{R}^{d+1} : (x_1 = Q(c_1), x_2 < \ell) \text{ or}$$

$$(Q(c_1) < x_1 < Q(d_1), x_2 = \ell)\}, \text{ or } (x_1 = Q(d_1), x_2 < \ell)\} ,$$

and the corresponding discrete structure $L = H \cap \mathbb{Z}^{d+1}$. It is clear that if ℓ is large enough

$$c_i, d_i \in L \text{ for } i = i = 1,\ldots,n \tag{8}$$

From this point on the reader is invited to verify that the proof of theorem 1 of [12] applies. We modified the definition of K_n in order to have H and L instead of the families H_i and L_i. The corresponding modification in the proof is straightforward and one should use (7) as in the proof of (a) above.

The basic argument is that on $\{\tau^A = \bar{\tau}^B = \infty\}$ either $(\xi_t^A, \bar{\xi}_t^B)$ or $(\bar{\xi}_t^B, \xi_t^A)$ hits K_n after an almost surely finite time. ℓ can be chosen then so large that (8) probably holds. Modify the processes (ξ_t^A) and $(\bar{\xi}_t^B)$ after this hitting time by letting them interact only on L (i.e., disregard the rest of $\mathbb{R}^{d+1}$) and use (7), Griffeath's lemma and the hyposthesis of the theorem as in the proof of (a).

Acknowledgement

We thank Joel Lebowitz for the warm hospitality with which he received us in Rutgers, where this work was done. This work was partially supported by CNPq (Brazil) and NSF (USA). We thank also the organizers of the meeting and the IMA staff for their help and warm hospitality.

References

1. R. Durrett (1980) - On the growth of the one dimensional contact processes. Ann. Probab. 8, 890-907

2. R. Durrett (1984) - Oriented percolation in two dimensions. Ann. Probab. 12, 999-1040

3. R. Durrett (1985) - Stochastic growth models: ten problems for the 80's (and 90's). Contemporary Mathematics 41, 87-99

4. R. Durrett, D. Griffeath (1982) - Contact processes in several dimensions. Z. Wahrsch. Verw. Gebiete 59, 535-552

5. D. Griffeath (1978) - Limit theorems for nonergodic set valued Markov processes. Ann. Probab. 6, 379-387.

6. D. Griffeath (1979) - Additive and Cancelative Interacting Particle Systems. Springer Lecture Notes in Mathematics, vol. 724.

7. D. Griffeath (1981) - The basic contact process. Stochastic Process Appl. 11, 151-186

8. T.E. Harris (1974) - Contact interactions on a lattice. Ann. Probab. 2, 969-988

9. T.E. Harris (1976) - On a class of set-valued Markov processes. Ann. Probab. 4, 175-194

10. T.M. Liggett (1978) - Attractive nearest-neighbor spin systems on the integers. Ann. Probab. 6, 629-636

11. T.M. Liggett (1985) - Interacting Particle Systems. Springer Verlag

12. R.H. Schonmann (1986a) - A new proof of the complete convergence theorem for contact processes in several dimensions with large infection parameter. Ann. Probab. (to appear)

13. R.H. Schonmann (1986b) - The asymmetric contact process. J. Statistical Physics (to appear)

FRACTAL AND MULTIFRACTAL APPROACHES TO PERCOLATION: SOME EXACT AND NOT-SO-EXACT RESULTS

H. Eugene Stanley

Center for Polymer Studies and Department of Physics
Boston University, Boston, MA 02215 USA

1. SCALE INVARIANCE AND POWER LAWS

I feel flattered to be included in such an impressive group of talks by mathematically gifted people. The last **rigorous** proof that was completely "my own" was in 1968 (Stanley 1968a) and this is now 1986. Hence in accepting Professor Kesten's kind invitation I assume that my role is to unveil a set of percolation results that—unlike those in "the book" (Kesten 1981)— has not yet reached the stage of completely rigorous proofs. Accordingly, I shall focus mainly on relatively recent developments in percolation theory, some of which almost certainly may be amenable to exact analysis.

Where appropriate, I'll use the language of fractal geometry. Fractal concepts were first explicitly introduced into percolation theory 10 years ago (Stanley et al 1976, Stanley 1977) when the various percolation exponents were related explicitly to fractal dimensions, thus providing an easily visualizable understanding of these rather abstract quantities. Consider, for the sake of specificity, an infinite square lattice exactly at the site percolation threshold $p = p_c \cong 0.593$.[*] This system has no built-in scale: there occur finite clusters of all sizes and shapes. Suppose we observe the system from a slowly rising balloon. As soon as the balloon is sufficiently high that we can no longer detect the lattice constant a_o, there is no possible measurement that can determine the dimensions of a cluster. There occur finite clusters of all sizes and shapes!!!

"Clusters of all sizes and shapes" is a truism, but all sizes do **not** occur democratically. Suppose we ask how many clusters will we find that are twice as big as the size we are looking at. The answer is neither "the same number" (democracy), nor "exponentially few" (totalitarianism). Rather it is $(1/2)^\tau$ where, as we shall see below, $\tau = 187/91 = 2.0549$. In fact, the exponent τ has *exactly the same value* for all two-dimensional systems, whether they be lattice or continuum, **even if the elements interact** (Geiger and Stanley, 1982b). We say that τ is "universal" in the sense that τ is independent of the details of the two-dimensional system under consideration. Hence the reduction factor $(1/2)^\tau = 0.24066$ is the same, regardless of what size we start with. Thus for *any* size of the cluster we examine, we know that the relative population of clusters twice the size will be smaller by exactly a factor 0.24066. Away from the percolation threshold $p = p_c$, we find exponential decay: the reduction factor becomes smaller and smaller as the cluster size under consideration increases.

---

[*] In site percolation, sites are randomly occupied with probability p. In bond percolation, bonds are randomly occupied with probability p.

Of course, there is nothing special about the number 2. In general, for any positive number λ, the number of s-site clusters (per lattice site) $n(s)$ obeys the functional equation

$$n(\lambda s) \sim \lambda^{-\tau} n(s). \tag{1.1a}$$

The tilde symbol is conventionally used to indicate that (1a) holds only asymptotically as $s \to \infty$. Algebraic equations (like $ax^2 + bx + c = 0$) have as their solution **numbers**, while functional equations have as their solution **functional forms**. The solution of (1a) is obtained by noting that if (1a) holds for any positive value of λ, it certainly must hold for $\lambda = 1/s$. Accordingly, the solution of (1a) is the "power law"

$$n(s) \sim s^{-\tau}. \tag{1.1b}$$

Since (1.1a) holds asymptotically, so does (1.1b). Because of (1.1b), double logarithmic graph paper is the only paper that need be stocked by anyone with percolation disease.

There are many related diseases that require of their patients no more than double logarithmic paper. Any disease for which the underlying physics is governed by a "scale symmetric" equation of the form of (1.1a) is one for which double-log paper suffices. A common question is "when do you know that you will find scale symmetry in nature?" This question was debated recently at the annual meeting of the Materials Research Society, motivated in part by the desire of materials scientists to be able to predict which sort of materials might exhibit scale-symmetric behavior. The point of view that I advocated is that scale symmetry is "the" symmetry obeyed by random phenomena.

The simplest random system is a random walk on a square lattice. Here the time for a drunk to sweep out a characteristic rms length scale R of visited bars obeys a functional equation analogous to (1.1a),

$$t(\lambda R) = \lambda^2 t(R). \tag{1.2a}$$

On setting $\lambda = 1/R$, we obtain the more familiar relation

$$t = R^2. \tag{1.2b}$$

This is one of those rare occasions in statistical physics where the governing functional equations hold for *all* values of λ, not just asymptotically as $R \to \infty$. For this reason, the customary tilde signs are replaced by equalities in (1.2a) and (1.2b). If our drunk is capable of taking, say, 10 steps in a straight line before randomly stepping North, East, South, or West, then on *short* time scales (1.2a) and (1.2b) will not hold since $t = R$. However *asymptotically* they will. One could imagine the reverse: the drunk is purely random for, say, 10^4 steps (30 steps/minute × 300 minutes) but asymptotically he (or she) becomes sober with $t \sim R$. Asymptotically as $t \to \infty$ (1.2a) and (1.2b) fail, but there are nonetheless 4 decades on $\log_{10}$ paper for which t is linear in R, with slope 2.

This second possibility illustrates a typical feature of scale symmetry: phenomena obey functional equations like (1.2a) over only some limited range of the independent variable. Although this feature may not be very appealing to a mathematician, it is immensely valuable to a scientist since it permits random phenomena to be *quantitatively* characterized—over this range. A recent example is shown in Fig. 1a, where a low viscosity liquid (water) is pushed under pressure into a liquid of much higher viscosity. At small times (or on short length scales), the structure of the "water viscous finger" is scale symmetric. At large times (or on large length scales), the structure feels the walls of the confining cell and becomes asymptotically one-dimensional. Recently Hentschel and Deutsch (1986) have discussed the details of the "crossover" from the $d = 2$ behavior to the asymptotic $d = 1$ behavior.

A radial cell of huge diameter has the advantage that the walls are not felt as soon, and the resulting scale symmetric pattern can be recognized more easily. Thus the pattern of Fig. 1b has the same quantitative scale invariance as the pattern of Fig. 1a. We say that both patterns have a fractal dimension d_f that is identical.

In order to define d_f, let us return to the paradigm random scale invariant system, percolation. In the limit as $p \to p_c^-$, one of the many clusters becomes infinite in extent. This cluster has been termed (Stanley 1977) the incipient infinite cluster (IIC), although its precise mathematical definition is not unanimously agreed on (L. Chayes, private communication). Suppose we ask about the "local density" ρ of the IIC. One way to answer this question is choose a point on the IIC and call it the origin. Then we can construct a family of square 'sandboxes' of edge L (in units of the lattice constant), centered on the origin. For each sandbox, we count the number of cluster points (or 'mass') and divide by the area L^2 of the sandbox. For dimension d,

$$\rho(L) \sim M(L)/L^d. \tag{1.3a}$$

If $L = 1$, we find mass 1 and hence $\rho = 1$. Statistically, the density of the IIC will *decrease* monotonically with L. The fractal dimension d_f measures the rate of this decrease:

$$\rho \sim L^{d_f - d}. \tag{1.3b}$$

The exponent $d - d_f$ is termed the co-dimension.

Geometrically speaking, the co-dimension tells us how the fraction of space "carved out" by a voracious fractal scales with L. Thus we can understand the trivial theorem that if two fractals with dimensions d'_f and d''_f intersect in a set of dimension $d_\cap$, then

$$d - d_\cap = (d - d'_f) + (d - d''_f). \tag{1.4}$$

The fraction of space $d - d_\cap$ carved out by *both* fractals is the product of the fraction for each, so the exponents add.

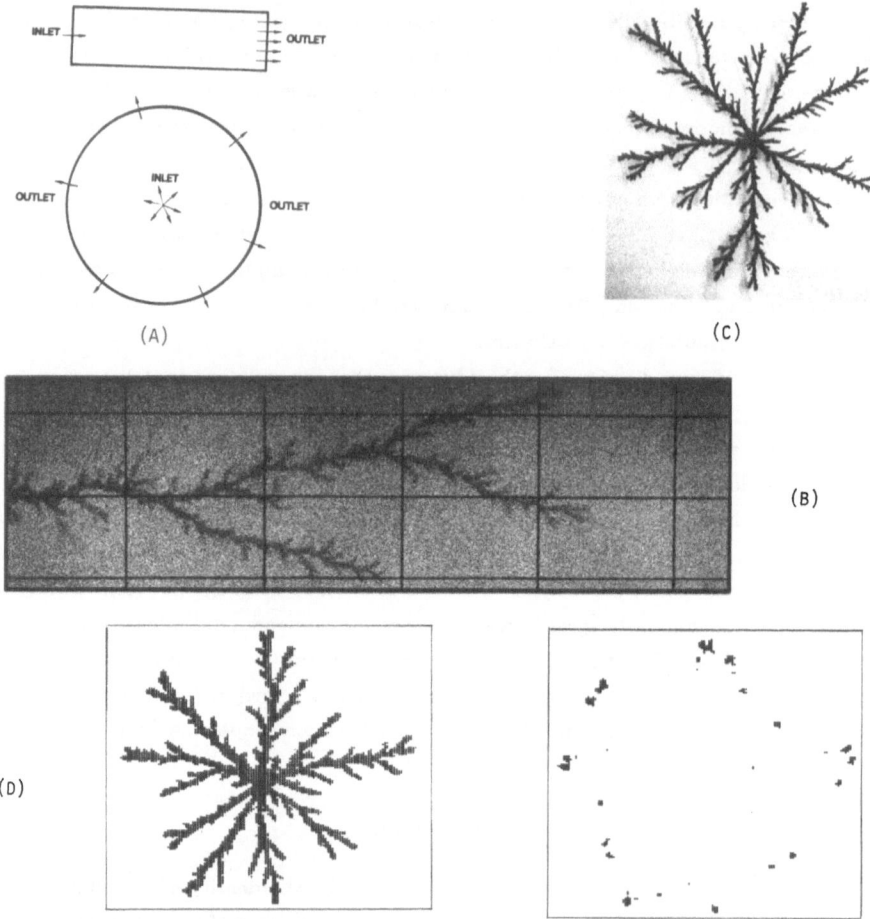

Fig. 1: (a) Schematic illustration of the lateral and radial Hele Shaw cells. (b) Typical *lateral* viscous finger [from Nittmann et al (1985)]. (c) Typical *radial* viscous finger [Daccord et al (1986)]; note that the east arm of the *radial* finger resembles the *lateral* finger, consistent with the idea that on length scales smaller than the width W_{HS} of the Hele Shaw cell in (b) the finger cannot know that asymptotically it is not 2-dimensional but rather 1-dimensional. (d) The "growth region" of a typical radial viscous finger, obtained by subtracting the images of the same finger photographed at slightly different times. The fact that the growth regions are mainly on the tips is a striking proof of the high degree of screening. The detailed analysis of the growth of viscous fingers (Nittmann et al 1987) confirms that viscous fingers have not only the same fractal properties as DLA (Amitrano et al 1986) but also the same "multifractal" properties. The relation of viscous fingering and dendritic growth is discussed in Nittmann and Stanley (1986,1987).

The generating function for the $n(s)$ is

$$G(h) \equiv \sum_{s=1}^{\infty} n(s)K^s \equiv \sum n(s)(1-h)^s. \qquad (1.5)$$

We equate the "fugacity" K to a parameter $1-h$. The reason is that h can be interpreted as the probability that a site is connected to a "ghost site." Then each term in the generating function is the number of *finite* clusters of s sites: the number of 1-site clusters per lattice site is $n(1)(1-h)$, the number of 2-site clusters is $n(2)(1-h)^2$, and so forth. From (1.1b) and (1.5), it follows that

$$G(h) \sim h^{\tau-1}. \qquad (1.6)$$

One can argue that $\tau - 1 = d/d_f$ (Stanley 1987), but this result is not as rigorous as it could be. "Volunteers?"

2. EXACT ENUMERATION APPROACHES

2.1 The functions $P(s,p)$ and $P(b,p)$

Consider, for the sake of specificity, random-site percolation on a square lattice. What is the probability $P(s,p) = sn(s)$ that a randomly-selected site is a member of an s-site cluster? We can answer this question exactly for s small. For $s = 1 - 3$, we find (Fig. 2a)

$$P(s = 1, p) = pq^4 \qquad (2.1a)$$

$$P(s = 2, p) = 4p^2q^6 \qquad (2.1b)$$

$$P(s = 3, p) = 3p^3[4q^7 + 2q^8]. \qquad (2.1c)$$

For bond percolation, it is customary to calculate $P(b,p)$, the probability that a randomly-chosen bond belongs to a b-bond cluster (Sykes et al. 1981 and refs therein). Thus (Fig. 2b)

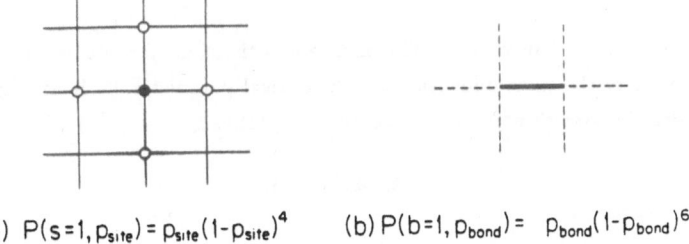

(a) $P(s=1, p_{site}) = p_{site}(1-p_{site})^4$ (b) $P(b=1, p_{bond}) = p_{bond}(1-p_{bond})^6$

Fig. 2: (a) Derivation for the square lattice of Eq. (2.1a), the probability that the origin belongs to a one-site cluster [site percolation, site counting]. (b) Derivation of Eq. (2.2a) [bond percolation, bond counting].

$$P(b = 1, p) = pq^6 \tag{2.2a}$$

$$P(b = 2, p) = 6p^2 q^8 \tag{2.2b}$$

$$P(b = 3, p) = 3p^3[2q^9 + 9q^{10}]. \tag{2.2c}$$

2.2 Lattice animals

The general form of $P(s, p)$ for site percolation is

$$P(s, p) = sp^s D(s, q). \tag{2.3a}$$

An analogous equation holds for $P(b, p)$ for bond percolation. The factor p^s arises from the s occupied elements (sites or bonds) while the "perimeter polynomial"

$$D(s, q) = \sum_t g_{st} q^t, \tag{2.3b}$$

arises from the unoccupied perimeter that bounds the cluster. Here g_{st} is the number of clusters or "lattice animals" with mass s and perimeter t. Since $D(s, q)$ gives the number of distinct lattice animals of mass s, grouped according to the number of perimeter elements,

$$A_s \equiv D(s, q = 1), \tag{2.4}$$

gives the total number of general s-element lattice animals regardless of perimeter.

The generating function for general lattice animals,

$$G_A(K) \equiv 1 + \sum_{s \geq 1} A_s K^s, \tag{2.5a}$$

displays, for both bond and site percolation, a singularity of the form

$$G_A(K) \sim |K - K_c|^{\theta - 1}. \tag{2.5b}$$

The exponent θ is called "universal." This means θ is the same for site and bond percolation, and for all lattices of the same dimension d. The critical parameter K_c is usually denoted $1/\lambda$, where λ is called the growth parameter because asymptotically

$$A_s / A_{s-1} \to \lambda. \tag{2.6}$$

Thus in an enumeration of the number of the animals of size s, one finds (asymptotically) λ times as many animals as at the previous order. This has the unhappy consequence that computer times increase by a factor of λ for each new order—i.e. "exponentially", as $\lambda^s = \exp(s \log \lambda)$.

2.3 Exact results: linear chain ($d = 1$) and Bethe lattice ("$d = \infty$")

Except for very special cases (such as a $d = 1$ linear chain lattice and a "$d = \infty$" Bethe lattice), we cannot evaluate $P(s,p)$ and $P(b,p)$ for s larger than $s_{\max}$, and b larger than $b_{\max}$. Here $s_{\max}$ and $b_{\max}$ depend on the complexity of the lattice and on the University computer budget (for the square lattice, $s_{\max} \cong 20$ and $b_{\max} \cong 15$).

For the linear chain, we see that for site percolation (Fig. 3a)

$$P(s,p) = sp^s q^2, \tag{2.7}$$

if we measure site content. For bond percolation (Fig. 3b)

$$P(b,p) = bp^b q^2, \tag{2.8}$$

if we measure bond size (cf. Table 1, which appears in Sec.5 below). Thus

$$A_s = 1 \tag{2.9}$$

for both site and bond lattice animals. Hence the generating function (2.5a) becomes

$$G_A = (1 - K)^{-1}, \tag{2.10}$$

so that the growth parameter is given by $\lambda = 1$, and the critical exponent $\theta - 1$ defined in (2.5b) is given by

$$\theta - 1 = -1. \tag{2.11}$$

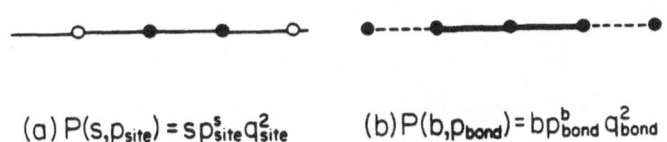

(a) $P(s, p_{\text{site}}) = s\, p_{\text{site}}^s\, q_{\text{site}}^2$ (b) $P(b, p_{\text{bond}}) = b\, p_{\text{bond}}^b\, q_{\text{bond}}^2$

Fig. 3: (a) Derivation for the linear chain of (2.7), the probability that the origin belongs to a s-site cluster [site percolation, site counting]. (b) Derivation of Eq. (2.8) [bond percolation, bond counting].

The A_s can *also* be evaluated exactly for the Bethe lattice, giving the result (Fisher and Essam 1961)

$$\theta - 1 = 3/2. \tag{2.12}$$

It is believed that θ takes on the value 5/2 for all systems with $d > 8$ (Isaacson and Lubensky 1980). For $1 < d < 8$, the A_s have been evaluated exactly for small values of s, and one finds that θ varies smoothly with d (Gaunt 1980). Parisi and Sourlas (1981) have argued that

$$\theta(d) - 1 = \sigma(d') + 1. \tag{2.13}$$

Here $\sigma(d') + 1$ is the exponent characterizing the singularity (Lee and Yang 1952) in the complex H plane of the Gibbs potential $G(H, T_o)$ for an Ising model of dimension $d' = d - 2$ at *any* temperature $T_o > T_c$ (Fig. 4).

2.4 Critical exponents

From the basic functions $P(s, p)$ one can calculate the analogs for percolation of the thermodynamic functions that are singular at normal critical points. For example, the role of the Gibbs potential $G(T)$ is played by $G(p)$, the mean number of clusters per site regardless of size; like $G(T)$, $G(p)$ pertains asymptotically as the system size becomes infinite. Clearly

$$G(p) = \sum_{s \geq 1} n(s, p), \tag{2.14}$$

where, as mentioned above,

$$n(s, p) = s^{-1} P(s, p) \tag{2.15}$$

is the number of s-site clusters per lattice site. Near p_c, one finds that G_{sing}, the singular part of $G(p)$, varies as

$$G_{\text{sing}} \sim |p - p_c|^{2-\alpha}, \tag{2.16}$$

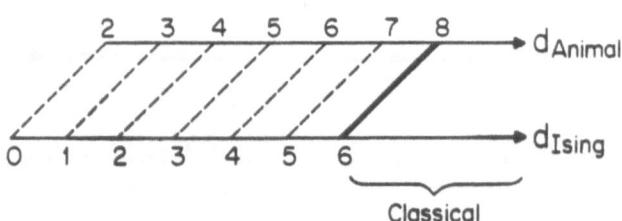

Fig. 4: Schematic illustration of the Parisi-Sourlas (1981) mapping between the lattice animal problem ($d_c = 8$) and the Lee-Yang singularity problem in two lower spatial dimensions ($d_c = 6$).

in complete analogy with the behavior of the singular part of the Gibbs potential $G(T)$ near the critical point T_c. For the linear chain, $n(s,p) = p^s q^2$ for $s \geq 0$ [cf. (2.7)]. Hence from (2.14),

$$G(p) = q. \tag{2.17}$$

That the mean number of clusters per site is the fraction of empty sites is obvious since for a *finite* system the number of clusters is equal to the number of empty sites. Since $p_c = 1$, from (2.16) we have $2 - \alpha = 1$.

The mean size of a finite cluster is defined as the first moment of $P(s,p)$

$$\chi(p) = \langle s \rangle = \frac{\sum_{s \geq 1} s P(s,p)}{\sum_{s \geq 1} P(s,p)}. \tag{2.18}$$

As $p \to p_c$, one finds

$$\chi(p) \sim |p - p_c|^{-\gamma}, \tag{2.19}$$

in analogy with the behavior of the isothermal susceptibility $\chi(T)$ for a magnet near its critical point. For the linear chain, (2.7) leads to $\gamma = 1$, since

$$\chi(p) = \frac{1+p}{1-p} = 1 + 2p + 2p^2 + \dots. \tag{2.20}$$

The probability of the origin belonging to a finite cluster of *any* size is given by the denominator of (2.18),

$$P_F(p) = \sum_{s \geq 1} P(s,p), \tag{2.21}$$

and hence the probability of the origin belonging to an infinite cluster is

$$P_\infty(p) = p - P_F(p). \tag{2.22}$$

As $p \to p_c$, one finds

$$P_\infty(p) \sim |p - p_c|^\beta, \tag{2.23}$$

in analogy with the behavior of the spontaneous magnetization M for a magnet near the critical point. For a linear chain, substituting (2.7) into (2.21) results in a series that converges for all $p < 1$, and gives $P_F(p) = p$. Hence by (2.22) $P_\infty(p) = 0$ except for $p = p_c = 1$. Thus $\beta = 0$ for $d = 1$.

The analog in percolation of the mean field or "classical" theory is the Bethe lattice, which predicts correct values of the critical exponents for $d > d_c$. In percolation, $d_c = 6$.

The "art" involved in the exact enumeration procedures is to devise tricks that permit one to *extrapolate* from the information contained in the functions $P(s,p)$ for $s \leq s_{\max}$ to obtain estimates of critical properties. One standard procedure is to notice from the general form of the function $P(s,p)$ that if we set $q = 1 - p$, we have an expansion that is *exact* through order $s_{\max}$.

This "low-density expansion," were it *not* truncated, would have a singularity on the positive real axis at p_c (Fig. 5a) which might be located using methods such as Padé approximants. Thus, e.g., by substituting (2.1) into (2.18), we obtain to order two the low-density expansion of the mean cluster size for site percolation on a square lattice,

$$\chi(p) = 1 + 4p + 12p^2 + \ldots \qquad (2.24)$$

Note that when we expand $P_F(p)$ about $p = 0$, we obtain

$$P_F(p) = p \qquad (2.25)$$

to all orders, as may be verified by substituting (2.1) into (2.21) and setting $q = 1 - p$.

We can also set $p = 1 - q$ in the functions $P(s, p)$ and thereby obtain a family of "high-density expansions" in the variable q (Fig. 5b). Thus, for example, on substituting (2.1) into (2.16), we obtain through order four

$$P_F(p) = q^4 + \ldots \qquad (2.26)$$

which is obvious since we must create four empty sites in order to obtain a one-site cluster in a nearly full lattice.

3. COMPUTER SIMULATION APPROACHES
3.1 Methods

Perhaps no physics problem is more tantalizing to the computer enthusiast than percolation. Two approaches spring immediately to mind, and each has been developed extensively in recent years.

In the first approach, we start by placing a site at the origin and then use a random number generator to assign a random number

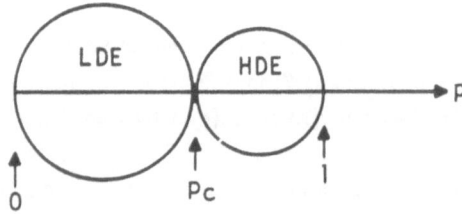

Fig. 5: Schematic illustration of the idealized circles of convergence in the complex p plane for low-density expansions (LDE) and high-density expansions (HDE).

to each of the z nearest neighbors of this site. If the random number is less than p, we join the neighbor to the site at the origin, while if the number is larger than p we do not. For p near zero, this process dies out after very few steps and we are left with a tiny cluster. We then repeat this process, say, 10^4 times to obtain good statistics. Among the advantages of this approach are (i) the fact that $P(s,p)$, not $n(s,p) = s^{-1}P(s,p)$, is calculated, and (ii) the lack of any "boundary." Both facts were emphasized by the work of Pike and Stanley (1981) on the minimum path exponent. A drawback is that it is difficult to obtain good statistics close to p_c, and finite size scaling cannot be used as easily.

A second approach is to begin with a finite section of a lattice with L^d sites. We assign a random number to every one of the sites, and then proceed to design an algorithm that recognizes clusters efficiently. This procedure gives the distribution $n(s,p)$ defined in (2.15), which means that to generate one 10,000-site cluster near p_c requires generating 10,000 times as many one-site clusters as required in the first approach which gives $P(s,p) = sn(s,p)$. A second disadvantage is that the predictions are hampered by "boundary effects," though this apparent "cloud" has a silver lining: one can study a sequence of different "box sizes" L and then use the theory of finite-size scaling to extract the exponent ν defined by

$$\xi \sim |p - p_c|^{-\nu}. \tag{3.1a}$$

Here the connectedness length ξ characterizes the exponential decay with $\mathbf{r}$ (near but not at p_c) of the *pair connectedness function* $C_2(r)$, which in turn gives the probability that a site at position $\mathbf{r}$ is occupied and connected to a site at the origin (Fig. 6). Now $C_2(\mathbf{r})$ and $\chi(p)$ are related in the same fashion for percolation as their thermodynamic analogs $\langle s_0 s_r \rangle$ and $\chi_T(T)$

$$\chi(p) = 1 + \sum_{r>0} C(\mathbf{r}). \tag{3.1b}$$

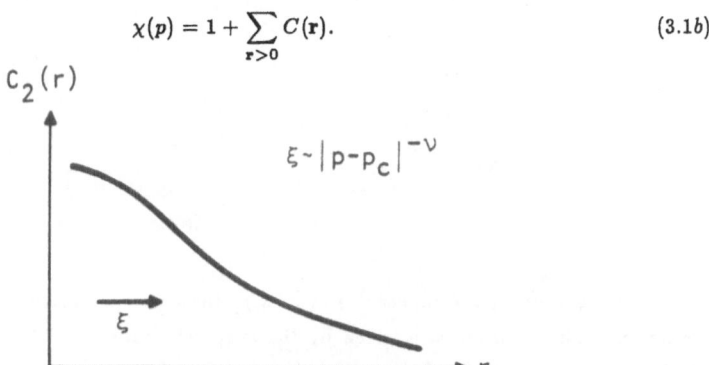

Fig. 6: Schematic illustration of the dependence on site separation r of the pair-connectedness function $C_2(r)$ (the probability that the origin is connected to a site at r). The decay with r is characterized by a length ξ termed the pair connectedness length, which is singular at the percolation threshold with exponent ν.

Thus, e.g., $C(1) = p$, $C(2) = 2p^2 + \ldots$ and $C(3) = p^2$ (since there are *two* paths whereby second neighbors can be connected and only one path for third neighbors). Hence for the square lattice,

$$\chi(p) = 1 + 4p + 4(2p^2 + p^2) + \cdots \tag{3.1c}$$

in agreement with (2.24).

Monte Carlo methods are complementary to exact enumeration procedures in some respects. For example, in the exact procedures, we obtain exact expressions for the cluster distribution function $P(s, p)$ for $s \leq s_{max}$ (Fig. 7), while in the direct computer simulation procedures we obtain highly approximate numerical estimates of $P(s, p)$ with s_{max} typically 2-3 orders of magnitude larger. Generally speaking, Monte Carlo methods have not led to higher accuracy in estimating critical exponents than, say, exact enumeration procedures. However they have aided our understanding of many features of the percolation problem, as well as giving a graphic illustration

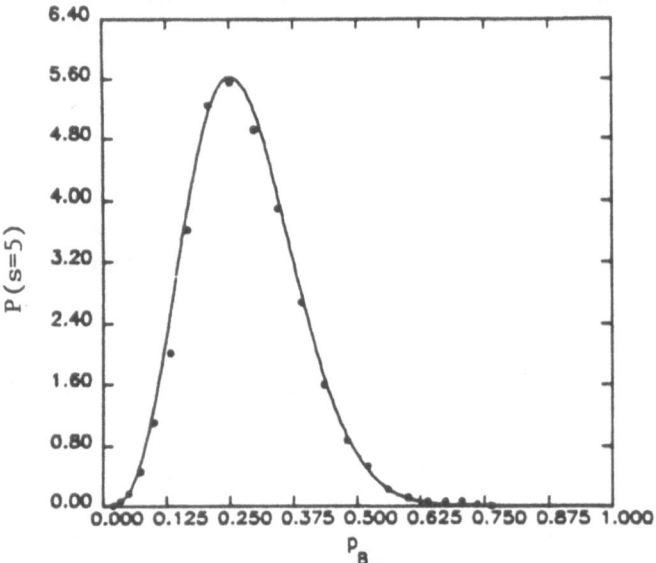

Fig. 7: Illustration of the function $P(s = 5, p)$ for bond percolation with *site* counting on an ice lattice. The solid curve is given by the exact (Blumberg et al 1984, Stanley and Teixeira 1980) expression $P(s = 5, p) = 455p^4(1 - p)^{12}$, while the points represent the connectivity analysis (Geiger and Stanley, 1982a,b) of the Rahman-Stillinger molecular dynamics tapes for a ST2 model interaction potential for liquid water. The agreement between exact calculations for a *lattice* and Monte Carlo calculations for an *interacting continuum* system is striking; no adjustable parameters have been used. Adapted from Stanley et al. (1981).

of what a "typical" million-site cluster really looks like ... and how one should describe it.

3.2 Scaling

The interpretation of the numerical results is so greatly aided by finite-size scaling that we shall take a short "detour" to explain this procedure. In order to explain finite-size scaling, it is necessary to recognize the empirical fact that the functions $P(s,p)$ appear (Stauffer 1985; Nakanishi and Stanley 1980,1981) to be *generalized homogeneous functions* in the variables s and $\epsilon \equiv (p_c - p)/p_c$. That is, asymptotically close to the "critical point" ($s = \infty$, $\epsilon = 0$), $P(s,\epsilon)$ obeys the functional equation (Hankey and Stanley 1972)

$$P(\lambda^{a_S} s, \lambda^{a_T} \epsilon) = \lambda P(s,\epsilon). \tag{3.2}$$

3.2.1 An Example: The Rushbrooke Scaling Law

Equation (3.2) has many implications, among which are a family of "scaling laws" relating the critical-point exponents in percolation. For example, in Sec. 2 we defined the exponents $2 - \alpha$, β, and $-\gamma$ that describe the singularities in $G(p)$, $P_\infty(p)$, and $\chi(p)$ respectively. These exponents can be related by the Rushbrooke scaling law (Stanley 1987)

$$\alpha + 2\beta + \gamma = 2. \tag{3.3}$$

To see this, we first note that the critical exponent of any function is simply the ratio of the scaling power of the function to the scaling power of the variable describing the path of approach to the critical point. Thus, e.g., the singular part of $P_F(p)$ obeys the functional equation

$$P_F(\lambda^{a_T} \epsilon) = \lambda^{1+a_S} P_F(\epsilon). \tag{3.4}$$

The scaling power is $1 + a_S$ since the summation in (2.21) is on the variable s. Since (3.4) is a functional equation valid for all positive λ, we can choose λ as we wish. Thus if we substitute

$$\lambda = \epsilon^{-1/a_T} \tag{3.5}$$

into (3.4), we obtain

$$P_F(\epsilon) \sim \epsilon^\beta \tag{3.6a}$$

with

$$\beta = \frac{1 + a_S}{a_T}. \tag{3.6b}$$

We attribute the same exponent to P_F and P_∞ since they are related by (2.22) and the variable p is itself not singular at p_c.

The exponent $-\gamma$ defined in (2.19) can also be evaluated in terms of the two scaling powers. From the definition (2.18), it follows that $\chi(\epsilon)$ satisfies the functional equation

$$\chi(\lambda^{a_T} \epsilon) = \lambda^{1+2a_S} \chi(\epsilon), \tag{3.7}$$

where the "scaling power" $1 + 2a_S$ arises from the fact that the summation in (2.18) is on the variable s, and the summand itself is sP_s. Substituting (3.5) into (3.7), we find

$$\chi(\epsilon) \sim |\epsilon|^{-\gamma} \tag{3.8a}$$

with

$$-\gamma = \frac{1 + 2a_S}{a_T}. \tag{3.8b}$$

Finally, we can evaluate the exponent $2 - \alpha$ defined in (2.16) in terms of a_S and a_T. From (2.14) it follows that the singular part of $G(p)$ obeys the functional equation

$$G_{\text{sing}}(\lambda^{a_T}\epsilon) = G_{\text{sing}}(\epsilon), \tag{3.9}$$

since the scaling power $-a_S$ arising from the definition of $n(s, p)$ in (2.15) is cancelled by the scaling power a_S arising from the summation in (2.9). Hence on substituting (3.5) into (3.9) we find (2.16) holds, with

$$2 - \alpha = \frac{1}{a_T}. \tag{3.10}$$

The Rushbrooke scaling law (3.3) can now be obtained trivially by eliminating the *two* unknown scaling powers a_S and a_T from the *three* equations (3.6), (3.8), and (3.10).

3.2.2 Finite-Size Scaling

Now we can introduce finite-size scaling. First we recall the functional equation (3.7) obeyed by the mean size function $\chi(\epsilon)$ for an infinite system. Suppose now that the system is confined to a box of edge L, where L is large compared to a lattice constant but not infinite. The essential *Ansatz* behind finite-size scaling is that $\chi(\epsilon, L)$ is a generalized homogeneous function in the variables ϵ and L. Thus finite-size scaling makes the assumption that

$$\chi(\lambda^{a_T}\epsilon, \lambda^{a_L}L) = \lambda^{a_\chi}\chi(\epsilon, L), \tag{3.11}$$

where $a_\chi = 1 + 2a_S$. Moreover, the new scaling power a_L can be argued to be identical to the scaling power a_ξ characterizing the pair connectedness length, ξ.

A typical prediction of finite-size scaling is the following. Let us make computer simulations of the system directly at p_c, so that $\epsilon = 0$. Choose λ so that the second argument in (3.11) becomes unity. Thus

$$S(\epsilon = 0, L) \sim L^x, \tag{3.12a}$$

where the exponent x is given by

$$x = a_\chi/a_L = (a_\chi/a_T)(a_T/a_\xi) = \gamma/\nu. \tag{3.12b}$$

Thus a plot of $\chi(\epsilon = 0, L)$ on double logarithmic paper should result, for large L, in a straight line of slope γ/ν (Fig. 8). In this fashion, we can obtain β/ν, $(2 - \alpha)/\nu$, and so forth.

3.3 Incipient Infinite Cluster (IIC)

As an example, we conclude this section by mentioning one particularly fascinating aspect of the "percolation disease" that shows signs of yielding to solution, in large part due to Monte Carlo studies. This is the question of how one describes the incipient infinite cluster (IIC) that appears as the percolation threshold is approached from below. Consider again the site-dilute random magnet. When the fraction p of magnetic sites is very small, the system consists of small disconnected clusters of magnetically correlated sites. As $p \to p_c^-$, the mean cluster size increases until at $p = p_c$ a single cluster (the IIC) spans the entire lattice. The percolation threshold p_c is a critical point. By studying the propagation of magnetic correlations through the incipient infinite cluster that appears in the dilute magnet at its percolation threshold, one can obtain information about order propagation near this critical point—a classic unsolved problem (Zernike 1940).

Since the incipient infinite cluster dominates the behavior of the system it is important to be able to describe its structure. If a cluster is considered as a network of wires carrying electrical current between two parallel bus bars, it can be decomposed into a conducting "backbone" and many "dangling ends" that do not contribute to the electrical conductivity (and hence order

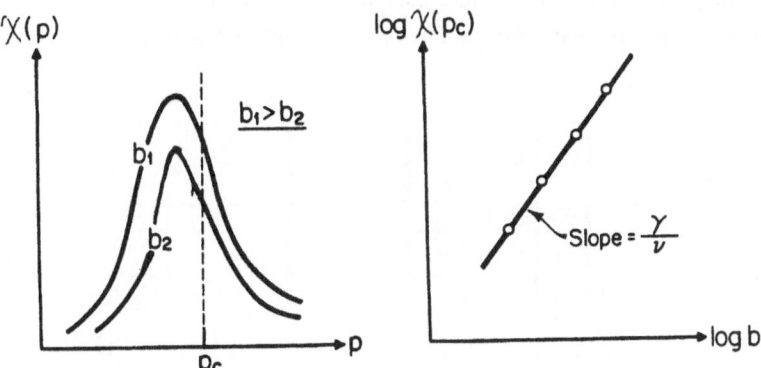

Fig. 8: Schematic illustration of finite-size scaling for $\chi(p)$, the mean size of a finite cluster, in a sample of linear dimension $L = b$. An actual plot that is remarkably linear for L larger than 20 is given in Fig. 3 of Gawlinski and Stanley (1981).

propagation) between the ends (Fig. 9). Describing the important features of this backbone is a problem that was first addressed in Stanley et al (1976), Stanley (1977), and Kirkpatrick (1978).

The backbone bonds may be divided into two classes (Stanley 1977), conveniently visualized as "red" and "blue." Red (blue) bonds are singly connected (multiply-connected); removing a single red (blue) bond breaks (does not break) the connection between two parallel bus bars touching the most eastern and most western bonds. Thus red bonds are "hottest": they must carry all the current. An example of this backbone decomposition is shown in Fig. 10, where "red" bonds are shown as heavy lines and "blue" bonds as light lines (the dangling ends are shown dashed).

The distinction between red and blue bonds seemed important (Stanley et al 1976) in connection with the dilute Ising magnet—e.g., "propagation of order" along a string of red bonds is analogous to propagation in a topologically one-dimensional system, while "propagation of order" in a "blue blob" is more like propagation in the d-dimensional magnet. For this reason, I posed the "red-blue" problem in a talk given at the annual Canadian Undergraduate Physics Association in Toronto in November 1977, and an undergraduate member of the audience, Rob Pike, came up afterward to announce that he thought he could solve the problem.

Pike indeed succeeded in formulating a computer algorithm that partitions bonds into three separate classes: red, blue, and "yellow" (the dangling ends). One worry that clouded the Pike project was the possibility that as $p \to p_c^-$, the mean number of red bonds χ_{red} would approach zero; this would occur, e.g., by the joining up of yellow dangling ends:

Fortunately, Pike's Monte Carlo data showed clearly that all three functions χ_{red}, χ_{blue}, and χ_{yellow} diverge. Pike's estimates for the corresponding critical exponents are

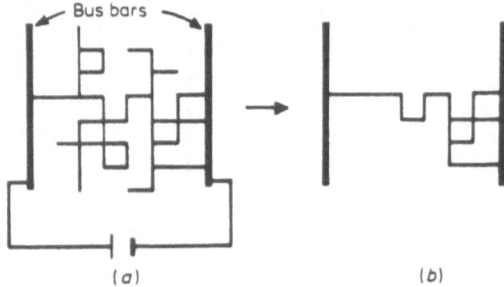

Fig. 9: Schematic illustration of the backbone of a cluster: bus bars are attached to the cluster extremities and a potential difference is applied in (a). Current flows in the backbone bonds shown in (b), but not to the "dangling ends" which are attached to the backbone at only one vertex. After Shlifer et al. (1979).

$$d_f(\text{red}) \cong 0.8 \qquad\qquad (3.13a)$$

$$d_f(\text{blue}) \cong 1.6 \qquad\qquad (3.13b)$$

$$d_f(\text{yellow}) \cong 1.9 \qquad\qquad (3.13c)$$

At the same time, Coniglio (1981) proved *rigorously* that

$$d_f(\text{red}) = 1/\nu = 3/4. \qquad\qquad (3.14)$$

Nienhuis has "proved" $d_f(\text{yellow}) = 91/48 = 1.895$. However $d_f(\text{blue})$ is still not exact, the best current estimate being 1.62 ± 0.02 (Herrmann and Stanley 1984)......"Volunteers?"

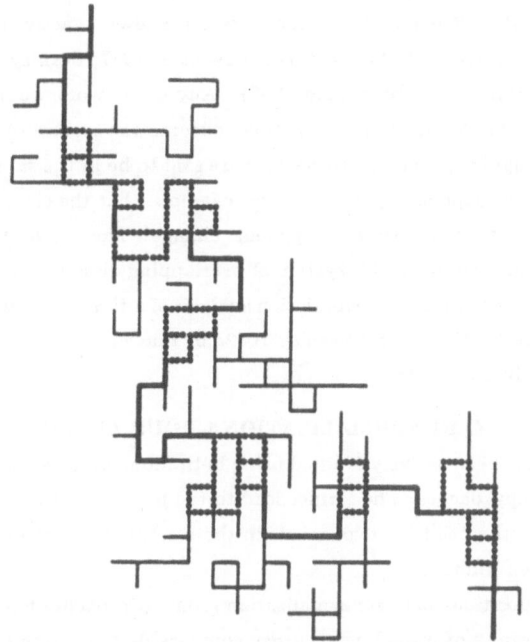

Fig. 10: A large machine-generated finite cluster just below the percolation threshold of the square lattice bond problem (Stanley 1977). The bonds of this cluster are of three sorts: dangling ends (light), singly-connected "red" bonds (heavy), and multiply-connected "blue" bonds (dotted). The red and blue bonds form the backbone, which (a) carries stress if the two ends of the cluster are pulled, (b) carries current if a battery is applied across the cluster, (c) carries fluid if the bottom of the cluster is oil and the top is my car [assuming the cluster represents a model of a randomly-porous material], (d) propagates spin order, if we view this as a large cluster of Ising spins just below p_c (see, e.g., Stanley et al 1976). This figure is taken from Stanley (1977).

The utility of the red-blue decomposition of the backbone to the classic problem of order propagation in the dilute ferromagnet has recently been demonstrated (Coniglio 1981,1982). However there remain interesting questions relating to cluster topology.

3.4 Some Open Questions

One important open question is the relation to cluster properties of the exponent μ describing the approach to zero as $p \to p_c^+$ of the electrical conductivity, $\Sigma_{el} \sim (p - p_c)^\mu$. The beautiful experiments of Deutcher (1981) and others should serve as a strong stimulus for a program designed to put the exponent μ on a firmer conceptual foundation.

Still another question concerns the new exponent defined through $\ell_{min} \sim L^{d_{min}}$, where ℓ_{min} is the *shortest* "cow path length" through the cluster backbone (from one bus bar in Fig. 8 to the other). Most $d = 2$ percolation exponents are known exactly, but not d_{min}. The latest estimates are $d_{min} \cong 1.10 - 1.15[d = 2]$ (Laidlaw et al 1987; Stanley and Herrmann 1987) and $1.3 - 1.4[d = 3]$(Stanley and Herrmann 1987). Any exact work would be timely and valuable indeed......Volunteers? Another open question concerns the nature of the "universality classes" for percolation (roughly speaking, two systems are said to be in the same universality class if they have the same critical exponents). Suppose, for example, that the elements are not constrained to the sites or bonds of a lattice. Do the exponents change? The answer to this question appears to be "no", at least for a simple $d = 2$ system of overlapping discs (see Gawlinski and Stanley 1981 and references therein). Computer simulation analysis of a *three-dimensional* system of *interacting* particles also suggests (Geiger and Stanley 1982a,b) that even the basic "microscopic" functions $P(s,p)$ are themselves unchanged.

4. RENORMALIZATION GROUP APPROACHES

Percolation has been actively studied using both momentum-space and position-space renormalization group approaches. The former lead to the prediction that the upper marginal dimension $d_c = 6$. Hence the utility of expansions in the variable $(d_c - d)$ for systems of the dimension $d = 2$ and 3 is often limited.

A variety of position-space renormalization group approaches have been developed recently, and some hold promise of providing accuracy comparable to or even exceeding that provided by the exact enumeration approaches. These have been recently reviewed elsewhere (Stanley et al. 1982), so I'll limit my remarks here to a brief description of one particular avenue.

As a simple example, partition a square lattice into cells of edge L. The cells play the role of "renormalized sites" (Fig. 11). If the sites are occupied with probability p, then the cells may be defined to be occupied with probability

$$p' = R(p), \tag{4.1}$$

where

$$R(p) = \sum_{state} \pi_{state} f_{state} \tag{4.2}$$

is the probability that the configuration spans. The summation is over all 2^N states of the system, where $N = L^2$,

$$\pi_{\text{state}} = p^s q^{N-s} \tag{4.3a}$$

is the probability of a particular state with s occupied sites, and

$$f_{\text{state}} = 1 \qquad \text{[if the state "spans"]}$$

and

$$f_{\text{state}} = 0 \qquad \text{[otherwise]}. \tag{4.3b}$$

The simplest example is $L = 2$; in this case there are 2^4 states, some of which are shown in Fig. 12. If "spanning" is defined from East-to-West, then (4.2) becomes simply

$$R(p) = p^4 + 4p^3 q + 2p^2 q^2. \tag{4.4}$$

The function $R(p)$ has been evaluated for $L = 3$ and 4 exactly, and by Monte Carlo methods for L up to 500. As L increases, $R(p)$ becomes sharper and approaches the step function sketched in Fig. 2b.

Equation (4.1) serves as a simple though highly approximate position-space renormalization group transformation with fixed points given by

$$p^* = R(p^*). \tag{4.5}$$

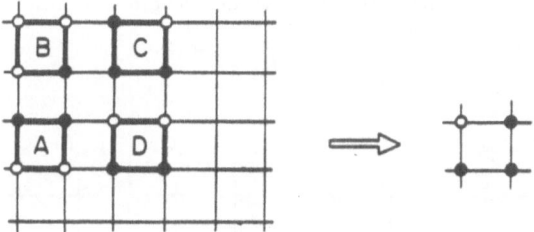

Fig. 11: Rescaling a lattice by forming cells out of groups of sites. Four-site cells on the square lattice are shown, with dots representing occupied sites. In this example, cells C and D are both "occupied" (each cell can be traversed), and therefore connected on the *cell* level. On the *site* level, however, we cannot connect them. On the other hand, cells A, B, and C are connected on the site level, though only the next-neighbor cells, A and C, are occupied. These incorrect connections force one to either introduce nearest-neighbor, further-neighbor, and multi-site probabilities, or to go to larger cells where the "interfacing" error plays a smaller relative role. After Reynolds et al. (1980).

When we substitute (4.5) into (3.1), we find two trivial fixed point at $p^* = 0, 1$ and also a critical fixed point at

$$p^*(L = 2) = 0.62. \tag{4.6a}$$

This estimate should be compared with those provided by the exact enumeration approaches,

$$p_c = 0.593 \pm 0.003. \tag{4.6b}$$

Although the agreement is not impressive, it is not terribly poor for a "single-shot" technique.

The strength of the position-space renormalization group approach is that one can systematically consider cells for larger and larger values of L. One finds that $p^*(L)$ varies smoothly and predictably with L, following the relation

$$p^*(L) - p_c(\infty) \sim L^{-1/\nu}, \tag{4.7}$$

suggested by finite-size scaling considerations. Extrapolating a sequence of estimates for $p^*(L)$ for a range of L from 2 to 500, we find the estimate (Fig. 13)

$$p^*(L = \infty) = 0.5931 \pm 0.006, \tag{4.8}$$

which is believed to be possibly even more accurate than the estimate (4.6b) from exact enumeration procedures.

To calculate the connectedness length exponent from the renormalization transformation (4.1), we note that all lengths in the rescaled system have been reduced by a factor of L from the lengths in the original system. Hence the connectedness length transforms as

$$\xi' = L^{-1}\xi. \tag{4.9}$$

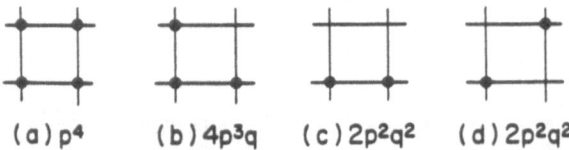

(a) p^4 (b) $4p^3q$ (c) $2p^2q^2$ (d) $2p^2q^2$

Fig. 12: Some of the 2^4 configurations that arise in the position-space renormalization group for site percolation for the square lattice using a $L = b = 2$ cell. Configurations (a) - (c) span from East to West, while configuration (d) does not.

Substituting (3.1a) into (4.9), we find

$$|p' - p^*|^{-\nu} \qquad L^{-1}|p - p^*|^{-\nu}. \tag{4.10}$$

Since the transformation $R(p)$ is analytic near p^*, when we subtract (4.5) from (4.1), we find

$$p' - p^* = R(p) - R(p^*) \cong \lambda(p - p^*), \tag{4.11}$$

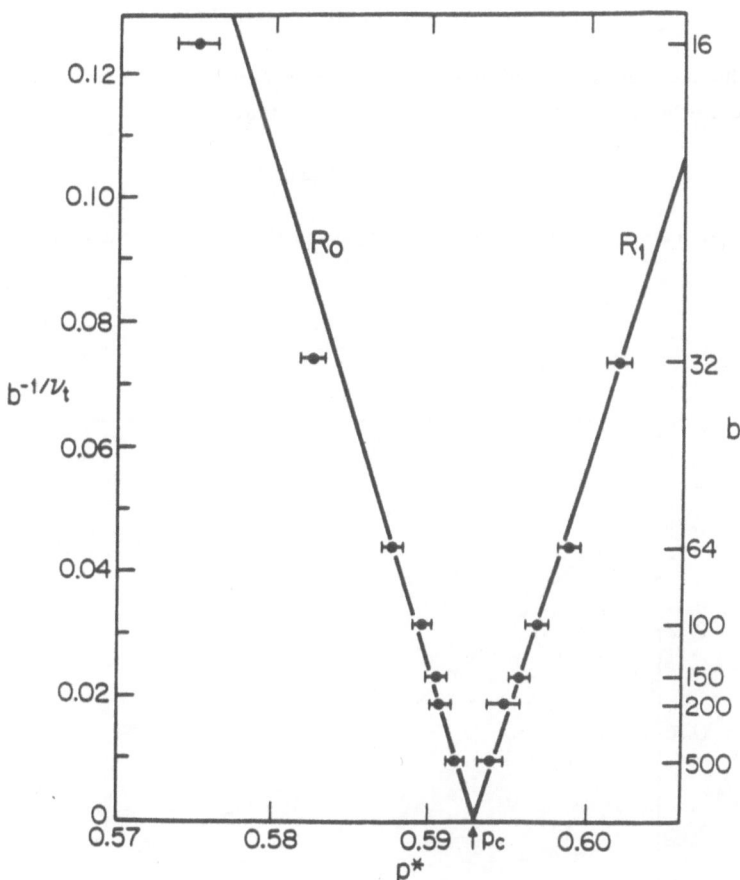

Fig. 13: Extrapolation of the sequence of fixed points $p^*(L)$ obtained from two different connectivity rules, R_o and R_1 using Eq. (4.7). We have chosen the trial value, $\nu_t = 4/3$, for this plot. Both curves approach approximately the same final estimate of p_c as $L \to \infty$. After Stanley et al (1982).

where

$$\lambda = (dR/dp)_{p^*}. \tag{4.12}$$

In order for (4.10) and (4.11) to be consistent, we must have $\lambda^{-\nu} = L^{-1}$, or

$$\nu(b) = \ln L/\ln \lambda. \tag{4.13}$$

Again, the error in $\nu(L)$ will decrease as L increases in a *predictable* fashion

$$\nu(L) - \nu(\infty) \sim A_1(\ln L)^{-1} + A_2(\ln L)^{-2}. \tag{4.14}$$

Figure 14 shows a sequence of estimates from $L = 5$ to $L = 500$. A least squares fit to these estimates results in the $L = \infty$ intercept (Eschbach et al 1981)

$$\nu(L = \infty) = 1.33 \pm 0.01, \tag{4.15}$$

which is in good agreement with the Nienhuis argument that $\nu = 4/3$ for $d = 2$.

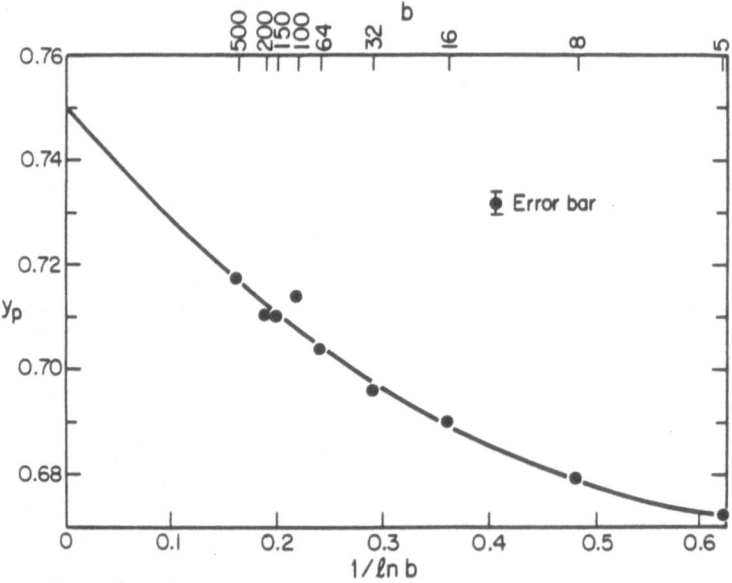

Fig. 14: *Quadratic* extrapolation against $1/\ln L$ for $y_p = 1/\nu$, using Eq. (4.14). The error bar shown is representative. After Reynolds et al. (1980).

5. BOND PERCOLATION AND THE "MONOCHROMATIC" POTTS MODEL

5.1 The Potts Model

Next we discuss the relation of bond percolation to a model of great recent interest, the Potts model, and the relation of the lattice animal problem to a particular generalization of the pure Potts model. These connections were first noted by Kasteleyn and Fortuin (1969) and by Harris and Lubensky (1981) respectively. The classic review is Wu (1982), and what follows is based on a collaboration with F. Y. Wu (Wu and Stanley 1983).

The Potts model is a natural generalization of the ordinary lattice-gas or Lenz-Ising model in which each spin variable s_i can exist in more (or less!) than two states (Fig. 15). We will first show how bond percolation arises as a limiting case of the Potts model. For concreteness, we shall illustrate several stages of the argumentation with a simple linear chain lattice with periodic boundary conditions, which has N sites and $E = N$ edges. With each site i, associate a variable $s_i = 1, 2, \ldots, Q$. When two neighboring spins s_i and s_j are in the same state, they have an "attraction" parametrized by the dimensionless coupling constant $K = J/kT$, while if they are in different states there is zero interaction energy (Fig. 16a). Hence the Boltzmann factor for a pair $\langle ij \rangle$ of neighboring sites is given by

$$\exp[K\delta(s_i, s_j)] = 1 + v\delta(s_i, s_j), \tag{5.1a}$$

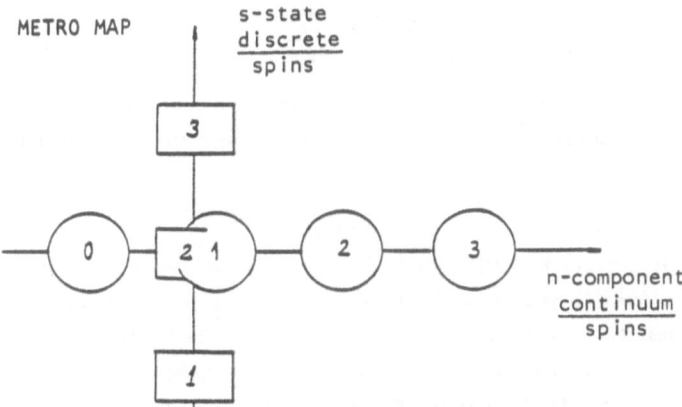

Fig. 15: Schematic illustration of a "metro map" showing how the Ising model has been generalized, first by allowing the two Ising spin states to become s discrete states (first proposed by Potts 1952), and next by allowing the two spin states to be replaced by a continuum of spin states confined to an n-dimensional spin space (first proposed by Stanley 1968b). The percolation problem corresponds to the $s = 1$ "station" on the North-South metro line. After Stanley (1981).

where

$$v \equiv e^K - 1. \tag{5.1b}$$

The *Zustandsumme* or partition function is the sum over all Q^N states of the system. Each state is weighted by the appropriate Boltzmann factor, which in turn is the product over all nearest-neighbor pairs $\langle ij \rangle$ of the Boltzmann factor (5.1a). Thus

$$Z_N(Q; N) = \sum_{s_1=1}^{Q} \cdots \sum_{s_N=1}^{Q} \Pi_{\langle ij \rangle} \left[(1 + v\delta(s_i, s_j)) \right]. \tag{5.2}$$

Since there are E edges in the lattice, there are 2^E terms that arise when the product in (5.2) is expanded. Each term becomes in 1 : 1 correspondence with a "graph" if we simply draw a line on the corresponding edge $\langle ij \rangle$ of the lattice for each factor $v\delta(s_i, s_j)$. For example, the term unity corresponds to a graph with no lines, and gives rise to a contribution of order v^o. There are E separate graphs, each with one line, giving rise to contributions of order v. In general, there are $\binom{E}{e}$ graphs giving rise to contributions of order v^e, where $e = e(G)$ is the number of lines or "edges".

The N-fold summation in (5.2) can be regarded as an operator that "filters out" a factor Q^n, where $n = n(G)$ is the number of clusters in the configuration. Hence the N-fold summation is replaced by a single summation over the 2^E graphs,

$$Z_N(Q; v) = \sum_{2^E \text{ graphs}} Q^n v^e. \tag{5.3}$$

Note that here we count clusters by their site content (cf. Table 1). Thus for the configuration with no lines, there are N clusters; for the E configurations with one intact line, there are $N-1$ clusters. The analog of Eq. (2.8) is $P(s, p) = sp^{s-1}q^2$.

TABLE 1

Exact results for $P(s, p)$ and $P(b, p)$ for site and bond counting for the $d = 1$ site and bond percolation problems (Reynolds et al. 1977). The symbol [KF] indicates that the Kasteleyn-Fortuin correspondence for bond percolation counts cluster size by their site content. The symbol [EE] indicates that exact enumeration procedures for site percolation utilize site counting, and for bond percolation utilize bond counting. For site percolation, p denotes site probability, while for bond percolation it denotes bond probability: $q = 1 - p$.

	SITE PERCOLATION	BOND PERCOLATION
SITE COUNTING	$P(s, p) = sp^s q^2$	$P(s, p) = sp^{s-1} q^2$
	[EE]	[KF]
BOND COUNTING	$P(b, p) = bp^{b-1} q^2$	$P(b, p) = bp^b q^2$
		[EE]

For the closed linear chain, n is simply $E - e$ for all values of e except $e = E$ (for which $n = 1$, not 0); the case $N = E = 4$ is illustrated in Fig. 17. Hence (5.3) may be trivially evaluated, with the result

$$Z_N(Q; v) = (Q + v)^N + (Q - 1)v^N. \tag{5.4}$$

Note that as $T \to \infty$, v approaches zero so that $Z_N(Q) = Q^N$. Also, for $Q = 2$, (5.4) reproduces the familiar result of Ernst Ising's Ph.D. thesis.

5.2 Bond percolation

The preceding formulation sounds tantalizingly similar to the bond percolation problem, which we now describe in similar terms. Imagine that each of the E bonds of our lattice is randomly intact with probability p, and define a cluster to be a set of sites joined by intact bonds. Clearly there are 2^E states of the system, and the probability of each state is

$$\pi_{\text{state}} = p^e q^{E-e}, \tag{5.5}$$

where now $e = e_{\text{state}}$ is the number of intact bonds in a given state.

In bond percolation the average $\langle A \rangle$ of any quantity A_{state} is

$$\langle A \rangle = \sum_{2^E \text{ states}} \pi_{\text{state}} A_{\text{state}}, \tag{5.6}$$

where the summation is over the 2^E states of the system. Substituting (5.5) into (5.6), we obtain

$$\langle A \rangle = q^E \sum_{2^E \text{ states}} A_{\text{state}}(p/q)^e. \tag{5.7}$$

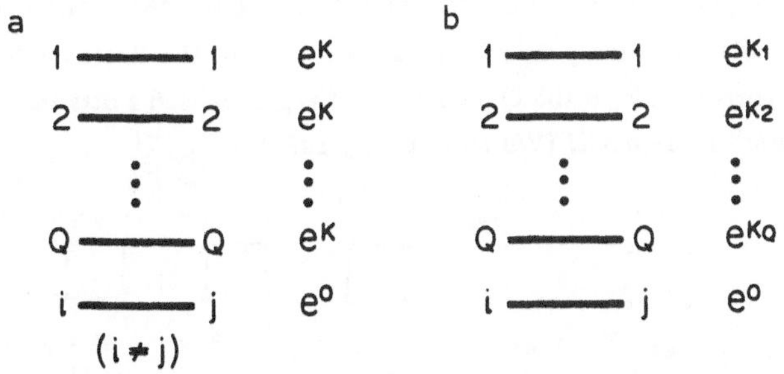

Fig. 16: Boltzmann factors for (a) the ordinary "monochromatic" Potts model, in which neighboring spins in the same spin state are weighted by an additional factor $\exp(K)$, and (b) the "polychromatic" Potts model, in which the additional factor $\exp(K_\alpha)$ depends upon which of the Q states the neighboring spins are in.

Motivated by the resemblance between (5.7) and (5.3), let us choose $A_{\text{state}} = Q^n$ where $n = n_{\text{state}}$ is the number of clusters in the state. Thus (5.7) becomes

$$\langle Q^n \rangle = q^E \sum_{2^E \text{ states}} Q^n (p/q)^e.$$

(5.8)

If we let the "high-temperature variable" v defined in (5.1b) be the ratio of intact to broken bond probabilities (Fig. 18), then (5.8) gives a direct relation between the average of Q^n in bond percolation and the Potts model partition function,

$$\langle Q^n \rangle = q^E Z(Q, v = p/q).$$

(5.9)

In percolation, the analog of the Gibbs potential is $\langle n \rangle$, the mean number of clusters regardless of size,

$$\langle n \rangle = (\partial / \partial Q)[\ln \langle Q^n \rangle]_{Q=1}.$$

(5.10)

Combining (5.9) and (5.10), we finally obtain the Kasteleyn-Fortuin result that $\langle n \rangle$ is the logarithmic derivative of the Potts model partition function, evaluated at $Q = 1$,

$$\langle n \rangle = (\partial / \partial Q)[\ln Z_N(Q, v = p/q)]_{Q=1}.$$

(5.11)

For the closed linear chain, the left hand side of (5.11) was evaluated directly in (2.17),

$$\langle n \rangle \propto Nq.$$

(5.12)

This result is also obtained by substituting (5.4) into the right-hand side of (5.11).

6. BOND LATTICE ANIMALS AND THE "POLYCHROMATIC" POTTS MODEL

Just as bond percolation corresponds to the $Q \to 1$ limit of the Potts model, so also the lattice animal problem corresponds to the $Q \to 1$ limit of a generalized Potts model that we shall call the "polychromatic" Potts model (Wu and Stanley 1983).

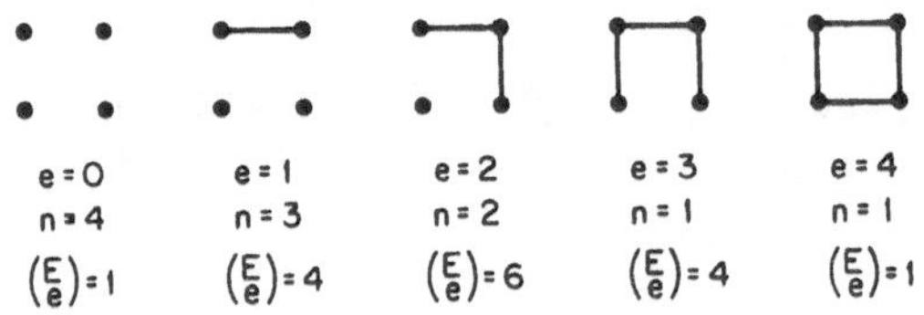

Fig. 17: Illustration of the use of the relation (5.3) to calculate the Potts model partition function for the linear chain with periodic boundary conditions. Here $N = E = 4$ for simplicity, but the argument leading to (5.4) holds for all N

In the ordinary or "monochromatic" Potts model, we place a black bond between two adjacent sites that are in the same spin state s_j. Clearly this can be generalized by placing Q different colors of bonds between neighboring spins that are in the same spin state. Thus when two neighboring spins s_i and s_j are in the same state, they have an interaction parametrized by the dimensionless coupling constant $K_\alpha = J_\alpha/kT$, where $\alpha = 1, 2, \ldots, Q$ denotes the state α (Fig. 16b). Equation (3.1a) for the Boltzmann factor is replaced by

$$\exp[K_\alpha \delta(s_i, s_j, \alpha)] = 1 + v_\alpha(s_i, s_j, \alpha). \tag{6.1a}$$

Here $\delta(s_i, s_j, \alpha) = 1$ if $s_i = s_j = \alpha$ and zero otherwise, and

$$v_\alpha = \exp(K_\alpha) - 1. \tag{6.1b}$$

In analogy to (5.2), the partition function is now

$$Z_N(Q, \{v_\alpha\}) = \sum_{s_1=1}^{Q} \cdots \sum_{s_N=1}^{Q} \Pi_{\langle ij \rangle}\Big[1 + \sum_{\alpha=1}^{Q} v(s_i, s_j, \alpha)\Big]. \tag{6.2}$$

Clearly (6.2) reduces to (5.2) if $K_\alpha = K$ for all states.

To obtain a graphical expansion for the polychromatic Potts model, we proceed exactly as in the monochromatic case by associating a graph of the lattice with each of the 2^E terms obtained by expanding the product in (6.2). There is no change for the zero-line graph, which gives rise to a contribution Q^N when operated on by the N-fold summation in (6.2). For the E

Fig. 18: Illustration of the correspondence between the "monochromatic" Potts model and the random bond percolation problem, showing that the variable v defined in Eq. (5.1b) corresponds to the ratio p/q of intact to broken bonds. In particular, low-temperature expansions (LTE) for the Potts model correspond to low-density expansions (LDE) for bond percolation, while high-temperature expansions (HTE) correspond to high-density expansions (HDE). Similarly, the critical parameter $v_c = Q$ corresponds to the ratio p_c/q_c; for the square lattice, $v_c = Q^{1/2} = 1$.

one-line graphs, the summation gives rise to the factor $[v_1 + \ldots + v_Q]Q^{N-2}$. Consider now the $\binom{E}{2}$ two-line graphs. If the two lines form two disconnected one-bond clusters, then we have a contribution $[v_1 + \ldots + v_Q]^2 Q^{N-4}$, while if the two lines form a single two-bond cluster, then there is a contribution $[v_1^2 + \ldots + v_Q^2]Q^{N-3}$. In general, we see that the appropriate generalization of (5.3) is

$$Z_N(Q; \{v_\alpha\}) = \sum_{2^E \text{ graphs}} \Pi_{\text{clusters}} [v_1^{b(c)} + \ldots + v_Q^{b(c)}]. \tag{6.3}$$

Here the product is over all clusters c appearing in a given graph, *including the one-site clusters*, and $b(c)$ is the number of bonds in cluster c.

We can see that (6.3) reduces to (5.3) if $K_\alpha = K$ for all states α. The Q terms in the square brackets of (6.3) become $Qv^{b(c)}$. When we form the product over all clusters of the factor Q, we obtain Q^n, where n is the total number of clusters in the graph. When we form the product over all clusters of $v^{b(c)}$, we obtain v^e since e, the total number of lines in the graph, is simply the summation over all clusters of the number of bonds in that cluster.

A special case of the "polychromatic" Potts model was introduced by Wu (1978) and shown by Harris and Lubensky (1981) to reduce to the bond animal generating function. This case is obtaining by setting $K_1 = 0$ so that two spins have no interaction if they are both in state $\alpha = 1$. The remaining $Q - 1$ states are treated on an equal footing by choosing $K_\alpha = K$ for $2 \le \alpha \le Q$. For this case, then, (6.3) reduces to

$$Z_N(Q; v) = \sum_{2^E \text{ graphs}} Q^P(Q-1)^{n-P} v^e, \tag{6.4}$$

where P denotes the number of one-site clusters (isolated "points").

Motivated by the form of (5.11), consider the logarithmic derivative of (6.4) evaluated at $Q = 1$. The presence of the factor $(Q - 1)^{n-P}$ in (6.4) means that when we form the derivative of Z_N with respect to Q, only those graphs with $n - P = 1$ make a contribution after setting $Q = 1$. Hence we can replace the summation in (6.4) over all 2^E graphs by a summation over all graphs consisting of only a single cluster. Thus if we form the logarithmic derivative evaluated at $Q = 1$, and use the fact that $Z_N(Q = 1; v) = 1$ from (6.4), then we find

$$(\partial/\partial Q)[\ln Z_N(Q; v)]_{Q=1} = \sum_{\text{graphs}}' v^e, \tag{6.5}$$

where the prime on the summation indicates that it is over only those graphs made up of a single cluster. We can group together all terms in (6.5) with the same number of bonds,

$$(\partial/\partial Q)[\ln Z_N(Q; v)]_{Q=1} = \sum_{b \ge 0} A_b v^b, \tag{6.6}$$

where A_b is the total number of b-bond "lattice animals". Thus (6.6) is the generating function for the bond animal problem.

Just as we illustrated the correspondence between the monochromatic Potts model and the bond percolation problem using the linear chain, so also we can illustrate the correspondence between the polychromatic Potts model and the lattice animal problem on the same lattice. We have already noted above that $A_b = 1$ for an infinite linear chain lattice, so that the bond animal generating function is simply

$$G_A(K) = \sum_{b=0}^{\infty} A_b K^b = (1 - K)^{-1}. \tag{6.7}$$

To calculate the partition function of the Potts model in the $N = \infty$ limit, we can appeal to the transfer matrix method. Thus in the large-N limit, we have

$$Z_N(Q; v) = \text{trace } \mathbf{T}^N \sim (\lambda_{\max})^N, \tag{6.8}$$

where $\lambda_{\max}$ is the largest eigenvalue of the transfer matrix $\mathbf{T}$. For example, for $Q = 3$, we have

$$\mathbf{T} = \begin{pmatrix} e^o & 1 & 1 \\ 1 & e^K & 1 \\ 1 & 1 & e^K \end{pmatrix}. \tag{6.9}$$

For all Q, we find that the largest eigenvalue is given by

$$2\lambda_{\max} = (Q + v) + [(Q + v)^2 - 4v]^{1/2}. \tag{6.10}$$

Forming the logarithmic derivative, we have for large N

$$(\partial/\partial Q)[N^{-1} \ln Z_N(Q; v)]_{Q=1} = (1 - v)^{-1}. \tag{6.11}$$

The agreement of (6.7) and (6.11) confirms the general result (6.6) for the linear chain.

7. SURFACES, INTERFACES AND SCREENING OF FRACTAL STRUCTURES

The next part of this talk is devoted to the subtle and fascinating subject of disordered surfaces. But what do we mean by "the" surface of a fractal object? In fact, we shall see that there are many different surfaces, depending on the physical process in question (Fig. 19). We shall discuss these roughly in order of increasing subtlety.

(a) External Perimeter ("Hull"): d_{hull}

The total number of *external* surface sites, or "hull," scales with the caliper diameter or radius of gyration L as

$$N_{\text{hull}} \sim L^{d_{\text{hull}}}. \tag{7.1}$$

For $d = 2$ percolation, d_{hull} appears to be about 1.74 ± 0.02 (Voss 1984, Sapoval et al 1985, Ziff et al 1984, Weinrib and Trugman 1985), thus motivating the conjecture

$$d_{\text{hull}} = 7/4. \tag{7.2}$$

It would be highly desirable to prove this conjecture. "Volunteers?" A possible clue is that (7.2) is equivalent to $d_{hull} - 1 = d_{red}$, where $d_{hull} - 1$ is the fractal dimension of a cut of the hull by a line [cf. Eq.(1.4)]. For $d = 3$, is $d_{\text{hull}} = d_f$, as might be conjectured; due to the extreme rarity of completely enclosed "three-dimensional lakes," it is difficult to imagine that the ocean front (hull) scales with a smaller exponent than the total perimeter.

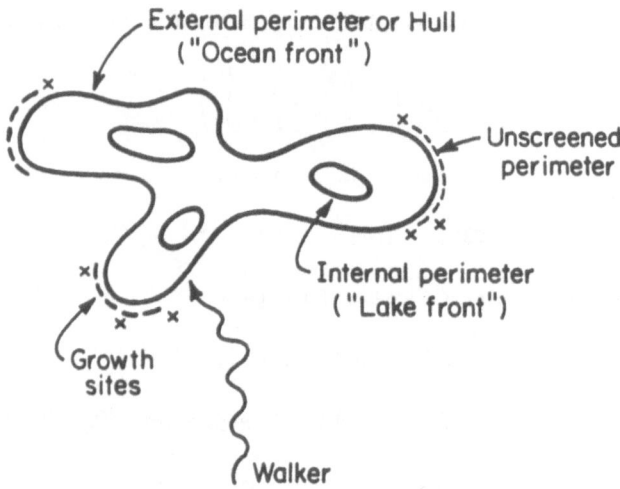

Fig. 19: Schematic illustration of four different fractal surfaces arising in the description of a percolation cluster. (a) The external "oceanfront" parameter or hull has a fractal dimension d_{hull}. (b) The total perimeter has a fractal dimension d_f, equal to that of the total bulk mass of the cluster. Since $d_f > d_{hull}$, it follows that the internal "lakefront" perimeter must have the same fractal dimension d_f of the total perimeter. (c) The unscreened perimeter where an incoming walker is more likely to hit has fractal dimension d_u (heavy solid lines). (d) The growth sites are those perimeter sites that form the living frontier of the cluster. These have fractal dimension d_g, but the nature of the G-site fractal depends on the actual mechanism of how the percolation cluster grows (see, e.g., Stanley et al 1984; Bunde et al 1985a,b; Herrmann and Stanley 1985).

(b) Total Perimeter: d_f

We know that for d-dimensional percolation the total number of perimeter sites $N_{\text{perimeter}}$ scales in the same fashion as the total number of cluster sites (Kunz and Souillard 1978),

$$N_{\text{perimeter}} \sim N_f \sim L^{d_f}. \tag{7.3}$$

For $d = 2$, $d_f = 91/48 = 1.896$. Hence the fact that $d_{\text{hull}} < d_f$ means that the ratio of *external* perimeter sites to *total* perimeter sites approaches zero at the percolation threshold. As the clusters get larger the *internal* perimeter sites ("lakefront" sites) completely swamp the *external* perimeter sites ("oceanfront" sites). For $d > 2$, internal perimeter sites are less commonly seen in finite computer simulations since it takes a lot of cluster sites to completely surround a "3-dimensional lake." An open question is the value of d_{hull} for $d > 2$: could it be that $d_{\text{hull}} = d_f$? Work is underway to test this possibility. Grossman and Aharony (1986) recently studied the "accessible perimeter," that subset of hull sites that can be reached by a random walker coming from outside the cluster. They found a fractal dimension $d_{\text{AP}} \approx 4/3$. The possibility that $d_{\text{AP}} = d_{\text{SAW}}$ (the fractal dimension of a self-avoiding random walk, known to be exactly 4/3 for $d = 2$) is under study currently. Particularly promising is recent work of Coniglio et al (1987) showing that the hull corresponds to the configuration of a dilute polymer chain at its theta point, so that a slight perturbation away from the theta point (a slight "repulsion") could cause the exponent to shift to its SAW value (Coniglio et al, unpublished).

(c) Unscreened Perimeter: d_u

Coniglio and Stanley (1984) introduced the concept of the "unscreened" perimeter to describe that portion of the hull that is effective in "termite" motion, which is the random walk responsible for conductivity Σ in the random superconducting network (RSN):

$$N_{\text{unscreened}} \sim L^{d_u}. \tag{7.4}$$

Moreover, they showed that the conductivity exponent for the RSN is simply related to d_u. Recalling the Nernst Einstein relation, we have

$$\sum \sim D \sim (R_{\text{cluster}})^2 \tau^{-1}. \tag{7.5}$$

Here the jump frequency τ^{-1} scales as the fraction of cluster sites belonging to the unscreened perimeter,

$$\tau^{-1} \sim [N_{\text{unscreened}}/N_f]. \tag{7.6}$$

Now substitute the Stauffer expression for the mean radius of the finite clusters, R_{cluster}, and the definitions for d_u and d_f into (7.6). We obtain

$$\sum \sim L^{2-(d-d_f)} L^{d_u - d_f}. \tag{7.7}$$

By definition $\sum \sim \epsilon^{-s} \sim \xi^{\tilde{s}}$. Hence

$$\tilde{s} \equiv \frac{s}{\nu} = d_u - (d - 2). \tag{7.8}$$

The conductance between two points scales as $L^{\tilde{\zeta}_{RSN}}$ where $\tilde{\zeta}_{RSN} = \tilde{s} + (d - 2)$. From (7.8) we find the extremely simple result that the conductance exponent is identical equal to the fractal dimension of the unscreened perimeter

$$\tilde{\zeta}_{RSN} = d_u. \tag{7.9}$$

Let us contrast now the RSN with the random resistor network (RRN), which is better understood. For the RRN, we can also relate the conductance exponent to fractal dimensions characterizing the substrate fractal. The Einstein relation holds, but we must set $n = P_\infty$, the probability that a randomly dropped ant will land on the incipient infinite cluster. As noted above, P_∞ scales as the co-dimension $(d - d_f)$, and D scales as $L^2/$time. Hence (Ben-Avraham and Havlin 1982; Gefen et al 1983)

$$\sum \sim P_\infty D \sim L^{d_f - d} L^{2 - d_w}. \tag{7.10}$$

By definition $\sum \sim \epsilon^\mu$ for the RRN limit, so we have

$$\tilde{\mu} = \mu/\nu = (d - 2) + (d_w - d_f). \tag{7.11}$$

The conductance between two point scales as $L^{\tilde{\zeta}_{RRN}}$, where $\tilde{\zeta}_{RRN} = \tilde{\mu} - (d - 2)$. Hence for the RRN problem, (7.9) is replaced by

$$\tilde{\zeta}_{RRN} = d_w - d_f, \tag{7.12}$$

but again an exponent is simply given in terms of fractal dimensions, **only**.

Fig. 20: One example of how a set of bonds or unit resistors between sites i and j is replaced by an equivalent number N_R of unit resistors placed in series (here $N_R = 1$). Thus by definition the resistance between i and j is equal to N_R, which in turn scales with the Pythagorean distance L between i and j with an exponent d_R (see Eq. (8.1)).

8. FRACTALS AND RANDOM RESISTOR/SUPERCONDUCTOR NETWORKS

It is convenient to think of the resistance between two points as the "mass" of 1-ohm resistors that we would place in series between the two points in order to have the same resistance (Fig. 20). In this way, we can interpret $\tilde{\varsigma}_{RRN}$ as a proper fractal dimension for some specific fractal object (the set of resistors)

$$N_R \sim L^{d_R}. \tag{8.1}$$

From (7.9) and (7.12) it then follows that

$$d_R = \begin{cases} -d_u & [\text{RSN}] \\ d_w - d_f & [\text{RRN}]. \end{cases} \tag{8.2}$$

From (8.2) we see that both the RSN and the RRN have resistance exponents d_R that are simply expressed in terms of fractal properties of the substrate. The expressions are completely different, of course, since the mechanism of transport is completely different (Fig. 21). From Fig. 21a, we see that the transport is determined by a "fisherman's net" structure, with a mesh size given by the connectedness length ξ. The strands of the net are made of singly-connected "red" bonds and multiply-connected "blue" bonds, the statistics of which will be described shortly. From Fig. 21b, we see that for the RSN just below p_c transport from one bus bar to the other is determined by the motion of charge carriers from one cluster to another—more precisely *out of* the unscreened

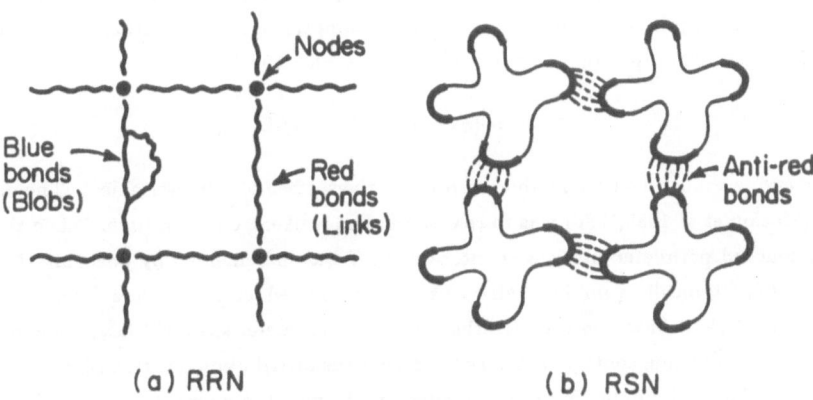

(a) RRN   (b) RSN

Fig. 21: Schematic illustration of the essential features of cluster structure in describing the conductivity of a general two-component random mixture in the limit of (a) the random resistor network, and (b) the random superconducting network. Adapted from Coniglio and Stanley (1984).

perimeter of one cluster and *into* the unscreened perimeter of the next (Hong et al 1985; Leyvraz et al 1986).

Thus the *clusters* in the RSN limit play the role of the *nodes* in the RRN limit. As one moves close to p_c in the RRN problem, the critical bonds are the singly-connected "red" bonds (the hottest). As one moves close to p_c in the RSN limit, the critical bonds are those bonds on the lattice which—if occupied—would connect two clusters. I call these pink (since they are "incipient" red bonds: once occupied they will become red); recently Stauffer suggested the term "anti-red" because in every sense they are the complement of the red bonds. Thus the red bonds control the physics of the RRN problem just above p_c, while the anti-red bonds control the physics of the RSN problem just below p_c. Coniglio (unpublished) has proved that the anti-red bonds have the same fractal dimension as the red bonds,

$$d_{\text{anti-red}} = 1/\nu. \tag{8.3}$$

Thus there is a certain symmetry between the RRN and RSN limits, which in some way should follow directly from the homogeneity theorems mentioned above. Work on this important topic is underway, and perhaps at this meeting some of you can help make progress along these lines.

9. MULTIFRACTALS: DLA AND PERCOLATION

Can we evaluate the fractal dimensions d_u and d_w appearing in (8.2) in terms of the fractal dimension d_f of the underlying substrate? Some progress along these lines has been made using arguments that require for their validity certain assumptions. In this section we will discuss the tests of a mean-field type argument (Coniglio and Stanley 1984) that

$$d_u = (d_f - 1) + (d - d_f)/d_w. \tag{9.1}$$

To this end, we must devise a method of probing the surface of a fractal object. The method we chose (Meakin et al 1985,1986) was to release random walkers, one at a time. When the random walker touched perimeter site i, a counter on site i was incremented by one unit (N_i becomes $N_i + 1$). After typically a million walkers have been released, statistics were done.

Our analysis is based on the idea that **all** of the perimeter sites will have some probability, however small, of being contacted. Hence to analyze the distribution function N_i ($i = 1, 2, \ldots, P$— where P is the total number of perimeter sites), we formed the moments μ_j defined through

$$[\mu_j]^j = \frac{N_T^{j+1}}{\sum_i N_i^{j+1}} = \frac{1}{\sum_i P_i^{j+1}} \sim [N_f]^{j\gamma_j}. \tag{9.2}$$

Here

$$N_T \equiv \sum_i N_i, \tag{9.3}$$

is the total number of incoming walkers, and

$$P_i = N_i/N_T, \tag{9.4}$$

is the probability that a given incoming walker will hit site i. The P_i are normalized to unity by virtue of (9.3).

We calculated the γ_j for $j = 1 - 3$, and found that the Coniglio-Stanley mean field relation holds to within the accuracy of our first calculations. We did notice a systematic dependence on j, so to test the possibility that the γ_j depend on j we extended the moment calculation to $j = 8$. The hinted dependence from $j = 1 - 3$ became much clearer (Fig. 22) and so we conclude that there is not a single exponent but rather an entire hierarchy of exponents (Meakin et al 1985,1986). This result has been independently obtained by Halsey et al (1986).

Why is the Coniglio-Stanley relation wrong? Presumably because it smears out the interface or "active zone" of the fractal into a band, and then assumes that there is an equal probability of capture for all surface sites within this band. In reality, there is a continuous gradation in "temperature," with the outermost tips being immensely hotter and the deepest invaginations

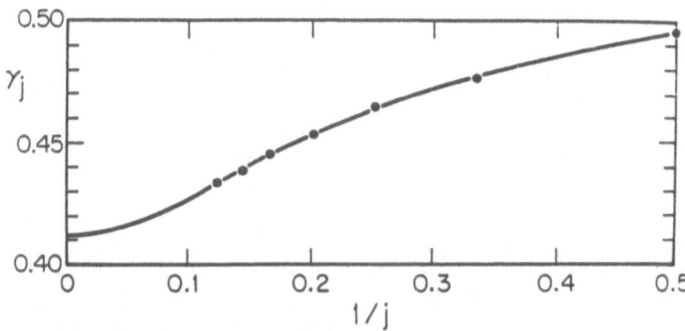

Fig. 22: The exponent γ_j characterizing the behavior of the jth moment of P_i. Here P_i, the basic quantity in surface growth, is the probability that perimeter site i is the next site at which the cluster will grow. In order to see if this apparent hierarchy or "spectrum" of exponents tends in the $j = \infty$ limit toward the expected limit, $1 - 1/d_f$, we have plotted these exponents against $1/j$. From Meakin et al (1986).

being extremely cold (Fig. 23). This situation is reminiscent of that found by De Arcangelis et al (1985,1986) for $N(V)$, the distribution of the number of bonds in the backbone across which the voltage drop is V. Here also there is a continuous gradation in temperature from the red bonds (the "hottest" in the sense that the full voltage drop of the entire cluster falls across each red bond) to the very cold bonds arising from the very long loops comprising the blobs.

This discovery of an infinite hierarchy of critical exponents—both in the voltage distribution of the percolation backbone and in DLA—is striking because normally one assumes that two exponents will suffice to describe a critical object. For example, y_h $(= d_f)$ and y_T $(= d_{red})$ are sufficient to describe percolation. However when we "do something" to the fractal, such as put

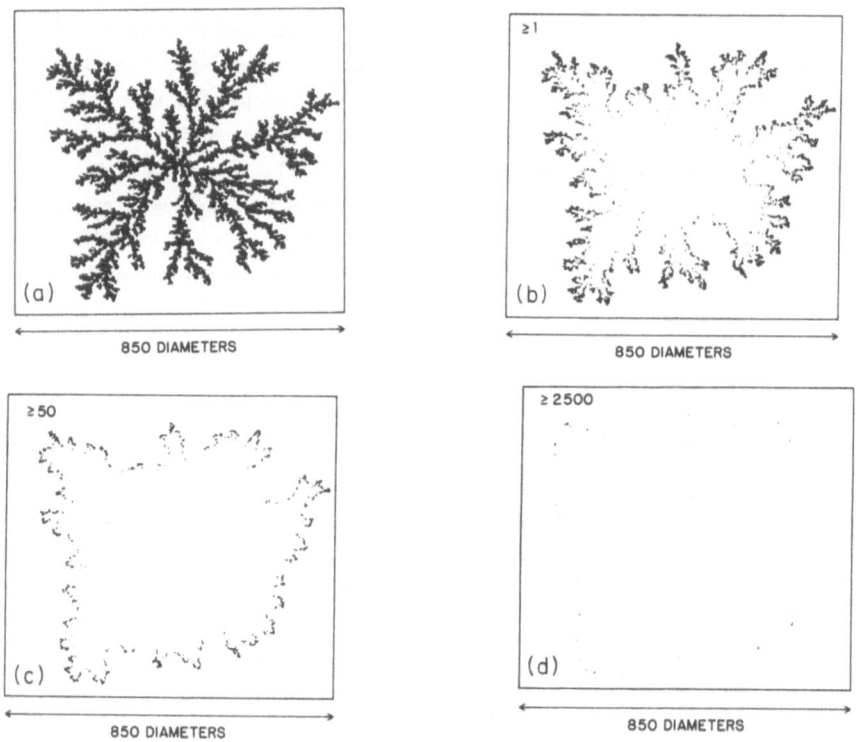

Fig. 23: This figure illustrates the harmonic measure for a 50,000 particle off-lattice 2d DLA aggregate. Figure 2a shows the cluster. Figure 2b shows all 6803 perimeter sites which have been contacted by at least one of 10^6 random walkers (following off-lattice trajectories). Figure 2c shows all of those perimeter sites which have been contacted 50 or more times and Fig. 2d shows those sites which have been contacted 2500 or more times. The maximum number of contacts for any perimeter site was 8197 so that $p_{max} = 8.2 \times 10^{-3}$. After Meakin et al (1986).

a battery across it or bombard it with random walkers, we introduce a new measure. Instead of each fractal site having weight 1, each site has a weight that depends on what we are doing to the fractal (e.g., each site has a voltage attached to it, or with each site we associate the number of hits on that site). Several groups (Boston University, Turkevich and Scher 1985, Halsey et al 1986) are currently seeking to understand the meaning of this new measure and what we can actually learn from this infinite hierarchy of exponents.

In retrospect, we might have anticipated this infinite hierarchy in advance. This is because for two extreme values of j, $j = -1$ and $j = \infty$, exponents differ by more than a factor of two: $\gamma(-1) = 1$ and $\gamma(\infty) = 1 - 1/d_f$. The first result follows immediately from the definition (9.2) and the fact that the total surface in DLA scales with exponent d_f just like the total mass. The second follows from the recent theorem (Leyvraz 1985) that P_{max} (the maximum value of all the P_i) scales with cluster mass to the exponent $1 - 1/d_f$. This prediction is confirmed by our calculations for $j = 1 - 8$. Our hierarchy or spectrum of surface fractal dimensions tends clearly toward a number fairly close to the predicted value $1 - 1/d_f$ (Fig. 22). By the way, the Leyvraz theorem can be "understood" as a special case of (9.1) where the "penetration term" is swamped.

10. FRACTAL GROWTH

How does a diseased or "damaged" region grow, and spread throughout a system? In the age of AIDS, this question of the growth or spread of a disease has taken on a new practical interest, in addition to the mathematical fascination that has existed for some time; for a recent work on the spread of damage through a cooperative system, see (Stanley et al 1987) and references therein.

Here we address not the general question of the growth of a damaged region, but the more specific question of how might we characterize the fashion in which the IIC, a percolation fractal, grows. It is important to state at the outset an obvious fact: completely different growth mechanisms can lead, eventually, to the same static fractal object (see, e.g., the discussion in Bunde et al 1985a).

For now, let us consider one of the simplest kinetic mechanisms of growth, the ant. Instead of dropping an "ant" (de Gennes 1976) onto a pre-formed fractal structure as we did before, we could instead drop the ant onto a Euclidean lattice but give her a set of rules with which she could form the fractal as she moves. This means that the ant would need not only the four-sided coin that all walkers carry (taking a square lattice for now), but she would also need a second coin providing some dynamical mechanism of generating the ultimate static fractal. For percolation fractals, this second coin is weighted so that each site can be open with probability p_c, and blocked with probability $1 - p_c$ (Bunde et al 1985a).

After some time has evolved, the ant is moving around in a rather interesting region of space that is characterized by 3 sorts of sites (the terminology in brackets suggests an epidemic interpretation of this entire growth problem):

(i) cluster sites already visited ["sick"]

(ii) sites already tested and blocked by the second coin ["immune"]

(iii) neighboring sites that have not been tested ["growth"].

Thus from the reference frame of the ant, only the growth sites are special in the sense that only these enlarge her territory.

Why are the growth sites interesting? Subjectively, they represent the "open frontier" (Rammal and Toulouse 1983) of the growing fractal. Objectively, this disconnected set of sites has a well-defined fractal dimension d_g in the sense that the number or "mass" of growth sites obeys a scaling relation of the form

$$N_g \sim L^{d_g}. \tag{10.1}$$

Here, as always, L represents the cluster diameter or radius of gyration.

The number of growth sites has been calculated as a function of cluster mass N_f. We predict for the "intrinsic" exponent

$$N_g \sim (N_f)^x \qquad \text{with} \qquad x = d_g/d_f. \tag{10.2}$$

Stanley et al (1984) obtained the first estimates of x, $x = 0.49$ [$d = 2$ percolation].

We now evaluate d_g for the two popular conjectured relations between d_w and d_f. The AO (Alexander-Orbach 1982) conjecture predicts $d_w = (3/2)d_f$, and the AS (Aharony-Stauffer 1984) conjecture $d_w = 1 + d_f$. Thus $d_g = 2d_f - d_w$ is

$$d_g = \begin{cases} d_f/2 = 91/96 = 0.9479 & \text{[AO]} \\ d_f - 1 = 43/48 = 0.8958 & \text{[AS]}. \end{cases} \tag{10.3}$$

For the intrinsic exponent x, we then predict

$$x = \begin{cases} 1/2 = 0.5000 & \text{[AO]} \\ 1 - 1/d_f = 0.4725 & \text{[AS]}. \end{cases} \tag{10.4}$$

Thus we see that the calculated values of x fall in between the AO and AS predictions.

Leyvraz and Stanley (1983) considered the conditions under which the AO conjecture might hold, focussing on the need for complete statistical independence of the increments in N_g. They noted that this statistical independence certainly occurs for the Cayley tree, since it is impossible to have correlations on a loopless fractal. In this fashion, they understand the numerical result that AO holds for the Cayley tree ($d_w = 6$, $d_f = 4$). In the asymptotic limit of a truly huge cluster they imagined that correlations in growth sites would begin to vanish even though loops exist. To the extent that these correlations eventually drop off, we can expect that in the asymptotic limit the increments in N_g will be statistically independent and hence the distribution will be normal, with $x = d_g/d_f$ exactly $1/2$ [AO].

Recently, the subject of growth sites has arisen in various contexts. One concerns a family of epidemic growth models, all of which eventually grow static percolation clusters (Bunde et al 1985a,b; Herrmann and Stanley 1985). Depending on the growth rules, the *dynamic* critical exponent d_g can differ—even though the *static* critical exponent d_f is the same for all rules.

We considered already one such epidemic model above. A second model is a variation of the above model in which the "ant" visits *only* growth sites, never re-visiting cluster sites. Clearly this requires the ant to make long-range jumps. In the simplest version of this model, the ant randomly chooses her next move from among all the existing growth sites, weighting them equally. This model can be simulated very rapidly (at least 100 times faster than the "walking" ant).

Had we set $p = 1$, we would have obtained an Eden cluster, with fractal dimension $d_f = d$. When a fraction $1 - p$ of the growth sites are poisoned, our flying ant acts rather like a discriminating butterfly, choosing carefully to land only on a non-poisoned site. A typical cluster at $p = p_c$ is shown in Fig. 24.

11. FRACTAL GEOMETRY OF THE CRITICAL PATH: "VOLATILE FRACTALS"

How does oil flow through randomly porous material just at that point where it can "break through" and reach the surface? How does electricity flow through a random resistor network where the fraction of intact resistors is close to the percolation threshold p_c? How does one describe the shape of the incipient infinite cluster that appears near p_c when one considers the polyfunctional condensation of monomers? There are but three of a host of questions that one can ask that are of some practical interest and that require for their answers a clear specification of cluster structure.

As mentioned above, it was noted (Stanley 1977) that there are two kinds of bonds in the backbone:

(i) singly-connected ("red") bonds and

(ii) multiply-connected ("blue") bonds.

Coniglio (1981,1982) provided a rigorous argument that for all d,

$$y_T \equiv \frac{1}{\nu} = d_{\text{red}}, \tag{11.1}$$

where $\xi \sim |\epsilon|^{-\nu}$ with $\epsilon = (p_c - p)/p_c$ and y_T is the thermal-like scaling field in percolation. Since

$$y_h \equiv \frac{\beta + \gamma}{\nu} = d_f \tag{11.2}$$

(Stanley 1977), it follows that both scaling fields in percolation are equal to geometric properties of the incipient cluster!

What about the blue bonds? To what is this fractal dimension d_{blue} related? Is there direct evidence that the red/blue picture of the backbone is valid? These are the questions addressed recently by Herrmann and Stanley (1984). To this end, they studied the "blob size distribution

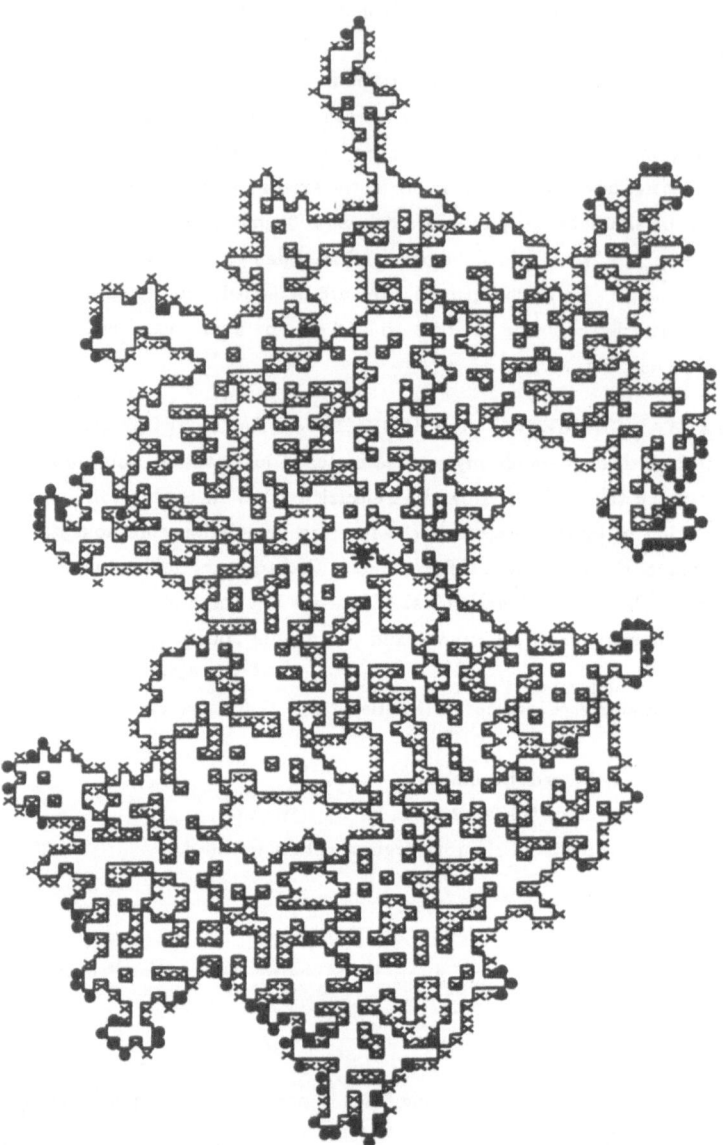

Fig. 24: Typical percolation cluster being grown by the "butterfly" mechanism for the case that each perimeter site has equal weight. This model is equivalent to the Eden model on a "diluted" lattice on which a fraction $1 - p_c$ of the sites have randomly been removed or "poisoned." After Bunde et al (1985a).

function" for *site*—as opposed to *bond*—percolation (cf. Figs. 25-27). They found—using an algorithm (Herrmann et al 1984) whereby the sites in the backbone is burned sequentially like trees in a forest—that the probability P_s that a randomly-chosen site in the backbone belongs to an s-site blob scales, for fixed box size, as

$$P_s = sn_s \sim s^{-d/d_f} \sim s^{-\tau+1}. \tag{11.3}$$

Now suppose we double L. What happens to the backbone? As Fig. 27a suggests, the actual identity of the backbone changes, as tiny strands of red sites become part of a big blob. In general, the little blobs cascade into big blobs and the entire distribution shifts "downward" (Fig. 27b). We say that the backbone is a **volatile fractal**, in that its identity depends on the box size.

Another volatile fractal is cluster-cluster aggregation. Here the time plays the same role as the box edge L. If we double t, then little clusters get eaten up, and become part of bigger clusters. The detailed "blob size distribution scaling analysis" for volatile fractals is presented by Herrmann and Stanley (1984) for the backbone case, and by Vicsek and Family (1984) for the "cluster size distribution scaling analysis" cluster-cluster aggregation case. The formalisms, though developed independently, are completely identical under the transformation

$$L \longleftrightarrow t. \tag{11.4}$$

From Fig. 25 it appears that the backbone is a pearl necklace, some of the pearls being swollen into a myriad of gargantuan shapes. Indeed, the backbone should be a string of

Fig. 25: Actual simulation of a backbone in site percolation just below p_c for a square lattice. The decomposition into blobs of all sizes from 1 to ∞ is apparent. After Herrmann and Stanley (1984).

pearls each of which is strung together by a drunken assembly line worker whose job is only to randomly choose a pearl from the pearl size distribution function P_s and attach it to the previously-chosen pearl in a linear "necklace" (the red sites are 1-site blobs). To test this appealingly simple possibility, we considered the statistical distribution of strings of "red sites". Specifically, we made a histogram of the number of strings with m sites plotted against m. We found that this histogram decays exponentially with m—one of the few examples when I've needed to use semi-log (as opposed to log-log) paper! The parameters entering this exponential decay are calculated in terms of the total number of red sites and the total number of blobs.

We conclude by making connection with the problem of anomalous diffusion and transport mentioned above. There we emphasized the fundamental importance of the Nernst-Einstein relation connecting a transport quantity, the *macroscopic* conductivity, to the diffusion constant which measures *microscopic* motion. Applied to the full cluster, we obtained the relation $d_R = d_w - d_f$. What happens if we chop off all the dangling ends? Clearly the conductivity cannot change, so that d_R is unaffected. However d_f is changed to d_{BB} and d_w is changed to d_w^{BB}, which

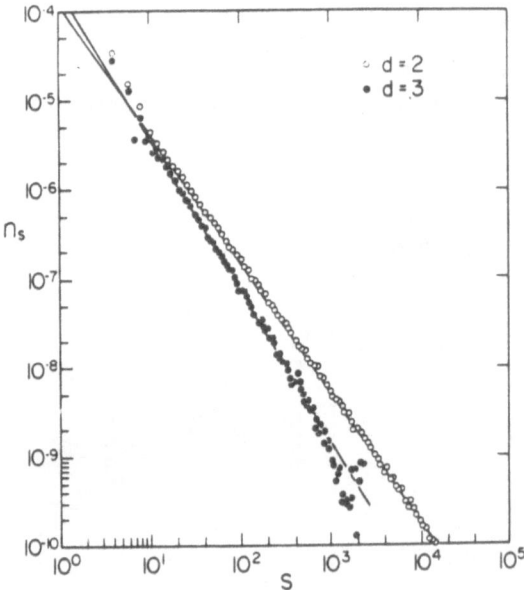

Fig. 26: Blob size distribution function for $d = 2$ and 3 for $L = 600$ ($d = 2$) and for $L = 60$ ($d = 3$). Note that the data are linear over many decades. After Herrmann and Stanley (1984).

is defined as the exponent connecting the number of steps *on the backbone* to the range of the walker: $N_w^{BB} \sim L^{d_w^{BB}}$. Hence the relation

$$d_R = d_w - d_f \tag{11.5}$$

becomes (Stanley and Coniglio 1984)

$$d_R = d_w^{BB} - d_f^{BB}, \tag{11.6}$$

Combining these two equations, we have the "invariance relation"

$$d_w - d_f = d_w^{BB} - d_f^{BB}. \tag{11.7}$$

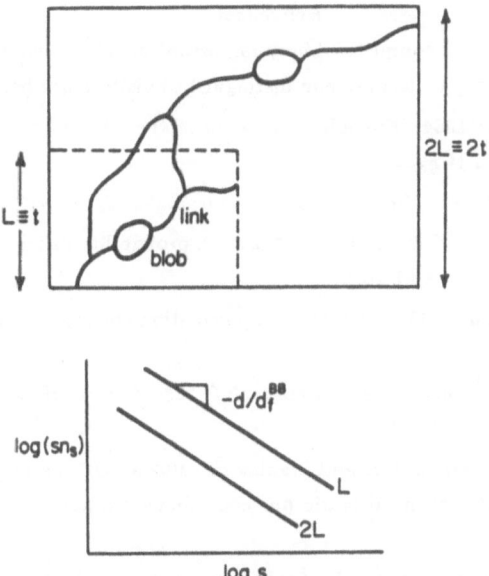

Fig. 27: (a) Schematic illustration of how small blobs become part of larger blobs when the box size L is doubled. (b) Schematic illustration of how the cluster size distribution $n_s(L)$ is uniformly depressed when L increases.

Acknowledgements

First and foremost, I wish to thank my recent collaborators. These include R.L.Blumberg [1], A. Bunde [49], A. Coniglio [39], G. Daccord [33], Z. Djordjevic [38], F. Family [98], A. Geiger [49], S. Havlin [972], H. J. Herrmann [57], D. C. Hong [82], N. Jan [809], J. Kertész [36], F. Leyvraz [41], I. Majid [91], A. Margolina [7], P. Meakin [44], H. Nakanishi [81], J. Nittmann [43], D. Stauffer [49], J. Teixeira [351], C. Tsallis [55], T. Witten [1], and F. Y. Wu [86]. I also wish to thank many without whose help the present understanding would not have arisen: A. Aharony [972], S. Alexander [972], Y. Gefen [972], P-G de Gennes [33], E. Guyon [33], B. Mandelbrot [48], and I. Procaccia [972]. I also wish to apologize that time and space considerations do not permit as coherent and complete a description of the subject as I had desired. Earlier work is reviewed elsewhere (Stanley et al 1980; Stanley 1981, 1982a,b,c, 1983, 1984a,b, 1985a,b, 1987; Stanley and Coniglio 1983). Last and not least, I wish to thank all my colleagues—especially my former graduate students—without whose insight I would not have arrived at my present appreciation for percolation.

References

This list is not intended to be complete. Many important articles have been omitted, and the work of my Boston University colleagues and distinguished visitors has been overemphasized.

Aharony A and Stauffer D 1984 "Possible breakdown of the Alexander-Orbach rule at low dimensionalities" Phys Rev Lett **52** 2368

Alexander S and Orbach R 1982 "Density of states on fractals: 'Fractons'" J de Physique **43** L625

Amitrano C, Coniglio A and diLiberto F 1986 "Growth probability distribution in kinetic aggregation processes" Phys Rev Lett **57** 1016

Ben-Avraham D and Havlin S 1982 "Diffusion on percolation clusters at criticality" J Phys A **15** L691-L697

Blumberg RL, Stanley HE, Geiger A and Mausbach P 1984 "Connectivity of hydrogen bonds in liquid water" J Chem Phys **80** 5230

Bunde A, Herrmann HJ, Margolina A and Stanley HE 1985a "On the universality of spreading phenomena: A new model with fixed static but continuously tunable kinetic exponents" Phys Rev Lett **55** 653

Bunde A, Herrmann HJ and Stanley HE 1985b "The shell model: A new growth model with continuously tunable forgotten growth sites" J Phys A **18** L523

Coniglio A 1981 "Thermal phase transition of the dilute s-state Potts and n-vector models at the percolation threshold" Phys Rev Lett **46** 250

Coniglio A 1982 "Cluster structure near the percolation threshold" J Phys A **15** 3829

Coniglio A and Stanley HE 1984 "Screening of deeply invaginated clusters and the critical behavior of the random superconducting network" Phys Rev Lett **52** 1068

Coniglio A, Jan N, Magid I and Stanley HE 1987 "New model embodying the physical mechanism of the coil-globule transition at the theta point of a linear polymer" Phys Rev B **35** xxx (1 March)

Coniglio A, Stanley HE and Klein W 1979 "Site-bond correlated percolation problem: A statistical mechanical model of polymer gelation" Phys Rev Lett **42** 518-522

Daccord G, Nittmann J and Stanley HE 1986 "Radial viscous fingers and DLA: Fractal dimension and growth sites" Phys Rev Lett **56** 336

deArcangelis L, Redner S and Coniglio A 1985 "Anomalous voltage distribution of random resistor networks and a new model for the backbone at the percolation threshold" Phys Rev B **31** 4725

deArcangelis L, Redner S and Coniglio A 1986 "Anomalous voltage distribution of random resistor networks" Phys Rev B **34** 4656

Deutscher G 1981 "Experimental relevance of percolation" in *Disordered Systems and Localization* (eds C Castellani, C DiCastro and L Peliti) Springer-Verlag Heidelberg p 26-40

Djordjevic ZV and Stanley HE 1987 "Scaling properties of the perimeter distribution for lattice animals, percolation and compact clusters," J Phys A **20** xxx

Eschbach PD, Stauffer D and Herrmann HJ 1981 "Correlation-length exponent in two-dimensional percolation and Potts model" Phys Rev B **23** 422

Fisher ME and Essam JW 1961 "Some cluster size and percolation problems" J Math Phys **2** 609-619

Gaunt DS 1980 "The critical dimension for lattice animals" J Phys A **13** L97-L101

Gawlinski ET and Stanley HE 1981 "Continuum percolation in two dimensions: Monte Carlo tests of scaling and universality for non-interacting discs" J Phys A **14** L291-L299

Gefen Y, Aharony A and Alexander S 1983 "Anomalous diffusion on percolating clusters" Phys Rev Lett **50** 77

Geiger A and Stanley HE 1982a "Low-density patches in the hydrogen-bonded network of liquid water: Evidence from molecular dynamics computer simulations" Phys Rev Lett **49** 1749

Geiger A and Stanley HE 1982b "Tests of universality of percolation exponents for a 3-dimensional continuum system" Phys Rev Lett **49** 1895

de Gennes PG 1976 "La percolation: un concept unificateur" La Recherche **7** 919

Grossman T and Aharony A 1986 "Structure and perimeters of percolation clusters" J Phys A **19** L745

Halsey TC, Meakin P and Procaccia I 1986 "Scaling structure of the surface layer of diffusion-limited aggregates" Phys Rev Lett **56** 854

Hankey A and Stanley HE 1972 "Systematic application of general homogeneous functions to static scaling and universality" Phys Rev B **6** 3515-3542

Harris AB and Lubensky TC 1981 "Generalized percolation" Phys Rev B **24** 2656-2670

Havlin S, Djordjevic Z, Majid I, Stanley HE and Weiss G "Relation between 'dynamic' transport properties and 'static' topological structure for branched polymers" Phys Rev Lett **53** 178-181

Hentschel GDE and Deutch JM 1986 preprint

Herrmann HJ, Hong D and Stanley HE 1984 "Backbone and elastic backbone of percolation clusters obtained by the new method of 'burning'" J Phys A **17** L261

Herrmann HJ and Stanley HE 1984 "Building blocks of percolation clusters: Volatile fractals" Phys Rev Lett **53** 1121

Herrmann HJ and Stanley HE 1985 "On the growth of percolation clusters: The effect of time correlations" Z Phys **60** 165

Hong DC, Stanley HE, Coniglio A and Bunde A 1985 "Random-walk approach to the two-component random-resistor mixture: Perturbing away from the perfect random resistor network and random superconducting-network limits" Phys Rev B **33** 4564

Isaacson J and Lubensky TC 1980 "Flory exponents for generalized polymer problems" J Phys Lett (Paris) **41** L469

Kasteleyn PW and Fortuin CM 1969 "Phase transitions in lattice systems with random local properties" J Phys Soc Japan **26S** 11

Kesten H 1982 *Percolation Theory for Mathematics* (Birkhäuser, Boston)

Kirkpatrick S 1978 "The geometry of the percolation threshold" AIP Conf Proc **40** 99

Kunz H and Souillard B 1978 "Essential singularity in percolation problems and asymptotic behavior of cluster size distribution" J Stat Phys **19** 77

Laidlaw D, MacKay G, and Jan N l967 "Some fractal properties of the percolating backbone in 2 dimensions" J Stat Phys (in press)

Lee TD and Yang CN 1952 "Statistical theory of equations of state and phase transtitions: II Lattice gas and Ising model" Phys Rev **87** 410-419

Leyvraz F 1985 "The 'active perimeter' in cluster growth models: A rigorous bound" J Phys A **18** L941

Leyvraz F and Stanley HE 1983 "To what class of fractals does the Alexander-Orbach conjecture apply?" Phys Rev Lett **51** 2048

Leyvraz F, Adler J, Aharony A, Bunde A, Coniglio A, Hong DC, Stanley HE and Stauffer D 1986 "The random normal superconductor mixture in one dimension" J Phys A **19** 3683-92

Meakin P, Stanley HE, Coniglio A and Witten TA 1985 "Surfaces, interfaces and screening of fractal structures" Phys Rev A **32** 2364

Meakin P, Coniglio A, Stanley HE and Witten TA 1986 "Scaling properties for the surfaces of fractal and non-fractal objects: An infinite hierarchy of critical exponents" Phys Rev A **34** 3325-3340

Nakanishi H and Stanley HE 1980 "Scaling studies of percolation phenomena in systems of dimension two to seven: Cluster numbers" Phys Rev B **22** 2466-2488

Nakanishi H and Stanley HE 1981 "Scaling studies of percolation phenomena in systems of dimensionality two to seven. II. Equation of state," J Phys A **14** 693-720

Nittmann J, Daccord G and Stanley HE 1985 "Fractal growth of viscous fingers: A quantitative characterization of a fluid instability phenomenon" Nature **314** 141

Nittmann J and Stanley HE 1986 "Tip splitting without interfacial tension and dendritic growth patterns arising from molecular anisotropy" Nature **321** 663

Nittmann J and Stanley HE 1987 "Non-deterministic approach to anisotropic growth patterns with continuously-tunable morphology" Phys Rev Lett (submitted)

Nittmann J, Stanley HE, Touboul E and Daccord G 1987 "Experimental evidence for multifractality" Phys Rev Lett **58** xxx

Parisi G and Sourlas N 1981 "Critical behavior of branched polymers and the Lee-Yang edge singularity" Phys Rev Lett **46** 871-874

Pike R and Stanley HE 1981 "Order propagation near the percolation threshold" J Phys A **14** L169-L177

Potts RB 1952 "Some generalized order-disorder transformations" Proc Cambridge Phil Soc **48** 106-109

Rammal R and Toulouse G 1983 "Random walks on fractal structures and percolation clusters" J de Physique **44** L13

Reynolds PJ, Stanley HE and Klein W 1977 "Ghost fields pair connectedness and scaling: Exact results in one-dimensional percolation" J Phys A **10** L203-L209

Reynolds PJ, Stanley HE and Klein W 1980 "Large-cell Monte Carlo renormalization group for percolation" Phys Rev B **21** 1223-1245

Sapoval B, Rosso M and Gouyet JF 1985 "Fractal nature of a diffusion front and relation to percolation" J Physique Lett **46** L149

Shlifer G, Klein W, Reynolds PJ and Stanley HE 1979 "Large-cell renormalization group for the backbone problem in percolation" J Phys A **12** L169-L174

Stanley HE 1968a "Spherical model as the limit of infinite spin dimensionality" Phys Rev **176** 718

Stanley HE 1968b "Dependence of critical properties on dimension of spins" Phys Rev Lett **20** 589-592

Stanley HE 1977 "Cluster shapes at the percolation threshold: An effective cluster dimension and its connection with critical-point exponents" J Phys A **10** L211-220

Stanley HE 1981 "New directions in percolation including some possible applications of connectivity concepts to the real world" in *Disordered systems and localization* eds C Castellani, C Di Castro and L Peliti (Springer-Verlag, Heidelberg) p 59-83

Stanley HE 1982a "Connectivity: A primer in phase transitions and critical phenomena for students of particle physics" in *Proc NATO Advanced Study Institute on Structural Elements in Statistical Mechanics and Particle Physics* eds K Fredenhagen and J Honerkamp (Plenum Press, New York)

Stanley HE 1982b "Geometric analogs of phase transitions: an essay in honor of Laszlo Tisza" in *Physics as Natural Philosophy: Festschrift in Honor of Laszlo Tisza* eds A Shimony and and H Feshbach (MIT Press, Cambridge)

Stanley HE 1982c "Renormalization group approach to polymer physics" Prog Physics (Beijing) **30** 95 [in Chinese]

Stanley HE 1983 "Aggregation phenomena: Models, applications and calculations" J Phys Soc Japan Suppl **52** 151

Stanley HE 1984a "Fractal concepts in aggregation and gelation: An introduction" in *Kinetics of Aggregation and Gelation* eds F Family and D Landau (North Holland, Amsterdam)

Stanley HE 1984b "Application of fractal concepts to polymer statistics and to anomalous transport in randomly porous media" J Stat Phys **36** 843

Stanley HE 1985a "Critical phenomena" in *Encyclopedia on Polymer Science* (Wiley, NY) vol 4

Stanley HE 1985b "Form: An introduction to self-similarity and fractal behavior" In *On Growth and Form: Fractal and Nonfractal Patterns in Physics* Proc 1985 Cargese NATO ASI Institute, eds HE Stanley and N Ostrowsky (Martinus Nijhoff Pub, Dordrecht) page 21

Stanley HE 1987 *Introduction to Fractal Phenomena* (Oxford Univ Press, London and New York)

Stanley HE and Teixeira J 1980 "Interpretation of the unusual behavior of H_2O and D_2O: Tests of a percolation model" J Chem Phys **73** 3404-3424

Stanley HE and Coniglio A 1983 "Fractal structure of the incipient infinite cluster in percolation" in *Percolation Structures and Processes* eds G Deutscher, R Zallen and J Adler (Adam Hilger, Bristol)

Stanley HE and Coniglio A 1984 "Flow in porous media: The backbone fractal at the percolation threshold" Phys Rev B **29** 522

Stanley HE and Herrmann HJ 1987 "The fractal dimension for the minimum path in two-dimensional and three-dimensional percolation" preprint

Stanley HE, Birgeneau RJ, Reynolds PJ and Nicoll JF 1976 "Thermally-driven phase transitions near the percolation threshold in two dimensions" J Phys C **9** L553-560

Stanley HE, Teixeira J, Geiger A and Blumberg RL 1981 "Interpretation of the unusual behavior of H_2O and D_2O at low temperature: Are concepts of percolation relevant to the puzzle of liquid water?" Physica **106A** 260-277

Stanley HE, Majid I, Margolina A and Bunde A 1984 "Direct tests of the Aharony-Stauffer argument" Phys Rev Lett **53** 1706

Stanley HE, Reynolds PJ, Redner S and Family F 1982 "Position-space renormalization group for models of linear polymers branched polymers and gels" in *Real-Space Renormalization* eds TW Burkhardt and JMJ van Leeuwen (Springer-Verlag, Heidelberg) Chap 7

Stanley HE, Coniglio A, Klein W, Nakanishi H, Redner S, Reynolds PJ and Shlifer G 1980 "Critical Phenomena: Past present and future" in *Proceedings of the International Symposium on Synergetics* ed H Haken (Springer-Verlag, Heidelberg) Chap 1

Stanley HE, Stauffer D, Kertész J and Herrmann HJ 1987 "Dynamics of spreading phenomena in cooperative models" Nature (submitted)

Stauffer D 1985 *Introduction to Percolation Theory* (Taylor and Francis, Philadelphia)

Sykes MF, Gaunt DS and Glen M 1981 "Perimeter polynomials for bond percolation processes" J Phys A **14** 287-292

Turkevich LA and Scher H 1985 " Occupancy-probability scaling in diffusion-limited aggregation" Phys Rev Lett **55** 1026

Vicsek T and Family F 1984 "Dynamic scaling for aggregation of clusters" Phys Rev Lett **52** 1669

Voss RF 1984 "The fractal dimension of percolation cluster hulls" J Phys A **17** L373

Weinrib A and Trugman SA 1985 "A new kinetic walk and percolation perimeters" Phys Rev B **31** 2993

Wu FY 1978 "Percolation and the Potts model" J Stat Phys **18** 115

Wu FY 1982 "The Potts model" Rev Mod Phys **54** 235

Wu FY and Stanley HE 1983 "Polychromatic Potts model: A new lattice statistical problem and some exact results" J Phys A **16** L751-L755

Zernike F 1940 "The propagation of order in cooperative phenomena" Physica **7** 565

Ziff RM, Cummings PT and Stell G 1984 "Generation of percolation cluster perimeters by a random walk" J Phys A **17** 3009

SURFACE SIMULATIONS FOR LARGE EDEN CLUSTERS

D. Stauffer

Institute for Theoretical Physics
Cologne University
5000 Köln 41, West Germany

Since the work with J.G. Zabolitzky on the Cray-2 at the University of Minnesota, from which I presented preliminary results at this conference, has been published in the meantime /1/, I now only given an overview, why simulations with millions of sites were needed to see the asymptotic behavior of the surface thickness of clusters in the Eden process. Also I add some remarks about still controversial points.

Introduction

Eden/2/ suggested a quarter of a century ago a simple model for cluster growth. For twenty years this work was cited only a few times a year, if at all. Recently about 2 papers appeared each month referring to this Eden paper. Why this sudden upturn of interest? One reason is the general interest in growth models /3/,/4/, of which the Eden model seemed to be the simplest case, similar perhaps to the Ising model in the field of thermal phase transitions, or percolation theory for disordered systems. The other more, specific, reason is the detection of nontrivial surface behavior /5/ by Plischke and Racz in 1984. I will ignore here the confusing history of that surface thickness /5-10/ and concentrate on our present knowledge.

In the usual version of the Eden model one starts with one site in the center of a lattice occupied. Then, at each time step, one adds one site by occupying one of the perimeter sites of this cluster. (Perimeter sites are empty neighbors of occupied sites.) The selection of the perimeter site to be occupied is completely random. In this way, out of the original single site one builds a cluster of s sites in s time steps.

Since at every time step each empty lattice site in the interior of the cluster has the same probability to be occupied, after some time it will be filled up. Thus deep in the interior of the Eden cluster, sites will be occupied with probability one: The cluster is fully compact. (See Leyvraz /12/ for mathematical proofs and literature, and Meakin's review in ref. 4 for a different version with holes even deep inside the cluster.) Thus the radius of a cluster in d dimensions increases asymptotically as $s^{1/d}$ and the perimeter as $s^{1 - 1/d}$. Thus we now concentrate on the thickness W of the surface of the cluster, defined through:

$$W^2 = <r^2> - <r>^2 \tag{1}$$

where we average over all perimeter sites their distance r (and its square) from the origin. Since there are no perimeter sites far away from the cluster or far in the cluster interior, it seems legitimate to use this definition for the thickness of the surface layer separating the cluster from the surrounding vacuum. We want to know how this thickness W depends on the cluster mass s.

II. Results

a) Round Clusters on Square Lattice

Simulations for s up to 1000 million sites gave three regions of behavior: For small s up to about a tenth of a million the thickness, plotted double-logarithmically versus s, gives a monotonically increasing curve, with monotonically decreasing slope, just as if W would be a power of log(s). For intermediate s between one and thirty million sites, the thickness increases as the square root of the radius, i.e. as $s^{1/4}$. For s between thirty and thousand million sites, the thickness W is about 7 % of the radius, i.e. it increases as $s^{1/2}$. Needless to say, a billion sites cannot be stored on a microcomputer, and the huge memory of the Cray-2 comes in handy.

Should we thus expect a fourth or fifth region of behavior with future increases of computational power or efficiency. I claim: No! There are good reasons to believe we have reached the asymptotic behavior. The first region is nearly normal for most power laws in Statistical Physics: For small sizes the asymptotic behavior is not yet reached. In the present case these corrections to the asymptotic power law behavior are unusually large, like $(1+18/r-...)$, thus making the first nonasymptotic regime also usually large. The second regime is the asymptotic regime, provided lattice anisotropy is neglected; simple capillary wave theory as well as more sophisticated approaches /13/ give a surface roughening proportional to the square-root of the length (radius), just as observed. The third regime, finally, is the same one were lattice anisotropies dominate /9-11/, as explained now.

The growth speed in the direction of the lattice axes is slightly faster than in the diagonal direction since the number of perimeter sites per unit length are different /9,11/. Thus very large clusters are not exactly circular but slightly distorted in the direction of diamond shape. This slight anisotropy explains why very large clusters have a surface width proportional to the radius, in the third region of behavior: Eq(1) averages over all perimeter sites, on the lattice axes, on the diagonal and in between. If the distance r on the axes is always a few percent larger than r on the lattice diagonal, then the difference $\langle r^2 \rangle - \langle r \rangle^2$ is loosing its meaning of a fluctuation and simply becomes a small fraction of $\langle r \rangle$. The transition from the second region, $W \propto s^{1/4}$, to the third regime, $W \propto s^{1/2}$, is thus reached if the thickness W due to this anisotropy alone is about as large as the thickness caused by the surface fluctuations even without lattice anisotropy. Since anisotropy of clusters was about the same from $s = 10^4$ to $s = 10^9$, one should not expect it to change for larger sizes: Therefore presumably the third regime for round clusters is indeed the asymptotic one.

b)   Flat Surfaces on Square Lattice

One way to get rid of anisotropies is to simulate flat surfaces in the "Eden deposit" algorithm /7/, instead of a round cluster. Thus one occupies initially the base line of width L in an LxH square lattice, with periodic boundary conditions only horizontally. Then perimeter sites are again selected randomly to be occupied. One may imagine this flat surface to be part of the surface of a huge cluster, with radius much larger than L. Growth occurs now only in the direction of a lattice axis, apart from fluctuations, and the results are simpler than for round clusters: In a first region for small L, the corrections to the asymptotic scaling behavior make life complicated, but in the second region from L near 500 to the maximum L=8000, the thickness W increased as the square root of L, as expected theoretically /13/.

Moreover, for round clusters in the first and second regime (below 30 million sites), the thickness W of a cluster with radius r is about the same as the thickness of a flat surface with a width L equal to this radius r. Thus the flat surface seems indeed to simulate reasonably well the round clusters, except that it avoids the problems from anisotropy.

In the simulation of flat surfaces of width L, after L growth steps the height h of the surface layer has increased by one lattice unit. One may therefore interpret this height also as time. The preceding results referred to equilibrium thicknesses $W(h=\infty)$ in the sense that the time h is sufficiently large to make W time-independent.

Instead one may ask how does W grow with time:

$$W \propto h^\beta \tag{2}$$

for

$$1 \ll W(h) \ll W(\infty) \tag{3}.$$

The effective exponent $\beta = d(\log W)/d(\log h)$ is found /1/ to be oscillating about $1/2$ for h up to 3, to decrease to about $1/4$ between h=3 and h=30, and to increase again for longer times h, until it levels off near $\beta=1/3$ for h above 3000. (These simulations used L up to 16 million, occupying up to $8 \cdot 10^9$ sites but storing only the surface region.) This asymptotic result $\beta = 1/3$ in two dimensions was predicted by Kardar et al /13/.

In summary, two dimensions seem well understood in these aspects: For radii r, widths L or time h larger than about 100 the trend towards asymptotic behavior seems to set in.

c) Simple Cubic Lattice and Higher Dimensions

Round clusters in three dimensions /10/ seem somewhat difficult to analyse, and thus most work in higher dimensions /5,7,2/ used flat surfaces. The behavior was similar to the square lattice: One needs linear dimensions aboved 100 to see the onset of the asymptotic (behavior). But that is quite difficult to reach now: In four dimensions, 100^4 is already quite large.

Moreover, for small L the thickness W increases much weaker with L in 3 and 4 dimensions compared to 2 dimensions, and thus condition (3) is much more difficult to satisfy. Thus for the equilibrium results we only know that the trends observed /7/ for small L in 3 and 4 dimensions are not continuued for L above 100; it is not clear numerically what this new behavior really is. Similarly, for the dynamic exponent β one finds a minimum near time h = 100, but it is not clear to what value β will increase for much longer times in more than 2 dimensions. The data do neither confirm nor contradict the prediction /13/ that the asymptotic exponents describing the surface thickness are independent of dimensionality in this Eden model.

III. Questions

a) Choice of Model

The above type of Eden clusters are called /7/ model A. In model C one first choses randomly an occupied site which has at least one empty neighbor, and then selects randomly one of these empty neighbors to be occupied. Refs. 7 and 9 claimed that model C is numerically better in the sense that the asymptotic behavior is reached for much smaller clusters. In particular, for model C, ref. 9 found $\beta = 0.307 \pm 0.007$ on the square lattice, which is in bad agreement with the theoretical /13/ prediction $\beta=1/3$ and the model A result above. However, inspection of fig. 3b in ref. 9 for the largest $L = 8000$ indicates that for fixed L as function of time h the effective dynamical exponent β first decreases, then increases, similar to model A above. Thus there is no reason to mistrust here the general universality principle that details of the models seldomly change the asymptotic exponents. For the equilibrium behavior of flat surfaces, on the other hand, model C indeed seems to get to its asymptotic regime earlier than model A, and ref. 9 could confirm on the square lattice the square-root law for the equilibrium width with less than 1 % error in the exponent.

Other possibilities to decrease the effects of too small cluster sizes may be Fourier transforms of the surface fluctuations /8/ or noise reduction /14/.

b) Round versus Flat

We mentioned above that numerically the surface thickness for round clusters or radius r is about the same as for flat surfaces of width $L=r$. This simple relation is questionable /7-9/: For flat surfaces one may postulate /15/

$$W(L,h) = L^{1/2} f(h/L^{3/2}) \tag{4}$$

in two dimensions, using $\beta=1/3$. For small arguments $f(y)$ varies as $y^{1/3}$. If

this law is applied also to round clusters, identifying h and L with the
cluster radius r, it gives

$$W \propto r^{1/3} \propto s^{1/6} \tag{5}$$

in two dimensions. Such reasoning neglects the problems of anisotropy; even if
one circumvents these problems by looking only at the surface width along the
lattice axes /10/ the numerical agreement with eq.(5) is not good for the
largest clusters. Moreover, the growth of a round cluster is known /10/ to give
a somewhat different surface thickness than that of a quadrant only; even more can
the growth results differ between round clusters and flat surfaces. Therefore
eq.(5) is not well founded, though it might still be correct.

c) Surface Tension

Kardar et al /13/ distinguish between usual Eden models and other surface
growth problems. In Eden models, according to Kardar et al one should not take
into account a surface tension, whereas generally one has a nonzero surface
tension. This distinction is crucial for their predictions in, for example, four
dimensions. Indeed, the randomness with which the next perimeter site is selected
to be occupied shows that microscopically there is no surface energy.

However, a nonzero surface tension might still be needed for a field theory
of the type used by Karkar et al. For example, a usual Ising model has a
microscopic energy quadratic in the spins, without a quartic term. Nevertheless,
the thermodynamic free energy, expanded in powers of the magnetization M (above
the Curie temperature), does have a nonzero term proportional to M^4; and in
renormalization group treatments of the Ising model one uses field theories
containing both the quadratic and a quartic term. Thus the distinction /13/
between this simple Eden model without interactions and other surface models with
interactions is not at all obvious.

d) Continuum Simulation

Simulating Eden clusters on a continuum, not on a lattice, lets the disturbing effects of the lattice anisotropy vanish. However, the continuum Eden model simulated by Botet /16/ has other properties which distinguish it even stronger from the usual case: Its interior density is no longer unity, as for most other Eden models (see Meakin in ref. 4 for another exception). Whereas on a lattice, all interior holes have at least one unit cell as volume, on a continuum the interior holes can be arbitrarily small. Very small holes will never be filled. Thus on a lattice finally all holes are filled up, whereas on a continuum some small holes remain. As a result, one should no longer identify the interior holes as part of the surface. Similar to percolation theory on a lattice /17/, the perimeter is now proportional to the cluster mass, and not a good measure of a physical surface.

Instead one should for continuum Eden models look ad the gradient of the density profiles /5,6/, and not at eq.(1), to define a surface thickness. It would be interesting to see if then the surface exponents are different from those for the lattice version.

Summary

Large scale simulations were able to clarify most problems with Eden clusters on the square lattice; but for higher dimensions the asymptotic region has not yet been simulated. In spite of its simplicity in the definition, the Eden model is numerically exremely complex.

I thank R. Botet, H.J. Herrmann, M. Kardar, P. Meakin, M. Plischke and D. Wolf for helpful discussions.

References

1. J.G. Zabolitzky and D. Stauffer, Phys. Rev. A **34**, 1523 (1986) and Phys. Rev. Letters **57**, 1809 (1986).

2. M. Eden, in : Proceedings of the Fourth Berkeley Symposium on Mathematical Statistics and Probability, edited by J. Neyman (University of California, Berkeley 1961), Vol. IV.

3. H.J. Herrmann, Phys. Repts. **136**, 153 (1986)

4   On Growth and Form, edited by H.E. Stanley and N. Ostrowsky, (Martinus Nijhoff, Dordrecht 1986).

5   M. Plischke and Z. Racz, Phys. Rev. Letters **53** 415 (1984); **54**, 2056 (1985); Phys. Rev. A **31**, 985 (1985).

6   H.P. Peters, D. Stauffer, H.P. Holters, and K. Loewenich, Z. Physik B **34**, 399 (1979).

7. R. Jullien and R. Botet, Phys. Rev. Letters **54**, 2055 (1986) and J. Phys. A **18**, 2279 (1985).

8. M. Plischke and Z. Racz, Phys. Rev. A **32**, 609 (1986).

9. P. Meakin, R. Jullien and R. Botet, Europhys. Lett. **1**, 609.

10. P. Freche, D. Stauffer, and H.E. Stanley, J. Phys. A. **18**, L 1163 (1985)

11. R. Hirsch and D. Wolf, J. Phys. A **19**, L 251 (1986); D. Wolf, J. Phys. A , in press.

12. F. Leyvraz, J. Phys. A **18**, L 941 (1985).

13 M. Kardar, G. Parisi and Y.C. Zhang, Phys. Rev. Letters **56**, 889 (1986) and **57**, 1810 (1986); see also D. Dhar, poster at StatPhys 16 Conference, Boston University (August 1986).

14 K. Kertesz and D. Wolf, priv. comm.

15. F. Family and T. Vicsek, J. Phys. A **18**, L 75 (1985).

16. R. Botet, preprint 1986).

17. D. Stauffer, Phys. Repts. **54**, 1 (1979).

DUALITY FOR k-DEGREE PERCOLATION
ON THE SQUARE LATTICE

John C. Wierman

Department of Mathematical Sciences
The Johns Hopkins University
Baltimore, Maryland 21218 U.S.A.

ABSTRACT

A generalization of the standard percolation problem, called k-degree percolation, is considered on the square lattice. In k-degree percolation, one investigates the probability of existence of an infinite path in which each vertex has k or more open edges incident to it.

One motivation for the study of k-degree percolation is to provide models which exhibit multiple phase transitions, since most substances exist in at least three possible phases. Quintas (1983) considered k-degree percolation on the ice lattice to estimate the volume function for water.

The duality approach to percolation problems on matching pairs of graphs is extended, establishing that the three standard critical probabilities -- p_H, p_T, and p_S -- are equal for k-degree percolation on the square lattice. Furthermore, the dual model for the 4-degree (or full-degree) model on the square lattice is shown to be equivalent to a standard percolation model on a nonplanar graph which is not a member of a matching pair of graphs in the sense of Sykes and Essam (1964), establishing that $p_H = p_T = p_S$ for this model. The approach may lead to the formulation of a broader class of graphs for which this equality holds.

1. Introduction

The classical percolation model consists of an infinite lattice graph G with an associated random mechanism which assigns one of two states -- *open* or *closed* -- to each vertex or edge. In a *bond percolation model*, each edge is open with probability p, $0 \leq p \leq 1$, independently of all other edges, with all vertices considered to be open. In a *site percolation model,* each vertex is open with probability p, $0 \le p \le 1$, independently of all other vertices, while an edge is open if and only if both its endpoints are open vertices (and closed if and only if both its endpoints are closed). A central concept in percolation theory is the *critical probability,* a threshold value of the parameter p above which there exists an infinite path of open edges and vertices with positive probability. The critical probability is analogous to a phase transition of a physical system represented by the percolation model.

In a site percolation model, an infinite path of open edges and vertices may be viewed as an infinite path of open vertices which have at least two open edges incident. This suggests a generalization of the standard percolation problem, called k-degree percolation, in which, for some integer k, one considers the existence of an infinite path in which each vertex has k or more open edges incident to it.

One motivation for the study of k-degree percolation is to provide models which exhibit multiple phase transitions. Most substances exist in at least three possible physical phases -- gas, liquid, and solid -- depending on the temperature and pressure of the physical system. As the parameters of temperature and pressure change gradually, the physical system's properties change gradually, for a broad range of values. However, the properties change dramatically at certain threshold values such as the condensation temperature and freezing temperature, where phase transition occurs. In a substance in the gaseous phase, molecular interactions lead to small clusters of molecules, with the average size varying slowly as temperature is lowered. At the condensation temperature, a giant cluster forms suddenly. With further cooling, the giant cluster becomes increasingly richly connected, untill at the freezing temperature it becomes a rigid crystalline solid. In the percolation model, these two phase transitions may be modeled by critical probabilities for k-degree percolation for two distinct values of k.

A k-degree bond percolation model on the ice lattice was considered by Quintas (1983) and Kennedy (1982) to estimate the volume function for water. The parameter p, considered a function of temperature, is interpreted as the probability of hydrogen bonding between water molecules. Physical assumptions relating volume to bond and cycle formation lead to a volume function which is qualitatively of the correct form. By matching estimated k-degree percolation critical probabilities with the phase transitions in water at 0 and 100 degrees Celsius, Quintas (1983) obtained a good fit to empirical data.

This paper initiates a theoretical investigation of k-degree percolation, beginning with the study of the model on the square lattice. The duality approach to the critical probability problem is extended to k-degree percolation in the square lattice site model, by minor technical modifications of standard methods. This establishes the result that the three standard versions of the critical probability -- p_T, p_H, and p_S -- are equal for k-degree percolation on the square lattice. Furthermore, the dual model for 4-degree (or *full-degree*) percolation in the square lattice site model is shown to be equivalent to a classical percolation model on a nonplanar lattice graph which is not a member of a matching pair of graphs [in the sense of Sykes and Essam (1964)]. Using the duality theory developed, we establish that $p_H = p_T = p_S$ for this model. The method, extended to a class of periodic graphs generalizing the matching graphs of Sykes and Essam, could substantially enlarge the set of graphs for which this equality is known to be valid.

Basic definitions and notation are presented in section 2. A description of the duality results is provided in section 3. Due to its length, the proof of the duality result is not included in this paper, but is available in a technical report [Wierman (1986)]. The relationship of the dual model of the full-degree model with a standard percolation model, is established in section 4. Discussion of proposed extensions, generalizations, and open problems is in section 5.

2. Definitions

The *square lattice S* is an infinite graph which may be imbedded in the plane so that each vertex $v = (v_1, v_2)$ has integer coordinates and a unit length edge connects $v = (v_1, v_2)$ and $w = (w_1, w_2)$ if and only if $||v - w|| \equiv [\, |v_1 - w_1|^2 + |v_2 - w_2|^2 \,]^{\frac{1}{2}} = 1$. The *matching graph S** of the square lattice has the same vertex set as S, but vertices v and w are connected by an edge of S^* if and only if $||v - w|| \leq 2^{1/2}$.

In the classical site percolation model, each site v is assigned the state "open" with probability p, $0 \leq p \leq 1$, and is assigned the state "closed" otherwise. An edge is open if and only if both its endpoint vertices are open, and closed if and only if both its endpoint vertices are closed. Thus, an edge may be neither open nor closed. Let P_p and E_p denote the probability measure and expectation operator parameterized by p, respectively.

For each $k = 0, 1, 2, 3, 4$, define a vertex v to be *k-degree open (closed)* if at least k edges incident to v are open (closed). A vertex v is *k-(+)degree open (closed)* if at least k edges incident to v are not closed (not open). [The term (+)degree refers to the degree in the graph of open edges plus the degree in the graph of edges which are neither open nor closed.] In the site percolation model, a vertex is k-degree open if and only if it is open and k or more adjacent vertices are open.

A path in S is a sequence of vertices $v_0, v_1, v_2, \ldots, v_n$ such that $||v_{i-1} - v_i|| = 1$ for each $i = 1, 2, 3, \ldots, n$. A path in S^*, or *-path, is a sequence of vertices $v_0, v_1, v_2, \ldots, v_n$ such that $||v_{i-1} - v_i|| \leq 2^{\frac{1}{2}}$ for each $i = 1, 2, 3, \ldots, n$. A path or *-path is *k-degree open (closed)* if every vertex in the path is k-degree open (closed), and is *k-(+)degree open (closed)* if every vertex is k-(+)degree open (closed). The *k-degree (k-(+)degree) open cluster* containing vertex v, denoted by $W_v(k)$ $(W_v(k+))$, is the set of all vertices which are connected to v by a k-degree (k-(+)degree) open path. Similarly, we define *-clusters, denoted by $W_v^*(k)$ $(W_v^*(k+))$, consisting of the set of vertices connected to v by k-degree (k-(+)degree) open *-paths. The cluster size, denoted by $\#W_v(k)$ $(\#W_v(k+))$, is the number of vertices in $W_v(k)$ $(W_v(k+))$. To denote the k-degree closed cluster or k-(+)degree closed cluster, we add a superscript $-$.

We now extend the three common definitions of critical probability to k-degree percolation. Define the *k-degree percolation probability* by

$$\theta(k; p) = P_p[\ \#W_v(k) = \infty\ ],$$

for some fixed $v \in S$. Note that by translation invariance of S, the k-degree percolation probability is independent of v, as lack of dependence in the notation reflects. The *k-degree cluster size critical probability* is defined by

$$p_H(k) = \inf \{\ p : \theta(k; p) > 0\ \},$$

and represents the threshold above which an infinite k-degree open cluster exists with positive probability. The *mean k-degree cluster size critical probability* is defined by

$$p_T(k) = \inf \{\ p : E_p(\ \#W_v(k)) = \infty\ \},$$

for any fixed $v \in S$ (but is independent of v as above). As in the classical percolation models, by definition, $p_T(k) \leq p_H(k)$.

Let $[m_1, m_2] \times [n_1, n_2]$. denote the subgraph of S contained in the region $\{\ (x, y) \in R^2 : m_1 \leq x \leq m_2, n_1 \leq y \leq n_2\ \}$. A path of vertices in $[m_1, m_2] \times [n_1, n_2]$ with initial vertex on the line $x = m_1$ and terminal vertex on the line $x = m_2$ is a *horizontal crossing* of

$[m_1,m_2] \times [n_1,n_2]$. Define

$$\tau(k;m,n,p) = P_p[\text{ there exists a } k\text{-degree open horizontal crossing of } [0,m] \times [0,n]\ ].$$

The *k-degree crossing critical probability* is

$$p_S(k) = \inf \{p: \lim_{n \to \infty} \sup \tau(k;n,n,p) > 0 \}.$$

[Note that this is different from the definition of Kesten (1982), which considers rectangles. For convenience here we follow the earlier proof of Russo (1978, 1981).] In the classical models, the p_S definition is the key to rigorous evaluation of the critical probability and the establishment of equality of the other versions of critical probability.

Crossing, percolation and critical probabilitites may be defined for k-(+)degree percolation, denoted by $\tau(k+;m,n,p)$, $\theta(k+;p)$, $p_H(k+)$, $p_T(k+)$, and $p_S(k+)$. Similar quantities may be defined for the matching lattice S^*, and all denoted by adding a superscript $*$ to the corresponding expressions for S.

Note that we need only consider k-degree percolation for $k \geq 2$. Every vertex has degree at least zero, so 0-degree percolation is trivial. In a path of 1-degree open vertices each vertex must be open and have two open neighbors, so the 1-degree open and 2-degree open critical probabilies are identical.

3. The Duality Result

We now state a result for the k-degree percolation model on the square lattice that is analogous to the principal result of Kesten (1982) for site percolation models on matching pairs of two-dimensional periodic graphs with one axis of symmetry.

Theorem 3.1: For the site percolation model on S,

$$p_H(k) + p_H^*((5-k)+) = 1 \qquad \text{for } k=3,4,$$

$$p_H(k) = p_T(k) = p_S(k) \qquad \text{for } k=3,4,$$

and

$$p_H(k+) = p_T(k+) = p_S(k+) \qquad \text{for } k=1,2.$$

Only brief comments on the proof of Theorem 3.1 are given here. A complete proof is presented in a technical report [Wierman (1986)], which is available from the the author upon request.

The result is a consequence of the duality properties of matching pairs of two-dimensional graphs. Every vertex of S is either k-degree open or (5-k)-(+)degree closed. Each configuration then generates a partition of the vertices into two sets, to which the topological lemmas of Kesten (1982) apply. Therefore, every rectangle $[0,m] x [0,n]$ is either crossed in one direction by a k-degree open path or in the other direction by a (5-k)-(+)degree closed path on the dual.

The first step in the proof follows the techniques discovered independently by Seymour and Welsh (1978) for the square lattice bond percolation model and Russo (1978) for the square lattice site percolation model, to obtain

$$p_H(k) + p_T^*((5-k)+) = 1,$$

$$p_T(k) = p_S(k)$$

and

$$p_T^*(k+) = p_S^*(k+).$$

Modifications must be made to take into account that the k-degree and k-(+)degree models are correlated percolation models, requiring attention in conditional independence arguments in the construction of circuits in annular regions. The FKG correlation inequality, which is valid for the events that rectangle crossing paths exist for both models, is used extensively, as in the classical setting.

The second step is to establish regularity properties of the percolation probability function. The proof follows the method of Russo (1978,1981) to show that θ is infinitely differentiable everywhere except at $p_H(k)$, and is continuous everywhere.

The final step uses Russo's formula [Russo (1981), section 4] and follows his adaptation of Kesten's (1980) path-cutting method to the site percolation model on the square lattice. Again, conditional probability arguments require modification due to the correlation betweeen the states of vertices.

This result establishes the equality of the three critical probabilities for a correlated percolation model. However, the result provides a new result about a classical percolation model, described in the following section, which is perhaps of greater significance since it may lead to a generalization of the equality to a broader class of graphs.

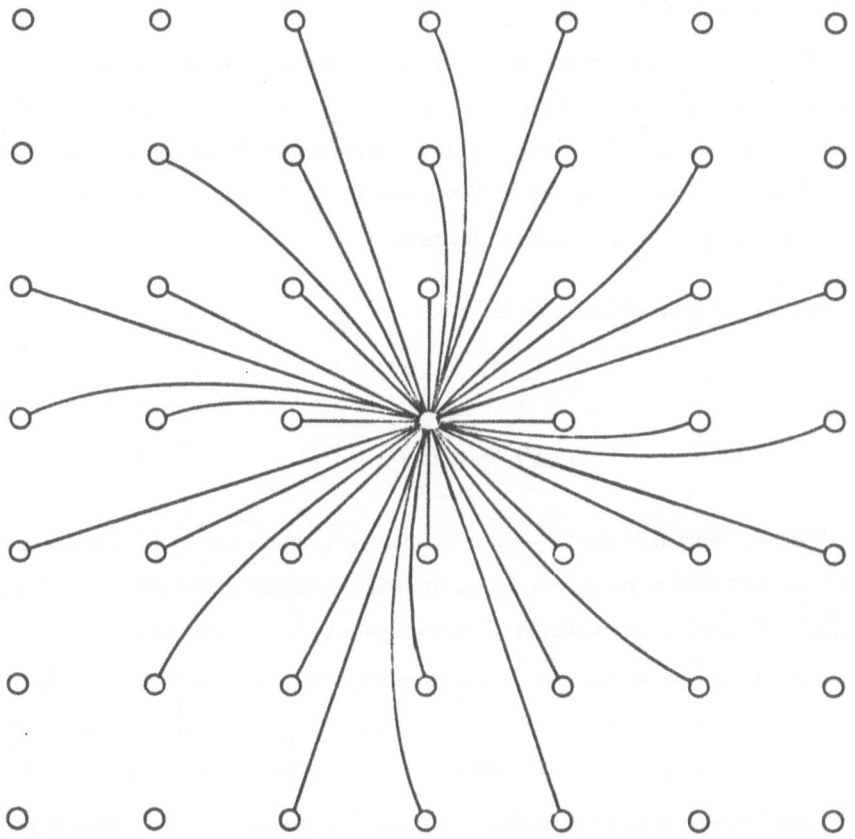

FIGURE 1.

Each vertex in D is adjacent to 36 other vertices which are within distance $2 + 2^{\frac{1}{2}}$. The 36 edges incident to a single vertex are illustrated above.

4. Equivalence of Models

Let D denote the graph imbedded in the plane with vertices at all points with integer coordinates, and edges between any pair of vertices v and w such that $||v-w|| \leq 2+2^{1/2}$. The edges of D are illustrated in Figure 1. The graph D is not in the class of matching graphs introduced by Sykes and Essam (1964), and does not satisfy the hypotheses of Kesten's (1982) principal theorem.

Proposition 4.1: For the site percolation model,

$$p_H(4) + p_H(D) = 1,$$

and

$$p_H(D) = p_T(D) = p_S(D).$$

Proof: We will show that the 1-(+)degree open percolation model is equivalent to the standard site percolation model on D, in the sense that the existence of a 1-(+)degree open path in S^* implies the existence of an open path in D and vice versa.

Consider a path $v_0, v_1, v_2, \ldots, v_n$ on S of vertices which are 1-(+)degree open in S. Then for each $i=0,1,2,\ldots,n$, there exists a vertex $v_i{}'$ which is open such that $||v_i - v_i{}'|| \leq 1$. [Note that v_i and $v_i{}'$ may be equal for some values of i.]

Suppose that two vertices v_i and v_{i+1} are 1-(+)degree open vertices which are adjacent in S^*. Then by the triangle inequality,

$$||v_i{}' - v_{i+1}{}'||$$

$$\leq ||v_i{}' - v_i|| + ||v_i - v_{i+1}|| + ||v_{i+1} - v_{i+1}{}'||$$

$$\leq 1 + 2^{1/2} + 1$$

$$= 2 + 2^{1/2},$$

so $v_i{}'$ and $v_{i+1}{}'$ are adjacent in D. Therefore, the path $v_0{}', v_1{}', v_2{}', \ldots, v_n{}'$ is an open path in D, so the existence of a 1-(+)degree open path in S^* implies the existence of an open path in D. This construction is illustrated in Figure 2.

Fix a vertex $a=(a_1, a_2)$ in D. We now show that for each vertex b which is adjacent to a in D there is a path in S^* from a to b such that if a and b are open, then all

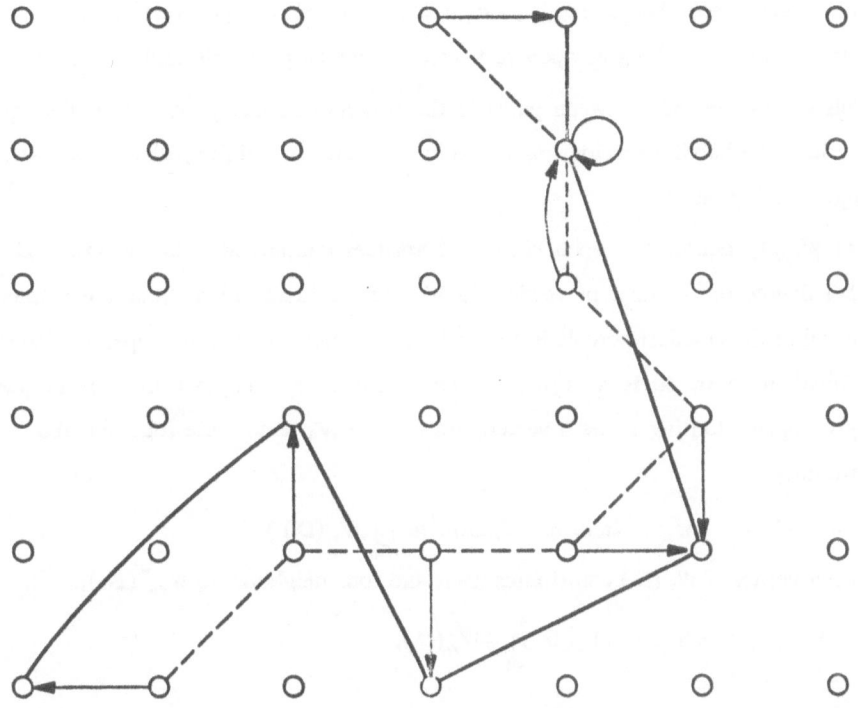

FIGURE 2.

Dashed lines represent edges of a 1-(+)degree open path in S*. From each vertex on the path, there is an arrow to the open vertex which implies that it is 1-(+)degree open. Solid lines indicate the open path in D corresponding to the original 1-(+)degree open path in S*.

vertices on the path are 1-(+)degree open. Note that by symmetry it is sufficient to consider only vertices b which lie in the sector bounded by the rays $y=a_2$ and $y=a_2+(x-a_1)$ where $x \geq a_1$. Figure 3 illustrates the 7 possible edges in this sector, showing an appropriate path in each case. Therefore, if $a_0, a_1, a_2, \ldots$ is a path of open vertices in D, we can construct a path of 1-(+)degree open vertices in S^* which passes through a_i for all i.

This correspondence between paths in the two models easily establishes the equality of the critical probability versions p_H based on existence of infinite paths and p_S based on rectangle crossing paths.

Let $W_v(D)$ denote the open cluster of vertices containing v in the graph D, and $\#W_v(D)$ denote the number of vertices in $W_v(D)$. Critical probabilities are defined for D as usual in the classical percolation model. Each vertex in D that is open has no closed edges incident to it, so is 4-(+)degree open, and every neighbor in S is at least 1-(+)degree open. Letting v_0 be a vertex and $v_1, \cdots, v_4$ denote the four neighbors of v_0 in S, we have

$$W_{v_0}^*(1+) = \{ w : w \text{ is in or adjacent to } \bigcup_{i=0}^{4} W_{v_i}(D) \}.$$

Since each vertex of $W_{v_i}(D)$ contributes itself and four neighbors to $W_{v_0}^*(1+)$,

$$\#W_{v_0}(D) \leq \#W_{v_0}^*(1+) \leq 5 \sum_{i=0}^{4} \#W_{v_i}(D),$$

so

$$E[\#W_v(D)] \leq E[\#W_v^*(1+)] \leq 25 E[\#W_v(D)]$$

for every vertex v. Thus, $E[\#W_v(D)]$ is infinite if and only if $E[\#W_v^*(1+)]$ is infinite. Therefore, the critical probability p_T is the same for the two models. $\square$

Proposition 4.1 can be used to verify that k-degree percolation on the square lattice site model does exhibit multiple phase transitions. Let D' denote the graph with the same vertex set as S in which each rectangle $[m, m+1] \times [2n, 2n+2]$ is close-packed. Using the containment principle and then Kesten's strict inequalities for critical probabilities of subgraphs, we obtain

$$p_H(D) \leq p_H(D') < p_H(S^*)$$

which by the duality results implies that

$$p_H(4) > p_H(2).$$

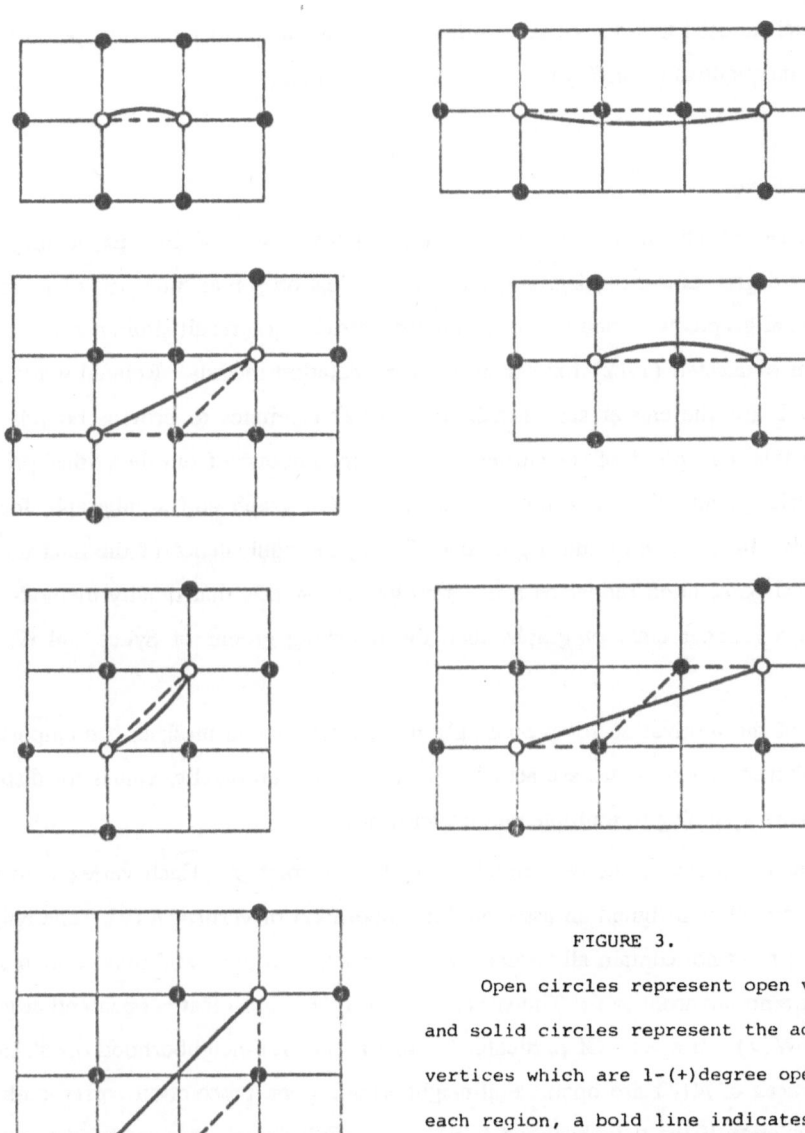

FIGURE 3.

Open circles represent open vertices and solid circles represent the additional vertices which are 1-(+)degree open. In each region, a bold line indicates an open edge in D, while dashed line segments represent a path of 1-(+)degree vertices between the same pair of vertices.

[Since the dual model for 3-degree percolation is not equivalent to a classical percolation model, strict inequalities for $p_H(3)$ have not been established.]

5. Generalizations

Preliminary investigations indicate that the duality result of this paper may be extended to k-degree and k-(+)degree percolation models on a matching pair of periodic two-dimensional graphs with one axis of symmetry, providing a result similar to the principal theorem of Kesten (1982) for classical site percolation models. Related work will provide bounds for the cluster size distribution, power estimates to provide bounds for critical exponents, analyticity of the clusters per site function away from the critical probability, and strict inequalities for critical probabilities of a graph and a subgraph, for k-degree models. In the case of full-degree percolation, the equivalence of the dual model with a classical percolation model on a different graph suggests that duality methods are applicable to a broader class of graphs than the matching graphs of Sykes and Essam (1964).

In view of the motivation for proposing k-degree percolation models, it is important to establish that in general there are actually distinct critical probability values for distinct values of k, corresponding to multiple phase transitions.

A further extension might be termed *k-neighbor* percolation. Each vertex v in the underlying graph G is assigned an associated neighborhood of vertices $N(v)$. The neighborhood $N(v)$ need not contain all vertices which are adjacent ot v, and may contain vertices which are not adjacent to v. The vertex v is *k-neighbor open* if it is open and at least k vertices of $N(v)$ are open. Of particular interest is the "full-neighborhood open" case, when all vertices of $N(v)$ are open. Full-neighborhood open percolation corresponds to classical percolation if the neighborhood $N(v) = \{v\}$. Full-degree percolation is obtained if the neighborhood of a vertex v consists of all vertices which are adjacent to v.

The extent of the class of neighborhoods for which the standard percolation results are valid for periodic two-dimensional graphs with one axis of symmetry is unclear. However, the equivalence of the dual model with a classical percolation model will permit the equality $p_H = p_T = p_S$ to be proved for classical site percolation on a class of much more richly connected graphs than the class of matching graphs of Sykes and Essam.

6. Acknowledgements

This research is supported in part by the U.S. National Science Foundation (Grant DMS-8303238) and the Institute for Mathematics and Its Applications at the University of Minnesota. The author thanks L. Quintas and J. W. Kennedy for stimulating conversations which introduced him to the k-degree percolation concept.

REFERENCES

Fortuin, C.M., Kasteleyn, J., and Ginibre, J. (1971) Correlation inequalities on some partially ordered sets. *Comm. Math. Phys. 22*, 89-103.

Harris, T.E. (1960) A lower bound for the critical probability in a certain percolation process. *Proc. Camb. Phil. Soc. 56*, 13-20.

Kennedy, J.W. (1982) The random-graph-like state of matter. *Proceedings of the 6th International Conference on Computers in Chemical Research and Education.*

Kesten, H. (1980) The critical probability of bond percolation on the square lattice equals 1/2. *Comm. Math. Phys. 74*, 41-59.

Kesten, H. (1982) *Percolation Theory for Mathematicians*. Birkhauser.

Quintas, L. (1983) A volume function for water based on a random lattice-graph model. *Chemical Applications of Topology and Graph Theory.*

Russo, L. (1978) A note on percolation. *Z.F.W. 43*, 39-48.

Russo, L. (1981) On critical percolation probabilities. *Z.F.W. 56*, 229-237.

Seymour, P.D., and Welsh, D.J.A. (1978) Percolation probabilities on the square lattice. *Ann. Discrete Math. 3*, 227-245.

Sykes, M.F. and Essam, J.W. (1964) Exact critical probabilities for site and bond problems in two dimensions. *J. Math. Phys. 5*, 1117-1127.

Wierman, J. C. (1986) Duality for k-degree percolation on the square lattice. Johns Hopkins University Mathematical Sciences Department Technical Report #467.